The History of the Future

BY

Kenneth J. M. MacLean

The History of the Future

Image Credits:

Book Cover: Image by Daniil Peshkov, purchased at 123rf.com

O'Neill cylinders: Images courtesy of Wikipedia.org/wiki. From NASA/Rick Guidice and NASA/Donald Davis.

Gene Editing and Gene Slicing, New York 3013 AD, Tower of the Overseers 3013 AD, and Abandoned Warehouse 2035 AD images are author created.

Contents

"The same revolutionary beliefs for which our forebears fought are still at issue around the globe—the belief that the rights of man come not from the generosity of the State but from the hand of God."

— John F. Kennedy[1]

"Faith gives us the fixed compass points to navigate by, to know who we are. Absent that we are going to believe the most attractive and seductive lie, or whatever generates in us the most fear."

— Father Robert McTeigue

"The price good men pay for indifference to public affairs is to be ruled by evil men."

— Anonymous

"Those who can make you believe absurdities can make you commit atrocities."

— Voltaire

"Science is but a perversion of itself unless it has, as its ultimate goal, the betterment of humanity."

— Nikola Tesla

Introduction

2045 AD

My name is Joe Courvall. I'm writing this book because I want to help humanity avoid a dystopian future. I'm not exaggerating. If nothing changes, the human race is sunk. If the people running this planet are allowed to continue on their heedless path, we aren't going to make it. I know this because I've traveled in time. I've seen the future and it's a dead end. Unless something changes.

This memoir is a warning. We can avoid a bad outcome if we understand the issues.[2]

Is it weird to call this a memoir when it's about things that haven't happened yet? Yeah, a little. But for me the future is the past. I've been to the future; I've seen it. The fluid and malleable nature of time is something you'll learn about when you read this. The...people...from the future taught my friends and me all about it.

I pieced this account together from extensive interviews of the participants, including the people from the future.

I've tried to make this memoir as entertaining as I can. It wasn't hard. The truth is a thousand times stranger and more interesting than fiction.

(Note: If you are reading this before 2045, you will find out at the end how this book got into your hands.)

Part I

CHAPTER **1**

2035 AD

It all started in the year 2035, when I was 24. (2035 may be the future or the past for you, the reader. Doesn't make any difference to me because it's all in my past).

I was born on September 11, 2011, on the tenth anniversary of the Towers coming down in New York. I am 34 years old as I write this.

I grew up in Midland, Illinois, a university town, in a subdivision dominated by academics from Carleton University. My dad is a professor of philosophy at Carleton and my mom is a medical doctor. They only believe what they read in their textbooks and academic journals, so they don't play a part in this story.

When I was ten I had an out-of-body experience that opened my eyes to an awareness beyond the material world, but that's all I'll say about it. My parents thought I was weird. They got "worried about my mental health" and sent me to a psychologist. That made me a little sensitive about opening up to people.

I won't bore you with my life story, other than to say that I have always felt a little different from the people around me. All of my friends feel the same way.

In high school I decided to do something practical that could make me some money. So I graduated from Carleton University with an engineering degree. Despite my unusual childhood experience I have a practical mind. I was good with engineering. I didn't want to do anything my parents did.

My life would probably have been uneventful if it weren't for my autistic friend, Germaine Robinson. Ger is a genius. He became associated with a (formerly) classified research project on the Carleton University campus and dragged me into it. Carleton is a private college in Midland, Illinois, with an international reputation. Ger is an inventor; he is the African American version of Nikola Tesla.

I work at Phoenix, a small engineering firm that does lab tests to certify parts for the electronics industry. One cold day in April Germaine showed up at my desk at Phoenix just after five, looking very excited. "What are you doing here, Ger?"

Germaine is tall and lanky, with frizzy hair. "I want to talk to you."

I knew that look. Ger is always coming up with new ideas. "What is it this time? A remote control can opener?"

Ger smiled. "You got some time?"

"Sure."

"Then come over to my place."

"Ok, but you have to walk with me." I walk to work because my apartment is only a mile away from Phoenix. Ger lives two blocks from me. I spend most of my time at the job in a chair, and at home in a chair, so I need some exercise every day. I'll walk even in the coldest weather, or when it's raining or snowing.

"OK, let's go."

We walked out of the five-story commercial building at the edge of downtown. Germaine's steps were half again as big as mine. "Slow down!"

Germaine stopped and let me catch up. He was obviously excited about something.

"Have you ever heard of Project Looking Glass?"

I groaned. When Ger talks in that conspiratorial tone I know he's going to go wacko on me. "Stargates and time travel. A bunch of nonsense."

I could tell Ger was offended. "Until last month I would have agreed with you. I'm an independent research analyst, among other things. I get hired by the university, and private research firms funded by venture capitalists, to evaluate the projects these guys are funding. In the course of my work I was in the Taubman Research Building a while ago. I walked into an old lab and found two legacy devices stuck in the back, gathering dust. I got curious because they looked like something I'd read about. In non-peer-reviewed journals, of course."

Ger was trying to make a joke. "Sure."

"I dusted them off and began to work with them."

I knew I was in for a lecture, but I have learned to be patient with Ger even though he can be tedious sometimes. I tried to sound enthusiastic. "Sounds subversive!"

I was only listening with half an ear as we walked. The early April sun illuminated a wintry landscape, with an inch of snow still on the ground. Climate change is real but global warming turned out to be fake. In 2035 the planet had cooled. It was cold eight months of the year up here in Midland, Illinois.[3] We're only fifty miles south of Chicago.

"There are two devices, Joe. One is a rectangular tank that contains a white computational substance. The other is a collection of metallic rings that move perpendicular to each other, surrounded by arrays of magnets. It appears that they were a part of an attempt, now abandoned, to research time."

We had been walking faster than I wanted, so I stopped. I knew Ger wasn't going to let go of this until I acknowledged him. "OK, Project Looking Glass. Supposedly a classified project run out of the S-4 facility at Area 51, an attempt to monitor the timeline and predict the future. What's it doing here at Carleton?"

"Good question. It looks like an attempt to recreate the original. When I walked back there I saw that the thingy with the concentric rings hadn't been completed. Parts were lying around it, as if the project had been abandoned."

"Or it ran out of funding. The Taubman Research Building is privately owned."

"When I mentioned it to Tanya, the receptionist, she knew all about it. Everybody knows about it, she said. It was what they call a boondoggle. Tanya even looked it up for me."

"'Project TRB-LG-3013, experimental research into the nature of time,' she said. 'The dates are 2027-2028. No record of why it was scrapped.'"

"'Any reason why the equipment was just left there to rot?' I asked her."

"'No, but it's off-limits for some reason. No one is allowed to remove the stuff, so it just gathers dust. No one works on it anymore. That's all I know.'"

"'Surely someone must be paying Taubman to keep that lab, he's an old skinflint,' I said to her."

"Tanya checked her records. 'Look at this Germaine!' she said. 'The lab *is* being funded, by a mysterious donor who calls himself Caesar 13. No one knows anything about him.'"

"Caesar 13?" I said to Ger. "Sounds like a dog's name, or a horse. It's a joke, Ger."

"Not so you'd notice. Every six months funding arrives for it, Tanya says. Somebody wants it there."

"A tax write-off maybe. Some people have too much money."

Ger shrugged. "Tanya said it was OK for me to look at it, and no one else seems to mind. It's a mystery, and I don't like mysteries."

We reached Ger's house and went inside. "I'll put the coffee pot on," I said.

Ger kept talking. "I found copies of some old schematics for the devices in that lab. They're from the early 1990s."

As we drank Ger and I looked at the drawings.

I couldn't make sense out of the schematics. "It looks like a fancy erector set. The only way to understand this thing is to build it and test it."

Ger was excited. "Correct! I knew you'd get it, Joe. That's why I wanted to talk to you. I need a good engineer to help me on this project. I think it has possibilities. I've been working on it every day for at least an hour during lunchtime. So far Tanya has given me clearance."

"I'll think about it." Ger is a little flaky sometimes, and I didn't want to talk about it anymore so I changed the subject. "Is your mom home? I want to get her recipe for that cherry cake she makes."

"You and your baking."

Ger lived in his parents' duplex, on the top floor. His mom lived downstairs, but she was almost never home, spending most of her time with her friends after Ger's father died. I like to bake, especially when I'm stressed. Ger's mom makes a killer cherry cake with royal icing on top.

Suddenly Ger's face crumpled. Tears began rolling down his face.

Shit! I should have kept my mouth shut. "I'm really sorry about your dad."

Ger nodded. "When dad got diagnosed with cancer last year I was hopeful. But he just went down the tubes so fast. The same thing happened to Lori."

"Yeah." Ger's father got pancreatic cancer and was gone in a month. Lori was Ger's aunt. She got cancer and died a year before Ger's father.

Ger gave me a look of despair and hopelessness. "It sucks. I was sad when dad died, but Lori was my favorite."

I looked at my friend with sympathy. We both had parents we didn't understand, and had confided in other family members. Lori for Ger, my Uncle Ken for me. Unlike Ger, my family has always experienced robust health. My grandpa is still going strong at 87, my grandma is 86. Dad is in his mid-fifties and looks twenty years younger (that's what other people his age say anyway). My mom looks like a contestant at a beauty pageant and she doesn't have to work at it.

I didn't want Ger to go into one of his depressions, but I didn't know what to do for him. Fortunately he snapped out of it. "Let's get out of here and get something to eat."

At the restaurant Ger made me promise to stop by the Taubman Building at lunch the next day to see his lab.

The next day I had to walk a mile during lunchtime to central campus to the Taubman Research Building. The TRB is notorious on campus for housing classified research projects for the military. It has been the focus of student demonstrations.

I saw Ger waiting outside the building by the front door. "Do you want to show me that can opener you're working on?"

Ger gave me an excited look, tinged with anxiety. "I hope they'll let you in."

"I have a security clearance. Remember, Phoenix got a military contract last year to test modules for the Navy's new F-35g fighter plane."

"Let's go in then."

We walked to the front desk. Two armed security guards stood on either side of the receptionist. "Going back in," Ger said to a woman with the nametag 'Tanya Harrison.' "Tanya, this is my friend Joe from Phoenix Corp. He worked on the F-35g project for the Air Force."

The receptionist spoke into her mic and looked at her display. "Joe Courvall?"

"Yes ma'am, that's me."

She looked me over carefully. Then she spoke into her headset and printed out a guest pass with my name on it. "OK Germaine. You're not to go anywhere except Lab B-103, understood?"

Ger nodded and we walked through a door that one of the security guards opened for us. "They say that every time," Ger remarked as we walked down a hallway to a bank of elevators and got in. When the elevator doors opened Ger checked in with another security guard. I handed him my pass. "OK Robinson, you and your friend are clear down here until 2 p.m., lab B-103. Don't make me come get you."

Ger nodded. We walked down a corridor painted black. Every ten feet a recessed light provided illumination. At the first door on the right (also painted black) Ger took his passcard and inserted it into the door, and we walked into a laboratory. In front, several lab tables and a bunch of chairs were arranged haphazardly. There was a rumpled bed against the right wall. We walked past a big partition to the back of the lab. On a lab table I saw a tank that had a white swirling gas in it. Off to the right I saw four thin concentric metal rings, each about six feet in diameter, surrounded by several circular arrays of what looked like magnets, above a platform about ten feet in diameter. Ger walked up to the control console and began fiddling. The rings began to move very slowly. I noticed that the space above the platform was slightly disturbed.

I turned my attention to the table. The tank was simply a transparent cube about two feet on each side. Inside the cube was a white, roiling fog. It was very unusual looking.

I looked to Ger. "This looks like a top-secret project. Why would I get clearance to be here?"

My friend grinned. "It used to be top secret." He indicated the equipment. "This is just Project TRB-GR-2035 now, named after me. These are mere curiosities, or at least they were before I got my hands on them. I've made some changes to the original designs I showed you yesterday."

Ger stared at the setup, a look of intense excitement on his face.

"Watch this." Ger went to the console and turned the power up. The rings began to rotate a little faster. The device gave an appearance of latent, untapped power. The space above the platform became more disturbed as the rings turned. A very faint blue energy began to surround the platform. Ger put on a headset. Suddenly the roiling white mist in the cube on the lab table began to take form.

Do you remember that old Star Trek show, the first series where Spock and Kirk go chasing McCoy, who stepped through a time portal? Well, that's what this thing was. There were fleeting images flowing through it now. Then I saw myself and Ger walking toward the TRB building, entering, and everything that happened until now.

"Notice the blue energy surrounding the platform. My theory is that this device is accessing temporal energy. The cube is the display."

My engineering mind was baffled. Ger operates in a different brainspace than I do.

Ger grinned and took off the headset. "My experiments show that this display cube, and the rotating rings, are connected through this headset." He pointed to the headset, which was a very thin silver band that went around the head and had two extensions that went over the temples.

"I've never seen anything like that," I said.

"They're quite common, I understand, in frontier physics research. Let me show you what it can do."

Ger put the headset back on. "Yesterday I made several trials with the device after I got it working. Looking Glass was designed to work with time, so on my last attempt I decided to find out what the earth would look like one thousand years into the future."

"How did you do that?"

Ger went over to the console and turned the power up. "By increasing the rotational speed of the rings, along with a focused, directed thought into the headset. I'm pretty sure the headset is using consciousness assisted technology. CAT is absolutely the cutting edge on the planet."

"That's way out of my league. I'm just an engineer."

"That's what I like about you Joe," Ger said playfully. "You know your place."

"OK, so show me what you've got."

Ger put the headset back on.

In the display tank, a city appeared. Ger took out his mobile and recorded it as we both stared into the tank. Hovercraft moved slowly back and forth in the skies under a pale sun. Several buildings were placed along what looked like the remains of city streets. People dressed in blue and black uniforms walked slowly, single file, on narrow footpaths that had been worn into the concrete. The remnants of a city grid system was barely visible, most of which consisted of sickly looking shrubbery, small pencil trees with a few leaves on them, and brown, half-dead grass. A few ancient crumbling buildings stood on city lots, neglected and unused.

The place looked vaguely familiar to me.

"As near as I can tell, this is New York, 3013 AD," Ger said.

"You're kidding."

"Watch."

Someone got out of line and began to walk toward what looked like a small hut. These were placed every couple of blocks. One of the hovercraft stopped its movement and turned toward the offending pedestrian, who got quickly back in line.

I looked at Ger. "Hopefully this is just the product of your warped imagination."

Ger smiled bleakly. "I do too. In the original Looking Glass project, the future displayed in the tank depended on the consciousness of the observer. A gloomy person would see a dark future. A positive person a brighter one." Ger sighed. "My dad was born in New York and we have family there. Maybe this device picked up on that."

"Let me try then. I'm pretty optimistic."

Ger took off the headset and the images in the tank disappeared. I put the headset on. "What do I do?"

"Clear your mind, think about what the future looks like a thousand years from now."

"OK. I'll keep a positive frame of mind." To my astonishment, images began to appear in the tank. The tank showed a similar, depressing scene, of a city. People were walking around like zombies; hovercraft looked to be monitoring everyone from the air, the remnants of crumbled buildings were everywhere. The images were sharp and lifelike.

"No future I want to be in," I said.

"Me neither."

I glanced over at the rotating rings. The blue energy was stronger now, and the space above the platform was disturbed, like what you'd see on a hot day driving in the desert. "What happens if you make the rings turn even faster?"

Ger blanched a little and looked at me. "I don't know, Joe. I'm too scared to try it right now. I think this thing is meant to be a temporal transporter."

We stood there for a second, staring at each other.

"In Project Rainbow, sailors on the USS Eldrige were supposedly transported in time and space," I said.

"Yeah." Ger pointed to the four rings. The biggest ring didn't move. The other three rotated perpendicular to each other, inside the large one. "My working theory is that the stationary ring is time, a scalar. It doesn't move. The other three represent the three spatial dimensions: length, width, and height. This device accesses the three dimensions of space moving through time."

I looked again at the blue energy surrounding the platform. The space around the platform was shimmering like the transporter on the old Star Trek.

"Turn that thing off," I said nervously.

The concentric rings slowly stopped turning. The blue energy disappeared, and the space around the platform returned to normal. The images in the display tank were gone. The display showed a random white mist, like a fog.

Ger was very pleased with himself. He replayed his time recording from the New York of 3013 from his mobile and put it on a display panel. We both stared at it again.

"I haven't told anyone anything I've done in here. Not even Tanya."

I looked up at the walls and ceiling. "I'd imagine that everything we're doing is being recorded, so good luck with that. Old Man Taubman is paranoid about security."

Ger grinned. "Actually, no. The cameras have all been deactivated. The employees on this floor use this space for sexual liaisons. They don't want anyone seeing them bump uglies on work time. You saw that bed against wall, didn't you? This place is like a safe zone within the building. A very valuable space."

I laughed. It was typical human behavior. The bad feeling I had vanished.

At that moment two giggling employees walked into the lab. They must have heard us talking behind the partition. "C'mon Robinson, you and your friend take a hike."

Ger looked at me and laughed. "They want to fool around." He looked at his watch. "Shit! It's past two."

"And I'm late for work!"

We hurriedly walked out of the lab and into the corridor. The security guard watched us all the way down the hallway to the elevators. We made the lobby and checked back in with the receptionist.

"Thanks Tanya," Ger said to the coolly dressed and professional looking woman. "Can I have clearance to be back here tomorrow at noon?"

Tanya nodded.

Ger winked. "Tell those degenerates down there to piss off during lunchtime. I don't want to be interrupted."

Tanya smiled. "Sure thing Germaine. You wouldn't want to go for coffee would you? My shift ends at five tonight."

"Uh, not tonight Tanya." Ger looked uncomfortable.

The woman was obviously interested in Ger. "He's crazy," I told Tanya.

"A girl can only hope..."

We walked out of the building. "I have to get back to work," I told Ger.

At work I couldn't concentrate. I told myself that Ger was crazy and his recording of the future was a fantasy. Even though I had seen exactly the same thing. My engineering mind was troubled by a device whose operation was way beyond my understanding.

I went to bed that night and dreamed that I was one of those zombie walkers in the New York of 3013...

CHAPTER 2

2035 AD

At lunchtime the next day I saw Ger in the lab. Tanya gave me another guest pass and the security guard recognized me.

When I walked into Lab B–103 Ger looked excited and a little disturbed The rings were turning and a faint blue energy surrounded the platform.

"Check this out. I was able to make another recording of the future."

This one showed a man with perfect facial features, a Roman nose, and olive-colored skin, sitting at a raised console. He was wearing a blue and yellow uni and surrounded by twelve others, also at consoles arranged in a circle. These individuals all had faces that looked airbrushed perfect. "Prepare the Time Shifter..." a voice said. In the background was a much bigger version of Ger's concentric ring device.

Ger's body was practically vibrating with excitement. "That's all I got, but this is 3013 AD again. If so, there's a working version of this device somewhere up the timeline."

The images in the tank were crystal clear and lifelike. I was starting to feel nervous. "Ger, what are you going to do with this thing?"

"What do you think? I'm going to experiment until I understand how this device works. Then I want to see what it can do."

"That's what I'm afraid of."

"All new inventions seem scary at first." He regarded me with a stubborn look that I knew well.

"Remember what happened to those sailors on the USS Eldrige."

"That's just a conspiracy theory, Joe."

11

Ger was getting defensive now so I changed the subject. "Why did you turn down Tanya yesterday?"

Ger squirmed a little. "I'm not ready for a relationship."

I knew better than to go any further. Ger's autism was kicking in. If I continued he'd completely close up on me. "OK buddy, I'm sorry I doubted you. Keep me posted on your progress."

My friend breathed a sigh of relief.

I turned to go, then remembered something. "Are you coming to the get-together tonight at Angelo's?"

Ger smiled. "Wouldn't miss it."

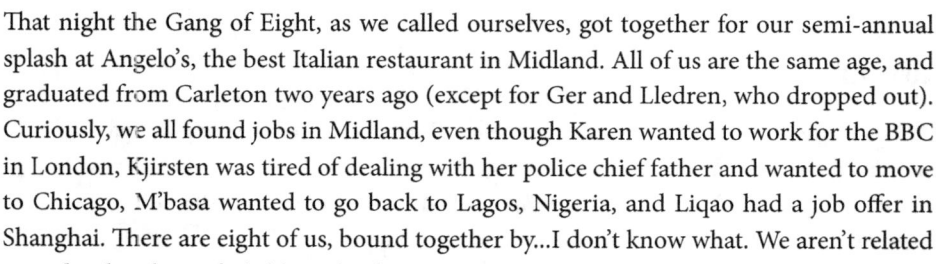

That night the Gang of Eight, as we called ourselves, got together for our semi-annual splash at Angelo's, the best Italian restaurant in Midland. All of us are the same age, and graduated from Carleton two years ago (except for Ger and Lledren, who dropped out). Curiously, we all found jobs in Midland, even though Karen wanted to work for the BBC in London, Kjirsten was tired of dealing with her police chief father and wanted to move to Chicago, M'basa wanted to go back to Lagos, Nigeria, and Liqao had a job offer in Shanghai. There are eight of us, bound together by...I don't know what. We aren't related to each other, but it feels like we're family. Kirra says it's a soul connection.

I got to Angelos early because I was starving, and loved their rosemary-potato focaccia rolls. If I ordered them with the others they'd be gone in a second.

The first to walk in was Kirra Bigbear. Kirra is small and dark, with flashing blue eyes and jet black hair. She always wears bright red lipstick, which makes her look really, really cool. She is Native American, an Ojibwa. Next to show was Kjirsten Chastaine, the daughter of Jack Chastaine, Midland's police chief. Kjirsten is a gloriously tall redhead with freckles around her nose. She was with her equally tall but soft-spoken Nigerian boyfriend, M'basa Ogunfatidime. Kjirsten had a habit of talking in a command voice like her father, which I find irritating sometimes. Ger walked in next and sat across from me at the big table, which I had reserved. Next came Liqao Chang, a smallish, nervous-looking guy with round, gold-rimmed glasses. Liqao has the thickest head of black hair I have ever seen. He's the IT guy for the Midland Police Department, and Jack Chastaine's right hand man in the department. He runs all their servers. When I saw Liqao I quickly gulped down the last two appetizers. Everybody laughed.

"What's so funny?" Liqao asked. He saw my empty plate. "You ordered those focaccia rolls and ate them all yourself!"

Kirra, Liqao's boyfriend, laughed. "Did you grow up in an orphanage of starving sharks?"

"I'm not that bad," Liqao complained.

"Yes you are," I said, licking the last few crumbs off the plate. "Order your own, you can afford it." Liqao is the son of a wealthy businessman, but he grew up with eight siblings. That might account for his acquisitiveness at the dinner table.

Lledren Cadwallader walked in next, still dressed in his soup kitchen clothes.

"Still haven't taken that marketing job with your old man," Ger joked.

"Uh, no," Lledren said, unruffled. "I'd rather work in the kitchen for minimum wage, helping people."

Everyone accepted this. Lledren was a maverick, and his lifestyle was respected by everyone in the group. Tonight he looked hungry so I ordered him a plate of focaccia rolls. When the plate came I had to slap Liqao's hand away, to general laughter.

The last to come was my fave, Karen Everard. Karen is a brown-eyed blonde with a certain...I don't know, inner force. I've always been just a little afraid of Karen. I don't think she understands how powerful she is.

"OK, let's get the alter-ego jokes out of the way," I said when everybody was seated. Amazingly, Ger and I have the same birthday, September 11, 2011. Today was April 13th, but someone would always start it up. A lot of jokes have been told about Ger being my dark side, and vice-versa.

We ordered, and the talk was comfortable and friendly. As I ate I soaked up the good vibes. Every time we get together there is an all-is-right-with-the-world feeling. I wondered why we only did this twice a year.

I looked at everybody, happily engaged in eating dessert, talking and joking. There was definitely a special bond between all of us even though we are completely different people with divergent interests.

Just then a guy walked into Angelos. He was of medium height and had brown curly hair. He walked over to our table, checking us out. He was holding some papers in his hand.

"Do you want something?" Ger asked.

He looked at Karen. "You're Karen Everard, the reporter for the *MIdland Chronicle*?"

"Yes. Say, you're Harry Kaine aren't you?"

"That's right Karen. You'll be interviewing me on Monday. I'll have some scientist friends along for our meeting." He handed her the printouts. "In order to discuss the subject intelligently, I've brought you some informational literature on Transhumanism."

Karen took the papers but was cool. "I'm doing my own research, Mr. Kaine. I make it a point before every interview."

Kaine flashed a smile. "Is that so?" His gaze was admiring and challenging.

To my surprise Karen flushed. The guy did nothing for me but apparently Karen was affected by that fake smile. I noticed that Kjirsten and Kirra were sitting up a little straighter.

"Just wanted to tell you about a slight change of plans. Our group will be meeting in Room 253 in the Life Sciences building at 7 p.m. on Monday. There's a conference table there. I'll have coffee and some rolls for everyone." Kaine smiled again.

"Uh, OK Harry," Karen said. "I'll see you Monday."

Kaine spoke confidently. "We could go out for coffee afterward."

Karen was recovering her composure. "It's a little early for that, Harry. Let's do the interview first."

Kaine laughed. "Sure thing, Karen. Looking forward to it."

The women were riveted. Ger and I exchanged 'I don't get it' glances.

After giving Kirra and Kjirsten a look, Kaine sauntered off. I could tell M'basa and Liqao were irritated too.

"What an asshole," Liqao remarked. Ger and I laughed.

"I thought he was nice," Karen said. "Very charming." The women nodded.

"What is Transhumanism?" I asked.

"Life extension via biological enhancement, nano-bio-technology, modifying the human body's genetics," Karen said.

Kirra looked alarmed. "Modifying the human body? What's that about?"

"I'm not sure, but I want to find out." Karen spoke a little smugly. "Harry called me personally about it."

M'basa looked at the three women. "Do you *like* that guy?"

"Well, he...he makes you feel like you're the only person in the room when he speaks to you," Karen said. "He gives you his undivided attention."

M'basa shrugged. "If you say so. What do you think, Kjirsten?"

"He's kind of magnetic." The women looked at each other and smiled.

Liqao, me, M'basa, and Ger looked at each other. "Watch that guy, he's trouble," Liqao said.

Kirra laughed at this. "More trouble than you, Liqao?"

This was a standing joke with the Gang. Liqao was obsessed with advancing his career and often got himself into scrapes because of his ambition.

As usual, Lledren sat quietly, observing us. Lledren is gay and a real loner. Out of all the Gang, I think Lledren is the most liked member.

After that the party broke up. I waited for Karen to get up and walked out with her. "Let's talk after you interview Kaine," I suggested as we walked out to the parking lot. "I think he's creepy."

"You're as bad as Ger when it comes to women," Karen said.

I have a sort of girlfriend, but I never found anyone I really liked. Except maybe Karen.

"Are you jealous of Harry?" she asked.

I sighed. "Maybe a little. I think he's too confident, too sure of himself."

"I don't think so."

"What kind of a guy asks you out when he's never even seen you before? I thought your interview was supposed to be as a professional."

Karen laughed and tossed her head back. "It is! But maybe I'll go out for coffee, you never know."

"He's arrogant." I spoke with some heat.

"That's a good sign, Joe, keep it up," she said, smiling.

I was confused.

She reached her vehicle. "I'll call you. Maybe Tuesday we can meet somewhere and I'll tell you how it went."

As I walked to my car I was irritated with myself. I don't like being angry, but I was angry at Harry Kaine. Why? I'd had relationships before but they were always casual, and I didn't think of Karen in that way. And why was that?

The day after the dinner with the Gang, Karen Everard was walking back to the office from lunch. As she was crossing the Diag on central campus she spotted me walking out of the First Street diner. She hailed me and I walked up to her. Behind us a class was getting out and students were walking toward them. An unusual looking woman with long red hair, dressed in mismatched winter clothing, was approaching them. The woman looked frightened and jumpy. She wore a nondescript brown jacket, a black pullover hat, athletic shoes, black slacks, and no gloves. She stopped, reached to the back of her head, and fiddled with something under her cap. Then she…disappeared for a second, and reappeared, holding a brown carryall bag.

"Did you see that?" someone behind them said.

"A glitch in the matrix!"

Karen felt me nudge her. "She wasn't carrying anything before she flickered," I said.

"You're right, Joe!"

The woman, obviously uncomfortable, turned around and began walking rapidly away from them.

Karen was intrigued and turned to the group of students. "Did you see that woman disappear or am I crazy?" She turned on her recorder.

"Yeah!" a shortish guy said. "She was gone. A second later she was back holding a carryall bag."

Karen interviewed the other students until everybody got too cold standing around. They all agreed about the details. Was there a story in this?

I was in complete shock. "Gotta tell Ger about this," I mumbled. That girl with her airbrushed face looked like the people in Ger's time viewer!

"I knew it," Karen said, watching me. "You and Germaine are up to something."

I smiled weakly. "Do you want to get some hot chocolate or a coffee?"

"Sure. I want to interview you about that woman anyway. Did you notice anything unusual about her?"

"As a matter of fact I did. Other than what the students said, I thought the woman's face was absolutely, perfectly symmetrical."

"You noticed it too! It was like she came from a...mold or something."

"Cosmetic surgery? Her skin was so perfect it looked airbrushed."

Karen looked at me curiously. "You're remarkably observant. A woman would notice that immediately, but I didn't know you paid that much attention."

I frowned. "I notice a lot. Especially that you and Kjirsten and Kirra have never looked at me."

"We never thought you were interested."

"I gotta be more interesting than Liqao."

Karen laughed. The wind was picking up so we walked to the Mason Street Coffeehouse, about a quarter-mile from the Diag. Karen ordered coffee and I ordered a hot chocolate and a roll.

"Do you think I'm gay?" I said with a little heat as we sat down.

Karen was surprised. "You, gay? No, Lledren's gay. We thought you might have it for that Kesha woman at COSA."

Kesha is a girl I met through Ger, who attracts women but never makes connections with them. COSA is the Children of Sexual Abusers organization. I had gone out a couple of times with Kesha.

"Well yeah, I do like her. But it's nothing serious."

"You have a great smile. You should do it more often."

I was stirring my chocolate. I put down my spoon. "You know, all three of you are gorgeous."

"First time I ever heard it. You are a bit...reticent."

"I thought girls liked shy guys."

"Sometimes."

We looked silently at each other. I could see that Karen was telling me that if I wanted a relationship with her I had to ask for it. "I want to ask you out, but I'm seeing Kesha tonight."

Karen shrugged. "OK. Now, let's get down to business. I want you to tell me everything you saw from the moment it started. Pretend you're a reporter at the scene."

She got out her mobile and set it to record. I described what I saw, and my reactions to the flickering woman on the Diag. I didn't want to tell her about Ger and his crazy time project.

"Don't hold back on me, Joe. This has something to do with you and Germaine, right?"

"You can read me like a book. Can you do that with everyone?"

Karen tilted her head to the side and gave me a mysterious, feminine look. "Pretty much."

I couldn't tell if she was bs-ing me. "Let's just say that Germaine is working on something that hasn't proven out yet. If we tell anyone it could...hinder Ger's progress."

Karen accepted this. "OK, but when you work it out I want an exclusive."

I smiled, beaming at her.

"I could get used to that," she said.

I felt my nervousness evaporate. As I said, Karen has a powerful feminine presence that scares me a little, but I was over myself now. For some reason I began to tell her about my childhood. We ordered some food and I explained how my father and mother had sent me to a psychologist when I was ten.

"They thought I was weird and needed help. My parents are academics and don't believe in anything outside the textbooks they studied at university."

"You must have said something or done something to make them do that."

I told her about my out-of-body experience. "I was ten years old. I was walking down the sidewalk when I suddenly saw myself walking from about three feet over my head. It felt really good. Then I popped back into my head. It's never happened again, but I'll never forget it."

Karen's eyes widened but she accepted it, to my surprise.

"I didn't tell my parents that of course! But I've always thought that the earth is a crazy place and that I came from somewhere else. When I told them that, asking for advice, they sent me to the psychologist for 'evaluation.' After I saw the psychologist my parents thought I was cured of whatever they thought was wrong with me because I never mentioned it again. But I had just learned not to talk about things like that."

Karen nodded. "I have to watch what I say at work. Melanie Fuscaldo at the Interfaith Center, who I just interviewed, calls people who feel like that star seeds. She says that souls have traveled all around the galaxy, and some of them have come to earth to help humanity." She spoke in a reporter's voice and I wondered if she believed that.

"I'm not big on religion or metaphysics, but the idea that you have a soul has been around for thousands of years."

I stopped because I was feeling embarrassed. I had never opened myself up to anyone like this. Karen seemed interested, but maybe she thought I was a flake.

She looked at me curiously. "I always thought you were just a boring engineer."

"I have unprobed depths."

Karen laughed. "Melanie says that all human beings are connected spiritually to a higher power."

"Do you believe in a higher power?"

"Uh...yeah. I do."

I smiled. "I do too."

I had no idea Karen was so deep. She told me a little about herself.

"My editor at the *Chronicle* occasionally publishes well-researched articles about subjects important to the community. I don't like to do them because they are a lot of work, and often about subjects I have no interest in. Like Transhumanism. I'm actually interested in spiritual subjects because both my parents are practicing Buddhists."

"Really? My parents are just dull academics."

Karen smiled. "It's probably why you became a boring engineer."

"I think that's more a function of my boring personality."

Karen turned her head to the side slightly and looked into my eyes. "Are you boring, Joe?"

"I don't feel boring inside, but it's how you appear to others I guess."

I could tell Karen was about to say something witty.

I expressed mock horror. "Don't answer that! My ego is far too fragile to hold up against you."

"You've made me laugh twice now in five minutes," she said, laughing lightly.

I stirred my coffee and spoke as drolly as I could. "Unintentionally, I'm sure."

Karen's eyes were sparkling now. "No, I don't think you're boring, Joe. Underappreciated, perhaps."

I caught her subtle meaning: I was too self-deprecating. I sat back in my chair. "I'm really enjoying this conversation."

"So am I."

I changed the subject. "You were telling me about the interviews you are doing. Spiritualism and Transhumanism! Why those two subjects?"

"Controversy! What else? I'm just an inquisitive reporter, after all."

"You're a little more than that, Karen."

Karen's eyes were dancing. "Very good, Joe."

We both took a sip of our drinks and our eyes met. Then we started laughing together.

Karen started up the conversation again. "I've already done an interview with Melanie Fuscaldo at the Interfaith Center for the *Chronicle*. It was more interesting than I thought it would be."

"Yeah, I've heard of her. Is she a space cadet?"

"Actually, she's a former medical researcher."

"Really?"

"Yeah. Melanie and Harry Kaine are polar opposites. The two interviews will make good feature articles, and hopefully generate a lot of comment among the readership."

I was interested. "I'd like to read them."

"First I have to write them!"

I told her about my work at Phoenix. "I think you're right that I gravitated to engineering because of my parents. My dad's philosophy career is just...thickheaded academic nonsense. My mom's medical career gives her a totally materialist point of view."

"There's some bitterness there."

I was pleased by Karen's observation. "There is. I still can't believe they sent me to the psychologist, even though I can sort of understand it. It's weird to know that at the age of ten you are more self-aware than your parents, who are supposed to know more than you."

"It was a betrayal."

"You're right! That's what it feels like."

"But the reason your parents did it is because they don't know any better."

"That may be," I agreed, a little reluctantly. "Maybe they just did the best they could."

Karen smiled. "Well, we've solved all of your life problems. It's smooth sailing from now on."

I grinned. "That's sounds like something Ger would say. I have no excuses now."

Karen told me that her job at the *Midland Chronicle* was OK, but she was looking for something more. "Unfortunately my job application to the *Chicago Tribune* – one of the few remaining big city legacy papers – has gone nowhere. The BBC isn't hiring, at least not Americans. So I'm stuck in Midland for now."

Karen told me that the *Chronicle* was one of the few legacy publishers that had survived in the standard journalistic format of dailies and Sundays, and still even had a limited print edition. "It's not too bad though, career-wise. Midland has a lot of popular local personalities. The conservative City Council often clashes with activists from the student and university community, so there's no lack of things to write about. Jack Martins, the world famous cosmologist, is always good for an interview if you don't mind his abrasiveness."[4]

"Martins! That guy loves to stick it to the establishment academic community. I like him for that."

Karen realized she was supposed to be working and changed the subject. "There's nothing more you can tell me about that disappearing woman on the Diag?"

"I don't think so. But it happened, I'm sure of that."

"OK, thanks."

The conversation died so we picked up our stuff and went back to work. I had always thought of Karen as one of those hard-driving reporter types, but she was no lightweight.

———— • ✳ • ————

When Karen got back to the office it was almost four. She had wasted an hour of work time talking to Joe, but he was more interesting than he looked. And his description of the incident on the Diag was very detailed. He had noticed that the redhead was wearing mismatched clothing, and that underneath her coat she was wearing some blue fabric he didn't recognize. Maybe she could use this strange incident for one of those "Unexplained Mysteries" articles the paper occasionally ran.

Bob Guza, the paper's senior editor, seemed pleased with her work. "Write up your interview with the Fuscaldo woman," he ordered. "When is your interview with Harry Kaine?"

"Monday."

"Good. This Sunday I'll put in the Interfaith piece. Next Sunday the Kaine story."

Karen told her editor about the flickering redhead on the Diag.

Bob looked at her curiously. "You say ten other students saw it too?"

"Yup. I've got their statements. It was...strange and remarkable."

Bob smiled. "Let's put that one on the back burner for now," he said drolly.

Karen laughed. "I guess you had to be there." She didn't pursue the matter and left the office.

On the way home she thought about her upcoming interview with Harry Kaine. It would counterbalance the metaphysical views of the Interfaith people. She knew next to nothing about Transhumanism other than what she'd read on the WorldNet.

Melanie Fuscaldo had trashed the entire subject. "Transhumanists believe in altering human biology to 'improve' it. All of them believe that your consciousness dies when the body dies. Therefore they seek life extension methods, not understanding that the soul is already immortal." When Karen mentioned Harry Kaine Melanie said, "Kaine is the leading proponent of Transhumanism and life extension in the country, and he's right here in Midland."

Karen went out that evening with two of her colleagues from work and got home around 11, feeling tired and a little tipsy. As she got ready for bed she thought about her eventful day. The highlight of it was her conversation with Joe Courvall. She and Joe had shared a deeper understanding about themselves today. She wondered why the Gang of Eight liked each other so much. They were all so different!

She fell asleep thinking about Joe Courvall's smile.

———— • ✸ • ————

That night before bed I thought about my friend Karen Everard. The students behind them on the Diag had been excitable, and ready to chase down the flickering woman. Karen had calmly turned to them with her mobile and asked for interviews, preventing an incident. I was pleased about our conversation at the coffee shop. Karen was more than just a reporter. She was soulful and beautiful.

The next morning at work I had to test and redesign a board. I kept thinking that the flickering woman and Ger's time viewer were connected, but I couldn't see how. I couldn't concentrate so I left my desk and went out for a short walk. I decided not to think any more about yesterday's strange incident.

An hour later I got a call from Karen. She insisted that I meet her at the Mason Street Coffeehouse at 6.

When I got there I could see that she was agitated. The first thing she said was, "I just reviewed my notes and recordings from yesterday. One of the students had his mobile on and captured the event. Joe, what we saw really happened."

I groaned playfully. "Just after I had successfully put it all out of my mind!"

Karen laughed. She looked delightful. "I like it that you have a sense of humor."

"My sense of humor thanks you."

Karen got out her mobile and we looked at the recording of the incident.

"Maybe it *was* a glitch in the Matrix!" I said.

"It looks like a badly spliced scene in a movie," Karen said.

We sat silently for a minute, watching the ten-second recording.

"Joe, does the Diag incident have anything to do with Ger's project at the Taubman Building?"

I was startled because neither Ger nor I had told anyone about his time research. How much did Karen know?

Karen smiled. "Don't look surprised. You told me yesterday that you and Ger are working on something. I wondered if the two were connected."

"I would hate to try and keep a secret from you."

"It's my job to ferret out what other people are hiding."

I leaned over the table to inspect her face. "Your nose isn't long enough for ferreting."

We both laughed and sipped our coffees.

I didn't think Ger would want me to blab about his research, so I changed the subject. "You're still going to see Harry Kaine, aren't you?" I asked.

"Yes. I'm told he's magnetically attractive to women."

Karen was regarding me with a feminine look I couldn't decipher. "Are you going to see Kesha tonight?" she asked.

It was nothing serious, although I did find Kesha intriguing. "No way. I'm going to stay home, listen to music, and try not to think about disappearing women."

"Maybe we should get drunk and have sex."

So there it was...right out in the open. But Karen was special; I didn't think of her as a mere sex partner. As I said, I was a little afraid of her. She had a feminine power that was scary to my more subdued maleness. I made an excuse. "I'd love to, but Kesha you know...I have to keep my head on straight."

Karen smiled. "You're wise, Joe Courvall."

I wasn't so sure about that. I was going to have to speak to Ger, though. Had he talked to anyone else about Looking Glass and his time device?

CHAPTER **3**

3013 AD, Safe Zone, Old New York

"Pietra 23, report!" the voice in her implant commanded. The Guardians had located her, and she was out of her assigned sector.

"You are out of position," the voice reminded her. "Report to Section Seven immediately after your duty period."

She groaned. It would mean a grilling, and she would have to come up with a very good reason for abandoning her post. But the need was great.

On foot, Pietra 23 approached the intersection of what used to be Lexington and 24th street in Old New York, where an abandoned clothing factory had once made the distinctive, tight-fitting blue and black unis for all Guardian citizens. These uniforms had not changed for almost a thousand years. Just like the rest of human society, Pietra 23 thought.

The Old City, or Old Sector, was an area of abandoned buildings about two miles square between the old 23rd and 34th streets. No hovercraft or spycams were allowed. It was what the Guardians called a "rest area," an area free from the constant and ubiquitous surveillance. Rest areas existed all over the world – after it was discovered that without them the populations exploded into anger and violence.

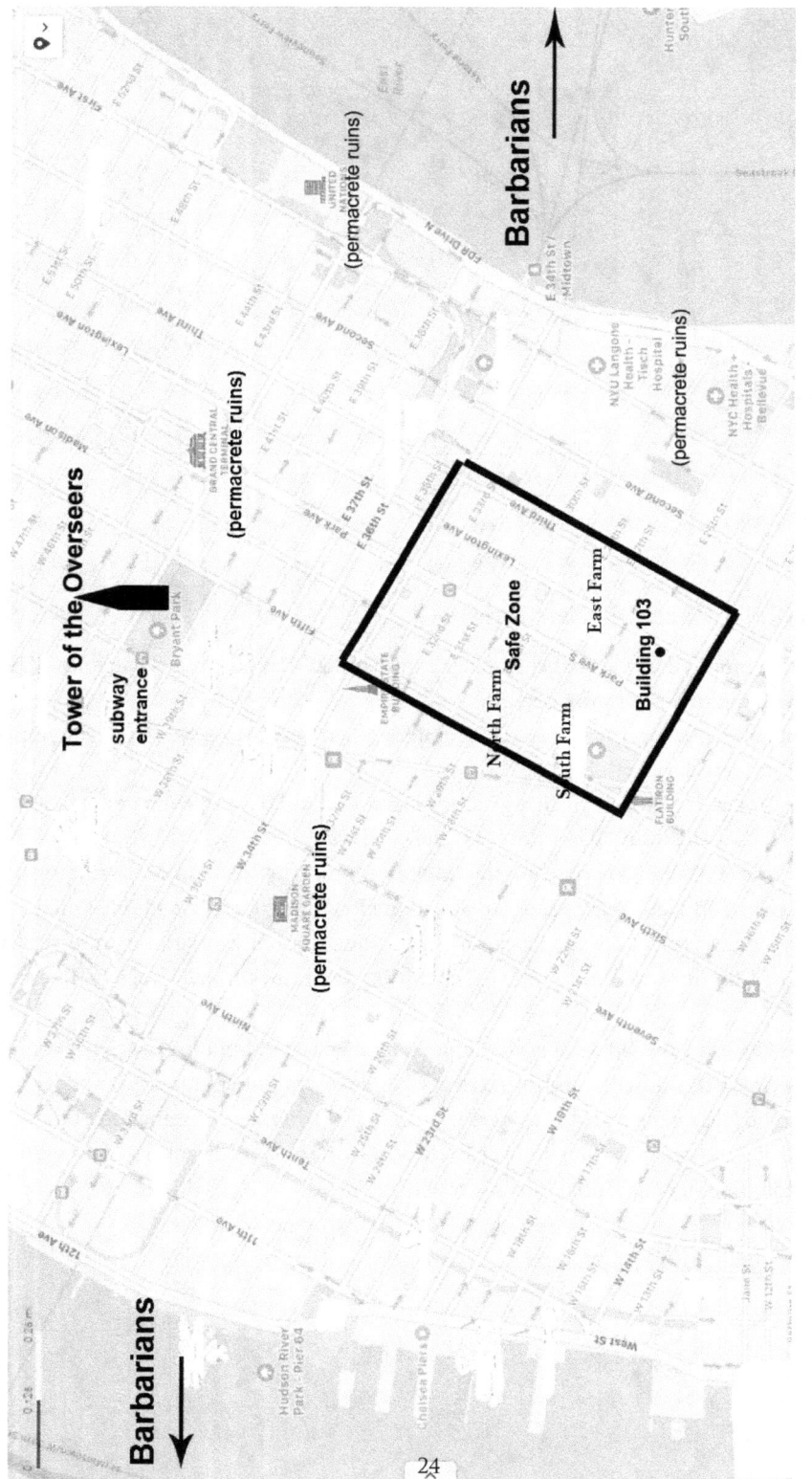

New York, 3013 AD

Old New York was like a museum from the 21st century, not that things had really changed that much in society since the establishment of the Guardians in 2123. Instead of mobiles there were brain implants. Instead of internal combustion aircraft, hovercraft flew overhead. Surface vehicles were forbidden by the Overseers. All travel on the streets was on foot.

Pietra 23 looked around at the shells of 21st century permacrete skyscrapers and marveled. Those were the days of freedom! Before Harry Kaine and his Planetary Referendum that had hopelessly changed the future of earth and humanity. Before human biology had been permanently altered in an attempt to prolong life. There was no sexual attraction in Guardian society. Over the generations since the Referendum, the human body had lost its ability to reproduce biologically.

The earth was slowly dying, and so was the human race. The population of the United States was now just over ten million, only thirty thousand or so left in the entire city. The "rest areas" were just an anachronism now. The Life Reading Index of the human race had slipped below 100, the minimum for race survival.

A few of the buildings in New York were still maintained, but most had been gutted long ago. Only their indestructible permacrete shells remained.

Pietra 23 approached the old clothing factory and went in a rusted shipping door that hung on only one hinge. She walked down a long dark permacrete corridor and approached another metal door, which was opened by a good-looking man about her height, with black hair, light brown skin, and a fine-featured face. "Nice to see you, Alan 133," Pietra 23 acknowledged.

"Hurry, we haven't much time," he said urgently, hustling her down a flight of dirty permacrete stairs and into the command and control area of the rebel clones. The room was dominated by a circular array of gigantic magnets enclosing four metallic rings, and surrounded by several large tanks containing a fine white computational substance. To the left was an operations room, and to the right, a bank of 13 monitoring consoles with computational devices and headsets. There were thirteen people in the room, representing four of the cloned human families.

Alan 133 approached one of the display tanks, his shoulders drooping. "This is our latest simulation for the future. I don't really want to show it to you, but—"

"Stop whining and run it," Pietra 23 said with a sort of angry fatalism. She already suspected what the temporal tank would show.

"In this simulation we showed what will happen if we started a biological enhancement program for our clones. The goal is unassisted biological reproduction." The tank showed a hopeful but short-term leap of progress that died out after the third generation.

Alan 133 looked up with tears in his eyes. "We've run simulations with every possible variant. The result is the death of the human race after twelve more clone generations. It s hopeless. We are somehow trapped in a temporal cul-de-sac that we can't get out of. "

Both of them looked over at the huge Time Shifter device.

"It's that or nothing," Pietra 23 said. "An intervention in time to save humanity."

Pietra 23 and Alan 133 walked over to the Shifter, as it was called. Surrounded by a bank of monitors and large voltage generators, the device was about a dozen feet in diameter, with circular arrays of permanent magnets surrounding a platform, above which was a series of four ten-foot counter-rotating metallic rings coming out of a huge coil. The entire device was surrounded by a molecule-thin transparent substance.

"How does this thing work?" Pietra 23 asked.

Alan 133 shrugged. "That's the province of the Lis," he said, indicating three of the dozen or so technicians seated at their tank consoles.

One of them stood up and offered his hand to Pietra 23. "I'm Li 308, at your service."

Pietra 23 saw a man, just a little over 6 feet, representing one of the Asian family clones.

"Actions speak louder than words," Li 308 said. "We are just about ready to run an experiment, and we need you both to see this."

On the platform was a small cage with a rabbit inside, happily munching on a carrot and some greens. "We are about to transport this little guy, we hope, fifteen minutes into the future."

The concentric rings began to move faster and faster. A vortex of shimmering blue energy, gradually increasing in intensity, surrounded the platform.

The entire operation was completely silent, except for a rustle in the air caused by the thin counter-rotating metal rings. Pietra 23 could hear the crunching of rabbit teeth, the creature inside the field seemingly unconcerned. The concentric rings were moving so fast now they were a blur, until the rings came to a sudden stop. The platform was empty.

"Now what?" Pietra 23 asked.

"We wait," Li 308 said. "Spacetime is programmable, and it is possible to move physically forward and backward in time." He paused for a second. "We hope."

Pietra 23 leaned against one of the empty workstation desks and thought about the contradictions in Guardian society. On the one hand, constant surveillance from spycams and hovercraft were enervating and mind-dulling. On the other hand, nobody cared what you did on your free time, they just wanted to know what it was and when you did it. Humanity's Life Reading Index had been slowly declining over the centuries since the Planetary Referendum in 2098, when humanity whole-heartedly embraced life extension technologies. Almost a thousand years later society had become completely apathetic, leading to a sort of indifferent and listless life resembling ants in a hive.

"Sixty second countdown," a voice intoned. "Fifty-nine, fifty-eight . . ."

If Pietra 23 hadn't looked up a second before, she would have missed it. One instant the platform was empty, the next a rabbit and a rabbit cage had appeared.

A cheer went up around the room.

"Look!" Li 308 said. "The rabbit and the cage seem to be in the same pre-launch condition. It worked!"

"I don't understand," Pietra 23 said.

"No temporal discontinuities," a voice said from one of the workstations. "Smooth sailing."

"What Li 355 means," Li 308 said, "is that the rabbit was transported cleanly from one point in spacetime to another without interference or alteration."

"That's what I said."

"We've had, er, unfortunate consequences in some of our earlier runs," Li 308 explained. "One time the rabbit came back an oldster, another time the cage was just dust."

"Does this mean the system is safe now for human transport?" Pietra 23 asked.

Li 308 eyed Li 355 carefully. "Well..."

"Tell me!" Pietra 23 demanded.

"On one of our test runs we, uh, saw one of the Pietra's back in 2035, interacting with Harry Kaine," Li 57 said hesitantly. "We think it was you."

"You *think* it was me?" she asked.

"We're certain, Pietra 23. Because—"

"Because when we followed the timeline and the temporal signatures," Li 308 continued, "the point of origin was directly from this room."

"So I'm supposed go back in time and meet *the* Harry Kaine?"

"The Founder himself," Alan 133 said. "Director of the Transhumanist Movement, instigator of the Planetary Referendum."

Pietra 23 almost experienced excitement, an unusual emotion in Guardian society. Harry Kaine, the most important historical figure of the past 1,000 years! The man who gave the famous speech at the UN in 2048 that began society's intense interest in what the 21sters called Transhumanism. Creator of the Transformation Centers that had wrecked humanity's future. Everyone in Guardian society knew about Harry, who was idolized by the Overseers in all of the 21st century history tranches.

Pietra 23's brain implant began pulsing, demanding that she return to her Guardian duty station in Section Seven, and download her report. Each day she had to cover her grid outside the Safe Zone, observe carefully, and record all that occurred from her implanted data recorder. Hovercraft and spycams could never record all of the interpersonal, day-to-day interactions. Humans on the ground collected valuable information for the Guardian databanks.

"Time to go?" Alan 133 asked, observing her eyes suddenly open wider. It was the distinctive gesture of communication with the brain implant.

"Yes, I have to report to Commissar Christian 172," she said, "but I don't want to." This one, Pietra 23 noticed, was showing mild interest in her. Very unusual, she thought. She didn't know whether she liked it.

"All right," Alan 133 said. "Our calculations indicate that you will arrive back here in...37 minutes from now. That means you must have left this temporal position before then. We haven't pinpointed the exact time you left."

"What am I supposed to do when I get to wherever I'm going?" Pietra 23 didn't want to go back to the past for an ill-conceived result that might get her killed. Maybe that rabbit got lucky.

"Presumably, contact Harry Kaine. He's the key actor in the destruction of the human race."

"What happens if I don't go?"

Li 308 considered. "If you don't go at the appointed time, you wouldn't come back because you never left. Not leaving would void whatever you did in the past, because future events can affect past events."

"That makes no sense to me."

"That's because we think of time as linear. A straight line, always moving one way: from past to future. It isn't! Time is connected in a closed manifold...well, let's keep it simple and say that time is connected in a circle. So future events can affect past events even without a time machine – the temporal energy goes around the circle in both directions."

Pietra 23 was fascinated. "OK. Take the straight line and connect the ends. That makes a circle, so time – temporal energy, as you say – can flow both ways. Future affects past and past affects the future. But I still don't know whether I should go."

Li 308 grimaced. He was committed to the project but he wasn't the one who had to risk his life. The Shifter had never been tested on a human being. "The Shifter has only been up for a few weeks. All we've been able to do is examine the timeline in a haphazard manner. Theoretically, someone up the timeline from us could alter the past and affect the entire stream of time. But so far, the only action has been you returning in...28 minutes." Li 308 looked up at her. "You may have done something very important back there. Or, you might have done nothing significant. I don't think we should chance it."

Li 355 spoke. "If Harry Kaine can be diverted from his disastrous life extension project the worldline of humanity – or at least our portion of it – should evolve more favorably."

Everyone was silent. Pietra 23 debated silently with herself. "All right, I'll go. What am I supposed to tell Harry Kaine when I meet him?"

Alan 133 looked uncomfortable. "That will be up to you."

"Try to find out why he is so interested in altering the human body," Li 57 suggested.

"That's not much of a plan," Pietra 23 said.

Alan 133 shrugged. "An intervention in time is our last hope. The race is dying, we have to do something."

"Got him," Li 308 said. "I've located Harry Kaine on the timeline in 2035, when he was just beginning to promote life extension technologies."

"Where?"

"At Carleton University in Midland Illinois, just south of Chicago."

Li 308 went to his console. "Our records indicate that you go . . . went . . to Midland on a cold and snowy April day In 2035 the planet was undergoing a period of cold weather. You had better take some clothing appropriate for that period."

Li 57 grabbed Pietra 23's arm and ran her to a 3D printer at the back of the room. "Stand here," she said, and put on a headset. "I've studied this period and know what you need."

In a few minutes Pietra's head was covered by a flexible black cotton cap that went around her ears. A brown winter coat was draped over her shoulders. She wore black pants that were too short and her feet were encased in a pair of flexible athletic shoes. "They actually wore stuff like this back then?" she asked.

"I'm afraid so," the tech said. "2035 was before the days of temperature-regulated films."

"What happens if I have to take this stuff off?" she said, looking at her tight-fitting Guardian uni underneath the clothing. "I wouldn't blend in."

No one spoke. Li 57 regarded her. "That looks fine. You won't look out of place." She paused. "The only thing left is a 2035 currency or credit card."

"Nothing can be done about that now," Alan 133 said, looking at Pietra 23. "The Shifter is ready to go, if you're leaving."

"Three minutes remaining in temporal launch window," Li 355 announced from his console.

She felt a sense of inevitability. "I feel as if I must go."

"Please step onto the platform," Alan 133 said. "We'll bring you back a second after you leave so you can make your Guardian report."

Pietra 23 walked over to the Time Shifter and stepped through the transparency to the platform. The circular metal loops were above her head. She looked up at the array of magnets and felt a sense of real excitement.

"I have entered the spacetime coordinates," Li 308 said, handing her a small device with a chronometer and a digital display. "This device is linked to the Shifter. When you are ready to come back, place it in your brain implant."

Pietra 23 pocketed the device in a pouch sewed onto her coat. "I'm ready," she said.

The circular rings began to move slowly and silently, then faster and faster. Pietra felt nothing and saw only a slight shimmering or undulation of the space around her. The rings were a blur now, and she saw a bluish haze build up around the platform.

Suddenly, she was standing in the freezing cold in front of an early 21st building with a sign that said, "Physics and Astronomy Building."

Now, she thought, how am I to find Harry Kaine?

CHAPTER 4

2035 AD

Pietra 23 looked around her. She had apparently materialized in this time right in front of the entrance to a building, but there was no foot traffic anywhere. The day was very cold. The winter jacket was an irritant over her Guardian thermofilm, but she decided to keep wearing it. Her bare hands were getting cold so she stuffed them into her coat pocket.

She looked around and saw "Carleton University, 1937" engraved on one of the building's cornerstones. So, she thought, I am at an ancient institution of learning. She began to walk toward a cluster of buildings close by, which would probably be the campus center. There she could hopefully find information about Harry Kaine. Her shoes crunched in the snow and her nose began to tingle with the cold. After five minutes or so of walking, people began exploding out of a number of the buildings, walking purposefully toward her. She panicked and ran, attracting a number of stares.

"What's wrong with her?" a voice said.

"She looked at you Jim, that's the problem."

Pietra 23 slowed down, trying to blend in with the crowd. She wondered whether her thoughts, amplified by her brain implant, would communicate to any of these ancient humans. She saw a boy holding a communication device, tapping away, and sent, "Where is Harry Kaine?"

Suddenly his head jerked up and he looked around, saying, "Harry Kaine? Who's Harry Kaine?" He was about to turn toward her and she lowered her head, not wanting anyone to see her face too closely.

She stopped and sat down on a nearby bench, adjacent to a road that led to a more crowded downtown area. For the first time she noticed vehicles moving along the roads.

These were automobiles! With internal combustion engines burning fossil fuels! For the first time she realized how far she was away from home, and felt a rising sense of panic. Her communication device was still in her pocket, reassuring her. She felt a little lightheaded, and used her implant to determine that the oxygen content of the atmosphere was higher here in 2035. She would have to move slowly and control her breathing.

After she calmed down she wondered whether she could receive as well as transmit thought. She slowed her breathing even more, and tried to listen in to the thoughts of the passersby. After a few seconds her head literally exploded with information, most of it jumbled thoughts mixed-in with strong emotions and even sexual desire. Why would these humans be thinking so much about sex?

She got up and walked slowly to an intersection of two roadways. She noticed that a few outliers, instead of waiting for the green traffic signal, chose to walk across the street even though automobiles were approaching from right and left. That would never happen in Guardian society. As the light turned, people shoved off and crossed the street. Pietra 23 saw a number of people entering a building marked with a sign that said "Densinger's Pub." This was the place! A student bar. Li 355 told her when she was on the platform that Harry Kaine came here regularly. Pietra 23 felt a little better. She had found the gathering place by sheer luck. Perhaps this boded well for the success of her mission to the past.

She felt very uncomfortable when she walked into Densinger's. The place smelled of the pungent odor of alcohol, banned in Guardian society. Loud, unmodulated voices assaulted her eardrums. I can't stay much longer in this place, she thought. But she at least wanted to identify Harry Kaine. She found a table next to the front window and hunched down. A waitress came over and said, "Can I help you?"

Again Pietra 23 felt panic overtaking her. Could she eat any of this food? She remembered that she had no currency or credit card. Quickly, she rose and left the table. As she turned her head she saw a female with black hair, muttering to herself, at the corner table, and tuned into the girl's thoughts with her brain implant. Immediately she saw an image of Harry Kaine and another female named Jennifer. The thoughts were tainted with the emotion of sexual attraction and jealousy. This girl was named Kathy Robeson and she liked Harry Kaine, but Harry liked Jennifer better. The emotional content of the girl's thoughts was very strong! Kathy was expecting Harry to arrive momentarily. More good luck.

Just as she was about to wait for Harry Kaine outside in the cold, the girl looked up. "Harry!" she shouted through the front window glass as Kaine approached the entrance. "Over here!"

Just then another girl entered the pub behind Harry Kaine. All heads turned as Kaine walked in; he exchanged greetings with almost everyone there. Pietra 23 saw that Kaine was older than the students; probably in his early thirties. An instructor perhaps? Jennifer and Harry joined Kathy at her table. Pietra 23 heard Harry say, "I'm buying, ladies."

It was the perfect excuse for her to join the other three at Harry's table.

"Well well, who have we here?" the Founder said, looking her over. His gaze and mannerisms were overtly sexual. Pietra 23 almost panicked at the intense emotional contact. The two women stared at her coldly. Jennifer was a brown-eyed blonde. "The more the merrier!" Harry said. "Have I seen you before?"

"I hope not," Pietra 23 replied nervously, wondering if Harry had already been introduced to a Pietra from her future.

The others laughed at this, especially the two women.

As Harry looked her over she studied him as well. "A beautiful redhead," Kaine said, noticing the hair that came out of her cotton cap. She would have to remember not to take the cap off and expose her brain implant! "I love those freckles over your nose."

Kaine spoke as if she were an item on display for his amusement. Pietra 23 was baffled; she didn't know how to respond to this.

Jennifer and Kathy relaxed; clearly this stranger was no threat.

"She looks like she came out of a lab," Jennifer remarked.

"Oh, if you only knew," Pietra 23 said.

Kaine's eyes intensified their focus on her. Had she given herself away? But no, in this society cloning technology was still in its infancy. "I'm Pietra." She almost said "Pietra 23" but managed to stifle her speech. Kaine noticed this. He was very observant; she would have to be more careful.

Pietra 23 sat back and watched the human interactions as Harry and the two girls ordered. Pietra 23 didn't want to drink alcohol so she ordered coffee. Jennifer and Kathy smirked. Harry defused the tension. "Coffee! Why not on such a cold day?"

When they got their drinks Pietra 23 sipped the bitter liquid and observed how the two females vied for Kaine's attention. Kaine enjoyed it, soaking up the admiration of Jennifer and Kathy. As the conversation continued she saw that Kaine was vain and self-centered, but with an almost magnetic charm. A few men stopped by the table and Harry had a word for each of them. Harry Kaine was enormously popular it seemed.

Gradually Pietra 23 relaxed. She knew Harry would eventually become the founder of Guardian society, so she just blurted it out. "I hear you're interested in the electronic or artificial augmentation of the human body." The two women groaned as if they were familiar with the subject, but didn't care for Kaine's interest in it.

Kaine's eyes lit up. "You're familiar with Transhumanism?"

"You might say I'm an expert."

At this Pietra 23 noticed a slight furrowing of Kaine's eyebrows; then it passed and his face cleared. From then on she had Kaine's full attention. Jennifer and Kathy were ignored. After ten minutes both women left the table, scowling at her.

"Don't mind them," Harry said. "They just want to get in bed with me."

Pietra 23 didn't know what to do with that suggestion. She certainly had no intention of having sex with Harry Kaine. That sort of thing had gone out after the cloning program began, even though her body was still capable of the sex act. She had to get out of here before the conversation veered to dangerous subjects.

"Do you have a lab or a workspace you could show me?" Pietra 23 asked. "I'm very interested in how far you've advanced on the subject."

Kaine showed intense, animated interest. "Why don't you come over this evening at 7 to the Life Sciences building, Room 253? It's on the second floor. I'll be talking with a couple of researchers from the university, and a reporter from the local paper will be there."

Pietra 23 managed a smile. "That will be excellent." She stood away from the table, beginning to feel physically ill from the strong emotional content of her interactions with these 21sters.

"Are you all right?" Kaine asked, getting up to put his arm around her waist. The physical contact sent a wave of electricity through her body and she was afraid her coat would part, revealing her Guardian uni. "I'm...I'm OK. I'm just not used to this cold weather."

She could see genuine concern in Kaine's eyes as well as an undercurrent of sexual excitement. It was time to leave. "I'll try to be there right at 7," she said. As Pietra 23 walked out she felt Harry Kaine's eyes on her body.

As she hurried away panic rose up within her. She had never felt this strongly in her entire life! Shaking with unaccustomed emotion, she looked frantically for a place where she could insert her Shifter device into her brain implant without being observed. She entered an alley at the back of the pub, brought out her device, and shoved the lead into her implant. Nothing happened. She took out the lead and carefully re-inserted it, concentrating very hard. But there was no response.

Pietra 23 stood behind a garbage container, feeling miserable. Little piles of snow had piled up next to it, and she almost slipped and fell on an icy patch. She saw the girl Kathy walk by on the sidewalk, looking for her. Pietra crunched down and inched behind the container. The back door to the pub opened; someone came out holding a garbage bag. Pietra 23 walked away. She had to find someplace to disappear without being noticed, for the temporal displacement could occur at any time. What was taking them so long?

After another hour of hiding in the alley and signaling several more times, she knew she was in trouble. The cold was seeping through to her extremities. These 21st-er shoes were worthless! Her feet were freezing. She was hungry and thirsty. She had only taken a sip of the coffee, which was too bitter to drink. She would have to walk to keep as warm as possible.

CHAPTER 5

3013 AD

"Li 308. A signal." It was Li 57.

Li 308, along with Alan 133, quickly walked over to the Shifter's control console.

"Another signal, Li 308. Pietra 23 is in trouble."

Li 308 cursed. "We forgot that it takes an hour to prepare the Shifter for another run! She'll just have to wait."

"The temporal trace shows her coming back in . . . approximately 15 minutes," Li 355 said.

"You'll never get that thing up and running by then," Alan 133 said bluntly. He was angry, an emotion he rarely felt about anything in his mundane life. Boring and dull, that's how you'd describe life in this Guardian society. Perversely, he was almost glad about Pietra's situation. There was a general heightening of emotional intensity in the room. It felt good.

"Perhaps we can," Li 308 said grimly.

Alan 133 raised an eyebrow and Li 308 explained. "Using the Shifter creates a space-time transform, which means that space itself becomes...distorted. We have to wait until it completely relaxes before we try again."

"Do you Lis know what you are doing?"

Li 57 frowned. "We are doing the best we can."

Alan 133 was becoming more animated. They had almost had a confrontation! "Everything looks normal."

"How would you know what was normal?" Li 355 snapped. "We Lis built this device!"

Alan 133 grinned. They *were* having a confrontation! How far could it go? The last recorded act of violence had been several centuries ago. He pointed to the display tank connected to the Shifter console. "Tell me that what you observe here is not space-normal."

"There is a safety delta," Li 355 explained defensively. "But you are correct. We could theoretically pull her out now."

Alan 133 rubbed his hands together. A crisis! He decided to push the emotional envelope further. "There are 54 Pietra's. One less isn't a terrible loss. And besides, you said that your temporal storyline showed her coming back."

Even the subdued Li technicians were feeling it now, Alan 133 observed. He awaited their response, wondering whether their decision would be based on technical necessity, or a more emotional decision based upon the value of a life. He felt that they had reached a crucial decision point.

The Lis gathered together. "Alan 133 is correct," Li 57 said. "The loss of one clone is an acceptable risk."

Li 308 consulted his chronometer. "This debate will have to be short. In five minutes we will have to activate the Shifter or risk a temporal paradox."

"I have a fondness for Pietra 23," Li 355 said. "I say we wait until we're certain."

Alan 133 watched the debate closely. After their uncharacteristic display of emotion, the Lis calmed down. In three minutes they had made a decision, an almost clinical one. Alan 133 was disappointed.

"Alan 133, enter the coordinates," Li 308 announced. "We're bringing Pietra 23 back."

Pietra 23 was walking on a cement path, surrounded by campus buildings. Her feet and face and hands were freezing. People were walking out of one of the buildings toward her. Suddenly she felt a tug on her implant. In front of two dozen people, she vanished.

When Pietra 23 returned to the Shifter in 3013 she felt slightly nauseous, but otherwise unharmed.

Alan 133 looked her over anxiously. "What happened back there? You seem... more alive."

"It was appalling!" Pietra 23 said almost angrily. Then she smiled. "And exciting."

"Do you feel any different now than before you left?"

"I do. Whatever I experienced is wearing off though." Pietra 23 felt that something valuable was slipping away from her.

Alan 133 gestured to Sigmund 78, a clone series that were experts in human biology. "Please do a rigorous life reading scan. Try to determine if there are any biological

THE HISTORY OF THE FUTURE

changes." Of course Pietra 23 would have to be debriefed, but this clone looked...enhanced, more vital.

Pietra 23 was led to a biological scanner, which determined the state of her cellular structure. "Life Reading Index is 113.3," Sigmund 78 announced.

Excited chatter from all of the techs broke out. Li 355, whose own LRI was almost 100, cried out. "100 is the threshold of race viability!"

"We have to keep her here for observation!" Alan 49 cried.

"I'm sorry gentlemen," she said. "I must report immediately to the Guardian station for my grid report. I am late already."

The others nodded grimly. There were no excuses in Guardian society; soon Pietra's implant would be buzzing painfully. Alan 133 put a hand gently on her shoulder. "Please report back as soon as you can for another scan."

Pietra 23 nodded. "Oh, wait. I made an appointment with Harry Kaine at 7 to talk about Transhumanism in the Life Sciences building. That will have to be arranged."

More excited chatter from the assembled clones.

"Gotta go." Pietra 23 shed her 21st century clothing and hurriedly left the building.

After she left, everyone in the Shifter building experienced emotions they had never felt before. "Pietra 23's mere presence has ignited the room somehow," Alan 133 said. Li 355 felt a tinge of...jealousy? It was clear Alan 133 liked Pietra. He did too. Both clones looked at each other, amazed at what they were feeling.

———— • ✹ • ————

When Pietra 23 arrived at the Section Seven Guardian post – a low synthfab building manned by ten monitor clones – she was met by Commissar Christian 172. "You were off the grid for over an hour," he said, indicating the reporting booth, a cylinder approximately 7 feet high and 4 feet wide. Pietra 23 stepped in. Immediately a bright white light surrounded her, flashed once, and disappeared. Christian 172 put on a VR headset with a razor-thin visor and was able to view her day. As she expected, the commissar kept her waiting, looking carefully at her movements. When she entered the safe zone the recording stopped.

"I was investigating some suspicious activity," Pietra 23 said lamely.

"What occurs in the safe zones is not your responsibility. Others have been assigned to monitor that area."

Pietra 23 hung her head in submission, acknowledging the correctness of this statement. "You are docked one credit," said Christian 172 almost gleefully.

"Blast!" Pietra 2543 shouted angrily, startling the monitor clone, who frowned with disapproval. In Guardian society each citizen had 10 credits to spend on essentials per week, and the deduction of even one credit meant that she would have to skip two Syn-thPak meals. Christian 172 turned to his console. "I am going to keep a close eye on you, 23," he said. "You display unaccustomed emotion."

Pietra 23's heart beat wildly. If Christian 172 ordered the monitor clone to perform a Life Reading scan, her enhanced LRI reading would show a jump. An investigation could be launched, potentially compromising the entire Shifter operation! And she might even be deactivated; deemed a threat to society. Her thoughts turned again to the depressing nature of Guardian society. Anyone spirited was immediately suspect, and their conduct suppressed – the very conduct that might lead to the increased viability of the race.

We are doomed, she thought, there's no doubt about it.

Christian 172 saw a look of hopelessness on 23's face. His frown disappeared. "You may go, Pietra 23," he said.

Pietra 23 walked out of the monitor station to her Community apartment (#PT23, of course) and wiped a bead of sweat from her forehead, even though the evening was cool. As she walked underneath one of the hovercraft flying silently overhead, she still felt an inner thrill from her trip to the early 21st. On an impulse she turned back to the safe zone. Her shift was over and she could do whatever she pleased, but had to avoid being spotted by night shift contact tracing personnel as she made her way to the secret back entrance of the Shifter building. Although no spycams or hovercraft were allowed in the Safe Zones, human observation of suspicious activity still occurred.

Alan 133 greeted her with as much enthusiasm as was possible in a cloned society. Without speaking they both walked to a biological scanner. "107.5!" Sigmund 78 announced in a tone of disbelief.

She locked eyes with Alan 133, both expressing pleased surprise. His LRI was 99.3 and hers had been 99.7, both on the very high end; which was why they had both been recruited for the Shifter project.

"There seems to be a degree of permanence," Alan 133 said. "When is the Shifter available?"

All of the techs turned around. Li 355 said, "I wish to go next. We must collect more data on Harry Kaine before Pietra 23's meeting with him."

"As director, it is my decision to go," Alan 133 replied with a little heat. Pietra 23 felt a very slight sexual component emanating from the two men, and just a hint of jealousy. But nothing compared to the 21sters.

"No one but Pietra 23 should go to the 21st," Li 308 objected. "Temporal paradoxes must be avoided. As the 21sters say, 'Too many cooks spoil the broth.'"

"I agree," Li 57 said.

Despite their disagreement, Alan 133 and Li 355 both thought this was sensible.

"Pietra 23 must return to the 21st and her meeting with Kaine," Alan 133 ordered. He turned to Pietra 23. "It has been a long day for you, Pietra 23. Return home, eat, get some rest, report back tomorrow evening after your Guardian shift."

Pietra 23 nodded and left the building, understanding that the Shifter could place her at any point in the past, regardless of how much time went by in 3013. Hopefully she would have no problem making the meeting at 7 with Kaine.

Everyone went home. Alan 133 walked through the facility, ensuring everything was locked down and ready for tomorrow's work. While he programmed the Shifter for Pietra's re-insertion, he made a decision. After watching Pietra 23 in 2035, Alan 133 wasn't sure that they could change the timeline by diverting Harry Kaine. Alan 133's models always showed that time was flexible, but that once something major occurred, it was impossible to alter it by going back in time. Small things might change, but not the big stuff. 21sters called this the Mandela Effect. Even though he hoped for success, he wouldn't be disappointed if the Kaine Experiment turned out to be a failure.

Alan 133 decided to take a different line.

During the 21st century, the average LRI of a healthy person was about 170. He was determined to discover whether Pietra 23's immersion in the past could cause a permanent increase in her Life Reading Index. If so, how much exposure was needed? Could being around others with enhanced LRIs salvage the human race here in 3013? Perhaps these cloned bodies could regain their sexual function. Then the Guardian brain implants would not be necessary. Humanity could be free.

It was a long shot for sure. Yet for the first time in his life, Alan 133 experienced the emotion of hope.

The next day Pietra 23 worked her Guardian shift (which consisted of human monitoring of societal interactions) and reported to Christian 172. She had been careful to do nothing to arouse suspicion. She had missed a meal and was feeling tired, which seemed to please the Commissar. Then she carefully made her way to the warehouse that contained the Shifter. There were two entrances to the building. She entered through the hidden one. This was an ancient subway maintenance door on what used to be 24th street, which was just inside the perimeter of the Safe Zone. The streets of Old New York were now covered with plant growth, but the surface underneath was still visible because foot traffic had scuffed everything down to the permacrete road. The subway maintenance door was invisible, surrounded by a thick covering of scraggly bushes. She took a different route every day, careful not to leave a path that anyone would notice.

Pietra 23 made her way down the still usable permacrete stairs and turned right into a musty, filthy subway tunnel. A pale light came through several small openings in the ceiling, enough to light her way. She had traveled this path so many times she could do it in the dark. Fortunately the Overseers only haphazardly monitored these ancient tunnels, because no clone wanted to go down here. The tunnels had long ago been looted in previous centuries, and Guardian society discouraged curiosity.

Pietra 23 picked her way around broken permacrete and centuries-old remnants of an ancient subway train. After a couple minutes of careful walking she spotted a small opening to her right, about four feet wide. She walked up another flight of permacrete stairs to a heavy synthfab door. Pietra 23 tapped a code on the door, which opened instantly. Here was the Shifter area. This building, #103, was listed in the Guardian databanks as a now unused clothing storage facility. The Shifter team had secretly converted a room for their use in the Old Sector. Over a period of years they had stolen the equipment necessary to build the Shifter.

Within the hour she was going to meet Harry Kaine back in the 21st, along with some of his 21ster scientists. Last night she had stayed up late, accessing the history grids. It was all Guardian propaganda, but she was able to understand much more about what the 21sters called Transhumanism, why it became so popular (mostly because of the influence of Harry Kaine), and learn some technical jargon. She also studied 21st century customs and language.

Sigmund 78 measured Pietra 23's LRI as Alan 133 and Li 355 watched anxiously. "106.9, lower than yesterday's 107.5, " Sigmund 78 announced. "But still over 7 points higher than this clone's baseline."

Alan 133 and Li 355's eyes met. Again, a flicker of hope!

Pietra 23 removed her uni and changed into her 21st century clothing. Both Li 355 and Alan 133 regarded Pietra 23's cloned body with interest as she stripped down.

Was it because of this clone's higher LRI? Alan 133 wondered. Clones were still printed/grown as male and female, but talk was that the following clone generations would be androgynous.

"I do not wish to be cold again," Pietra 23 said as she undressed. She put on fur-lined boots and heavy mittens to keep her extremities from freezing.

"You must arrive in exactly the same clothing as when you left," Alan 133 said. "You weren't wearing those mittens, and those boots are different from the shoes you wore last time. Dozens of people saw you vanish."

Pietra 23 went to the back and retrieved a 21st century clothing container. She put in a sweater, another pair of shoes, a couple of blouses that had been printed for her, and her Guardian thermofilm uni. "If I get stuck in the past again I want to be comfortable," she explained.

"You were not holding a clothing bag when you left," Li 355 pointed out.

"Too bad," Pietra 23 replied, using a 21st Americanism. She felt emotionally and energetically stronger than these clones. She liked the feeling. "I bring a bag or I don't go."

Both of the male clones shrugged; they didn't have the energy to argue. Alan 133 checked the Shifter's temporal coordinates while the Lis monitored its operation. "We will try to insert you within 1 second of your previous temporal position."

Pietra 23's implant was checked and she stepped onto the platform. The Shifter was activated.

CHAPTER **6**

2035 AD

Pietra 23 felt a blast of cold air strike her face. Her feet crunched on hard snow that covered the pavement.

"Wow! Did you see that?" a woman in the crowd said.

"A glitch in the Matrix!" a man cried.

"Are you a holographic projection?" a young man asked, walking up to her. He touched her. "You're real enough. But you flickered."

"And you weren't holding that bag, and now you are!" the woman said.

Pietra 23 froze. She had to think of something! A thought occurred to her; nonsense words she had heard from a 21st history program about a famous old movie. She said, "Klatu Barada Nikto."

She had no idea what it meant, but everyone laughed. The tension was broken, so she walked away. Then she realized she didn't know where the Life Sciences building was. She turned to ask someone but they were talking among themselves and she didn't want to draw more attention to herself. A woman was interviewing those who had seen her immersion in time. Pietra 23 panicked. Would the authorities come to arrest her?

As she walked she calmed down; there were no Guardians here in 2035. She had several hours to kill before 7 o'clock because she had re-entered 2035 right after she left, and that was in the early afternoon. Where would she go? She had forgotten to ask for 21st cash, or a currency card. Fortunately she had eaten a synthpak in Building 103's little cafeteria before she stepped onto the Shifter platform.

Pietra 23 walked into one of the campus buildings and felt warmer. There was a public area with benches so she sat down. She was tired from the previous day and a night of study, and fell asleep. When she woke it was dark. What was the time? A chronometer on the wall showed 6:55. She would be late, and she still didn't know the location of the Life Sciences building.

For some reason she felt anxiety, another foreign emotion. She couldn't think straight. Her eyes searched the building and found a map hanging on the wall by the front door. There it was! The Life Sciences building was halfway across campus. She would have to walk.

Pietra 23 stumbled into Room 235, half an hour late. Harry Kaine was there, along with three others. How was she to divert Kaine from his disastrous obsession with biological enhancement? The first step was to get to know him better, but she wasn't sure she even liked him...

Part II

CHAPTER 7

Winnie Sauvage had to pick up a young boy who had run away from abusive foster parents after he was taken away from his mother by Child Protective Services. It was her job to take him a safe house run by COSA, the Children Of Sexual Abusers organization. As she walked down the sidewalk with the child she saw a very tall man with fuzzy hair coming toward them.

"What are you staring at?" Winnie asked. The man's mouth was open, he was looking slack-jawed at her like she was the double-jointed woman at a circus.

"Uh, I've never seen anyone like you. My name is Germaine. My friends call me Ger."

"I'm not your friend. Are you a pervert?"

Ger saw a beautiful woman wearing black boots, a black leather jacket, and black jeans. Her hair was jet black. As he looked into bright blue eyes he saw great wisdom there, and an almost overwhelming sadness. Normally nervous around women, he felt safe around this one. "You're beautiful. Is that your son?"

"Get lost, retard."

As Winnie walked away she saw Germaine's eyes on her. He never looked at the child so he probably wasn't a pedo. He looked harmless so he couldn't be someone sent by the Bosses. She was lonely and turned around. "555-555-2231," she said, and walked away.

Winnie got a call that evening from Germaine.

"Are you free to have coffee?"

She liked the sound of his voice. "OK. There's a little cafe over on Third and Mason."

"I know it."

"I'll be there in half an hour."

Winnie went into their bedroom and saw Kesha, her best friend and roommate. "I'm going out for coffee with a normie I never met before."

Kesha looked at Winnie sharply but said nothing. She played with her corn rolls, always a sign of nervousness. Winnie hoped Kesha wouldn't say anything because she was seeing Joe Courvall, another normie.

"He looks like a nerd. Harmless. Early twenties. Gorgeous."

"Say nothing," Kesha said firmly. "Be careful."

When Winnie got to the cafe Germaine was already there, seated in the back against the wall. That was good. She didn't like to sit by windows or at the front of any public place. Ger's eyes were on her as she walked back and sat across from him at the little table. Winnie checked him out carefully. His vibe was good. She could sense the sexual interest but it wasn't creepy. He seemed relaxed in her presence.

Winnie thought this was very unusual. Normie men always think I'm too strange. If they knew what I've been through they'd run away screaming.

"Thanks for coming," Germaine said.

The waitress arrived and they both ordered espressos.

Germaine must have sensed the walls she put up because she saw the impulse for small talk disappear.

"I don't know why I wanted to see you," he said. "You're beautiful and...unusual."

Winnie broke into a laugh. Unusual! That was an understatement. She kept laughing, feeling a burst of emotion leaving her body, almost like an electrical charge blowing off.

"I'm sorry," Germaine said stiffly. "I know I'm goofy looking and weird but I thought—" He moved to get up.

Winnie put up an arm and blocked the way, still shaking with laughter. "It's not you...please stay...I'm having some kind of release. My counselor calls it a line charge – release of pent up emotion."

Germaine's eyes widened but he sat down. Winnie got control of herself again. "You seem to have a positive influence on me."

Germaine smiled. "I'm a high-functioning autist. I know I have strange effects on people. My friend Joe says I'm a catalyst."

"I'm glad you don't think I'm too weird."

"I didn't say that! It's just...you're my kind of weird."

Uh oh. "Don't tell me. You're a perv who likes older women, right?"

Now it was Germaine's turn to laugh. "I don't do perv. I like older women, yes, but there's something else about you I really like."

Oh my God, Winnie thought. I know I'm physically attractive but the men who go for me are usually losers. Was Germaine a loser? "What do you do for a living?"

"I work as a consultant in the physics department at Carleton, and for various venture capitalists interested in scientific projects and new inventions."

Winnie stared. Not a loser! "But...what do you want with me?"

Germaine seemed amused by this. "When I interact with people they're thinking, in the back of their minds, that I'm not normal. You're the same as me."

That's right, he gets it. Winnie felt tears coming to her eyes. "If you knew my past you might not be so into me."

"Tell me."

She shouldn't. She knew she shouldn't say *anything* to a normie. But she was very lonely and this guy felt safe. Winnie gulped. This would be the first time she told her story to anyone except one of the network psychologists, or a safe house counselor. A part of it anyway.

"It's OK," Germaine said

Winnie saw intense, genuine interest in Germaine's eyes. She had never experienced this before from any man, normie or perv.

"During the day I work at COSA, the Children of Sexual Abusers," Winnie began. "In the evenings I'm a part-time counselor in our survivor network. We all counsel each other. So far I haven't gone crazy."

Germaine nodded. Winnie took a deep breath and held it for a moment. She plunged ahead, speaking softly so that no one else in the cafe could hear except Germaine. She looked down at the table as she spoke.

"My name is Winnie Sauvage. When I was thirteen I was raped by my father, who never understood why I won't forgive him."

She looked up and saw Germaine nod silently in acknowledgment. She felt encouraged to go ahead.

"OK Germaine, I'll just come out and say it. There are people who traffic in human beings, and they do it for money. They aren't even human. The users – the people who look at child porn and participate in child sexual abuse in the pedophile networks – are desperately sick individuals."

It was all coming out now, she couldn't stop it. Speaking this way in public to a stranger like Germaine was way more cathartic than telling a network psychologist in a safe room.

"Child trafficking is big business, Germaine. I know this because...because when I was 14 I ran away from home and got picked up by some guys who saw me on the street. These people...procure children, young adults, and even babies, for pedos. I've met some of them, and have been forced to sleep with a few of them. The worst are the private 'children's foundations' and NGOs that are actually covers for child trafficking."

Germaine's eyes widened. This woman's story had the ring of truth. "You were a sex slave?"

Winnie nodded. "At the very highest level. That's why I know so much about it."

"OK. Keep going if you want."

"I want. For some reason I found favor with the pervs who captured me. I got passed around from client to client, up their perverted social ladder. One time I even slept with the head of the CIA."

This is where the guy gets up and leaves, Winnie thought. Germaine just sat there, listening. She continued.

"I found out some horrible shit when I was forced to sleep with these monsters. The child trafficking operations are surrounded by fanatical secrecy, Germaine. The Bosses who run the children have set up rigorously monitored private networks. The networks are run by rogue state actors, including former intelligence operatives and mercenaries, who do the wet work on whistleblowers. For moving large cargoes, these rogue elements provide ships, vehicles, trains, and aircraft for all the 'fruit.' That's what they call the children who are trafficked. If you knew what I know you would literally be in a mental hospital and under sedation."

Somebody dropped a glass of water on the floor behind them and they heard a soft explosion. Germaine didn't react, which pleased her. Somebody came over with a mop and a broom and began sweeping up broken glass. He was all attention.

"I was lucky because I found favor. What happens to some of these kids who don't is...unimaginable."[5]

Germaine's face was white now and he looked sick to his stomach.

"You think this is a conspiracy theory?" she said angrily. "Or a fantasy? I lived it."

"OK."

Fuck this normie, Winnie thought. It was all coming out now and she had to release it. "I gotta tell you this. I remember being sent to the villa of a wealthy European corporate executive, who I was forced to sleep with. On this day a 'hunt' was organized. A dozen children were released on the grounds, which were extensive. The older children were almost maddened with fear, for they suspected what was coming. The 'guests' assured the kids that the hunt was just a little 'fun,' and not to worry. I was forced to watch the proceedings. As the children were released some of the older ones tried to help the smaller ones find hiding places. After ten minutes or so the 'guests' (the hunters) tried to locate the children. I heard terrified, agonizing screams coming from the grounds. After an hour or so the guests returned, some of them with blood on their clothing...."

Germaine looked ill now. I was starting to feel my emotions shut down, but I was glad Germaine didn't say anything stupid like, "Why didn't you leave?" I had been watched closer than an evangelist at a revival meeting by my handlers.

"There's more but I can't tell you or the compartment of my mind where I store these memories – in an attempt to retain my sanity – will open up. Imagine having to see this. Imagine having to experience it. No, don't. Think of something positive.

"I was now eighteen, and starting to get used up. That's what happens to girls and boys who are trafficked. We 'lose our shine,' I heard one of my 'clients' say at a posh party I had to attend as a consort to one of these so-called VIPs. But one day a miracle occurred.

"I was at a big house in northern Virginia, the home of a prominent politician whose name I can't say out loud. A ruckus occurred outside. My handlers rushed in to see if the 'shipment' had been damaged, leaving me alone for the first time in five years. Apparently they trusted me. I found my way out of the house and got into the shrubbery. I saw my chance and ran out to the street in back of the house. I had some money on me – I actually received 'gratuities' from some of my 'clients' – and found my way to Chicago, where I joined COSA. Eventually I wound up here in Midland. Since then I have lived in safe houses with others who share my experiences. The memories keep returning so I write them down when they come up. I see a counselor almost every day. We have our own networks, but we have to remain hidden. As long as we say nothing publicly the Bosses leave us alone. If we talk, we know we'll end up like Isaac Kappy."

Germaine's dark face was blanched now. "You got out."

"Yeah."

We finished drinking our espressos, not talking. Winnie felt better, and wondered whether this guy believed her.

"Thanks for seeing me," Germaine said, and stood up.

The guy was tall! At least six inches over six feet. He had a wide, expressive face with dark brown curly hair piled on top of his head.

"Can I call you again?" Germaine asked her.

"Seriously? Do you really want to after what I told you?"

"Yeah. I'm an inventor, Winnie. I've lived a pretty sheltered life. You're telling me shit I never knew existed."

"Do you think I'm disgusting?" Winnie said anxiously. She hadn't told him much about the pervs she had been forced to sleep with. High level guys who ran big corporations, state-level pervs who ran major government agencies...but that was classified info that she didn't even want to think about.

"If even half of what you told me is true...then you're a warrior. A hero. To survive all that? It's incredible."

Winnie looked up at Germaine's face. It was an honest face, a no-bullshit face. "OK. You can call me, but I'm not sure about having a relationship with you. You see, we're not supposed to get together with normies...I mean, people outside our networks."

"Let's try it and see what happens, if you want to."

Winnie felt a rush of relief. She felt accepted. "Call me after I get off my shift tomorrow evening. At seven."

"I will."

Germaine left the cafe, walking purposefully. It was clear he had something to do and she was forgotten. It bugged her a little that she could be dismissed so easily, but she liked that he hadn't rejected her.

When Winnie got back to the house her roommate Kesha was waiting. "Well? You look relieved."

"Yeah. I told Germaine something of my story."

Kesha's face showed alarm.

"Relax. I didn't tell him about the babies in cages or the woodpecker grids, and I didn't mention any names. Nothing about the Bosses or anything classified. Nothing you can't find in open source materials."

Kesha relaxed. "You look better."

"For a normie he's pretty cool. A man who can shut up and listen – I didn't think they existed."

Both women laughed at this.

CHAPTER **8**

Despite Winnie's checkered past, Ger was almost as fascinated with her as he was with his time project. He knew she was still withholding something from him, but he didn't want to pry. During their conversation he had heard her mention "the Bosses." Ger knew just who to ask about that.

The next day Ger had an appointment downtown with a software tech at Berglin Enterprises, a company that sold security software to well-connected players in the national security establishment. As he was leaving the meeting he saw Ralph Zimring, Max Berglin's seven-foot-tall security chief, striding down the hallway. "Hey shorty, what you up to?" Zimring joked as he walked by.

"The Bosses," Ger said softly.

Zimring turned around, angry.

Ger was afraid. He had said it to pique the big man's curiosity, but Ralph didn't seem amused. Zimring was a giant who walked like a ballet dancer. The stories about his military prowess were legendary at Berglin Enterprises. Ger knew he had killed guys in his military service and as a mercenary.

"Don't ever repeat that word in public," Ralph said, getting in his face. "Who did you hear it from?"

"A couple of women I know."

Something clicked in Ralph's brain. "Winnie Sauvage and Kesha Montgomery, is that right? They're at that safe house on Warrington on the North side."

Ger was astonished. "How do you know that?"

"Don't sweat it kid. I'm paid to know what goes on in this town."

"Yeah, but what's it about?"

"Child trafficking and pedophilia. Keep your mouth shut about Bosses, Germaine, or those two girls aren't long for this world. And neither will you be."

Ralph walked away, swearing. He knew about the pedo rings and the trafficking. Almost twenty years ago when he was still a merc he had to track down and kill Gordze Khachidze, a viscous psychopath who ran a human trafficking ring for the Solntsevskaya Bratva. That brutal, bitter winter in Chechnya was something he'd never forget. The kid better be careful.

As Ger walked out of the building he was in shock. Being confronted by Zimring was scary, but he was more worried about Winnie. What was she into?

When he got home he couldn't concentrate. Normally, all he thought about was work and his time project, but this woman was getting on his brain. He sat down on his couch upstairs and thought. Should he stay away from Winnie even though he really liked her?

Nah. She was too fascinating.

---------- • ✹ • ----------

The next day at 7 in the evening I got a call from Germaine. "I want you to meet someone." Ger started a group conversation.

I saw two women sitting at an old-fashioned wooden desk. One was an intense blue-eyed woman with totally black hair, the other had beautiful corn rows and delicate brown skin.

"This is Winnie Sauvage. You know Kesha Montgomery," Ger said.

Kesha was smiling.

"Who is this other guy?" Winnie said. She looked irritated.

"Sorry Winnie," Kesha said. "This is my friend Joe."

Winnie turned to Kesha. "This isn't good. Too many normies."

"I'm not a normie," I said.

Winnie looked at me with intense blue eyes. "Have you ever been involved in a special program when you were younger? Run by the military for gifted children?"

"No. I don't know what a special program is."

She nodded. "OK, just checking. You have the look of someone who has been in the programs."

She was awfully mysterious, but I noticed that Kesha was still smiling at me. I smiled back.

Kesha turned to Winnie. "Your friend Germaine looks all right."

"Do you want to have coffee tonight?" Ger asked Winnie.

"Sure."

I was becoming impatient. "Did you want to show me something Ger?"

"Yeah. I wanted you to meet Winnie."

I could see why Ger was fascinated. She was an older woman, mid-thirties, and had a hard edge about her. Her jet black hair, black leather jacket and black slacks gave her the look of a woman who had seen and done things.

Ger turned to Winnie. "I checked you guys out, both of you."

Kesha bristled. "With who?"

Ger grinned. "Max Berglin's seven-foot pal, Ralph Zimring."

Kesha groaned and put her hand on her head. She looked over at Winnie. "I told you. Never get involved with normies."

"You did," Winnie said, pointing at me.

"Joe is harmless. I'm not so sure about Germaine."

"Who are Berglin and Zimring?" Winnie asked.

"Max Berglin owns Berglin Enterprises," I said. "He worked at the Lockheed Skunk Works. His buddy Ralph Zimring is a mercenary who could kill all of us here in five seconds or less."

Winnie snorted. She said something in a whisper to Kesha but I caught it. "The Bosses would destroy him."

"The Bosses?" I repeated.

Winnie blanched and Kesha backed off, a fearful look in her eyes. "See what you've done now," Kesha said to Winnie.

The connection was broken. Ger was angry. "Why did you say that?"

"What? What did I do?"

Ger told me about his conversation with Zimring, and something about Winnie. "Those two girls, Kesha and Winnie, have both been sexually abused. Winnie is a victim of human trafficking. That house they live in is a safe house."

"Human trafficking? You mean child porn and child abuse?"

"Yeah. By the way, don't ever mention the Bosses again, or Zimring says those two girls will be in big trouble. They both work for COSA."

I thought quickly. If Zimring said it, it was probably true. "OK."

"I saw Winnie with a child yesterday," Ger said. "She was helping him."

I remembered something. "I saw an article in the paper about that. COSA is trying to get the Midland Police to investigate Child Protective Services. They say there's abuse there."

Ger shrugged. "I don't read the paper, except for Crain's Business Report and peer-reviewed journals." He looked at me hopefully. "Do you think we should stay away from those two? Winnie is very...unusual, to say the least. "

"My intuition tells me they're OK. But we have to be careful."

"Thanks Joe. I really like Winnie."

I liked Kesha too, but I couldn't make up my mind between her and Karen.

CHAPTER **9**

Pietra 23 was a little out of breath when she arrived half an hour late to Room 253 of the Life Sciences building. She remembered to keep her black cap on her head to cover her brain implant. A discussion was already in progress, but Harry Kaine greeted her enthusiastically.

"...Ladies and gentlemen this is Pietra, who shares our enthusiasm for transhuman technologies." Pietra 23 took a seat in the small conference room. Three men and two women were present.

"Pietra, this is Dr. Winthrop Sauvage, who works in the field of cryogenics. To my right is Dr. Mary Chen, an expert in gene therapy. Sitting across from her is Dr. Tom Mankiewitz, a computer engineer and an authority on MNT, or machine nanotechnolgy. To Mary's right is Dr. Armand Singh, who works in artificial machine intelligence." Kaine indicated a blonde woman sitting by herself at the end of the conference table. "This is Karen Everard, a reporter for the *Midland Chronicle.*"

Winthrop Sauvage, an older man with graying hair and beard, nodded to her. "As I was saying, the ultimate goal of Transhumanism is physical immortality. If we are to realize that possibility we must see the human being purely as a physical system composed of working parts that cooperate to make up the whole, some of which have the tendency to get old and break down."

Harry Kaine nodded his approval of this.

"I'm fascinated by the idea of mind uploading," Dr Singh said. He was a tall, bulky man who wore a turban on his head.

"The uploading of a human mind to a non-biological container," Kaine explained to Pietra.

"Mind uploading," Singh continued, "is based on the idea that cognitive processing can be implemented on substrates other than neurons."

Pietra 23 saw Kaine's eyes light up at this.

"We've had decades of successful results in neurophysiology," Singh explained. "The blue brain simulation project has morphed into an actual, working, artificial brain. It appears that our minds are defined more by the information pattern they embody than the particular hardware they are on."

"The biological brain, while good for its time, is becoming outdated," Kaine remarked.

Pietra 23 saw the reporter jump at this, a look of shock on her face. It was clear that Harry was the leader in this discussion. His personality dominated the room.

"Just think!" Singh said. "An actual mind transfer to a brain that does not age. And whose components can be replaced when they wear out."

"That's still a ways off yet," Winthrop Sauvage said. "But we're getting closer."

Kaine turned to Mankiewitz. "Dr. Mankiewitz's specialty is utility fogs. Tom, would you describe utility fogs and their potential application to Transhumanism."

Mankiewitz was a thin man with a head of thick white hair. He had an aristocratic face with high cheekbones and wore round, wire-rimmed glasses. Pietra 23 thought he looked like the archetype for a 21st century intellectual.

"A utility fog is a collection of tiny nano-robots that can replicate a physical structure. The exciting thing is that molecular nanotech might replace the need for biological bodies altogether."

Harry Kaine placed his hands on his torso. "By God! Replacing these meat bodies, which age rapidly and get sick so often, would be a godsend to humanity."

"Indeed!" Mankiewitz replied. "Utility fogs are also a useful peripheral that can perform physical engineering and maintenance tasks. For example, a utility fog could be manufactured en masse to occupy the entire atmosphere of a planet and replace any physical instrumentality necessary to human life. Buildings could be constructed and dismantled within moments, enabling the replacement of existing cities and roads with farms and gardens. Utility fogs are the machines of the future."

"Connecting utility fogs to a brain implant would enable an interface between a transformed human and a programmable fog," Singh offered. "The possibilities are exciting."

Pietra 23 lurched and almost fell out of her chair. This is the very thing that must be avoided!

"Is there something wrong Pietra?" Harry Kaine asked.

"Uh...no...I'm sorry. I'm just shocked at the possibility. Using fogs, an entire society could be controlled by just a few people." She was thinking of Guardian society.

Kaine became excited again. "Of course Pietra. Eliminate petty dictators and anti-social personalities from public life. A guided society mentored by the most enlightened minds!"

"Let's not get ahead of ourselves," Mary Chen said. "We can begin to upgrade the human body right now via gene therapy – replacing bad genes with good genes. RNA interference can selectively knock out gene expression. Together, they give us the ability to manipulate our own genetic code. Supercomputers have become much more powerful and can be used to simulate gene changes in extreme detail before we attempt them with actual human beings. This will make unwanted side effects quite unlikely for the typical case."

Kaine snorted. "Much to the dismay of trash authors and films that promote a totally distorted view of genetic engineering."

"But what about the atypical case?" the reporter asked.

Kaine's head shot forward. "Are you with the program or not?"

Karen Everard sat back in her seat. "I didn't know I was here to write a puff piece, Harry. I'm looking at both sides."

Pietra 23 saw Harry Kaine frown. This was a man who didn't like anyone questioning his opinions. Just like Christian 172, her Guardian handler, and every other official in Guardian society she'd ever met.

"How does gene therapy relate to cloning?" Pietra 23 asked. She wanted to know how far Harry Kaine and these 21sters had gotten along the path to cloning human beings. Kaine smiled and his angry look disappeared. He truly was a charmer, she thought.

"Now that's a good question! Mary, do you have any thoughts?"

"Cloning techniques have become more sophisticated over the past decade. The physical mapping of genes is the first step, then gene testing to identify the presence of certain genes with a person's DNA. This can be used to determine if a person has the genes that cause genetic disorders. Gene splicing involves cutting out an impaired section of the DNA in a gene and adding new DNA in its place. Gene silencing turns off defective genes. Both of these techniques can be used to build a much more healthy human body, and prolong life. DNA cloning is the preferred method. Even individual genes can be cloned in this manner."

"Wow!" Everard said. "That seems to be something worth doing."

Kaine smiled benevolently at the reporter.

So this was how Guardian society started, Pietra 23 thought. Fools who thought that every advance in science must have a benevolent outcome. Pietra 23 noticed a sexual component in the exchange between Harry and the reporter, just as in Harry's responses to the women at Densinger's bar. She also noticed that Mary Chen was frowning. The two women were competing for Kaine's attention.

"I'm very excited," Singh said, ignoring the social by-play. "But cloning may only be the beginning. In a few decades we might even be able to 3D print perfect human bodies."

At this Kaine jumped out of his chair, and beamed at Dr. Singh. "Remarkable. Excellent, Armand!"

Pietra 23 noticed how Armand Singh basked in the attention and support from Harry Kaine. Kaine seemed to be able to emotionally charge up another person just by the force of his personality. Male or female, it didn't make any difference. She tried to keep her face from expressing the shock she felt, knowing what had happened with the cloning/printing program that began during the middle of the 22nd century. She was a product of it! The cloned bodies *were* beautiful and free of disease. But eventually their biological vitality had lessened to the point where humans could no longer reproduce, leading to an apathetic, dead society.

She was about to express these concerns to Singh, but the energy in the group was high and everyone was excited, carrying the conversation on a wave of emotional momentum. Pietra 23 felt every cell in her body begin to tingle. It was an amazing sensation. She could literally feel her life force being enhanced, just being part of this group of 21sters.

Before she could say anything Mankiewitz began to speak animatedly "Utility fogs and robotic self-replication will enable space colonization. The embrace of Transhumanism will be necessary to colonize space."

"That's right!" Mary Chen said. "As the population continues to grow, humanity will need to find other planets to colonize, and/or build structures in space. But human beings aren't designed to live in space. Our physiological issues with it are obvious. Even a few weeks in space can lead to deteriorating muscle mass. On the surface of Venus we would melt. On the surface of Mars, we'd freeze. The only reasonable solution is to upgrade our bodies."

Mary Chen paused. "We don't need to terraform the cosmos, but cosmosform ourselves."

At this Kaine became intensely animated and hugged Mary Chen. Pietra 23 could see the almost orgasmic expression on the woman's face.

"That's right Mary! Enhanced, transformed humans can colonize the universe. Using molecular nanotechnology combined with robotic self-replication, we can create megascale habitats for humans anywhere in space."

Pietra 23 was confused by this idea. Nothing like that existed in 3013. Humanity was confined to the earth. But the energy in the group was so high now that she began to feel uncomfortable. In Guardian society, gatherings of humans were carefully monitored. You learned quickly to contain whatever emotions you still had.

The reporter spoke up. "What if space is already occupied?"

Harry Kaine looked shocked, as did everyone in the room.

"Consider it! If intelligent life has evolved on other planets, space may be already filled up."

Kaine dismissed this with a wave of his hand. "That is absurd."

Pietra 23 noticed again that Harry Kaine didn't want to hear opinions that disagreed with his own.

"We're forgetting something," Dr. Singh said, changing the subject. "Artificial General Intelligence. AGI is the intelligence of a machine that can understand or learn any intellectual task that a human being can. It has almost been achieved. During the last 15 years we have made enormous progress. The next step is Artificial Super Intelligence, an entirely new form of life with intelligence far beyond even an enhanced or transformed human. ASI would render humanity obsolete, for its consciousness would be able to handle information on a universal scale. It's the ultimate Transhuman evolution.[6] Humanity itself may be a mere stepping stone toward an entirely new form of life."[7]

Kaine frowned at this. "Perhaps, Armand. But ASI applies to the more distant future. We have to first take the obvious steps of transforming humans. The current condition of human civilization is appalling. Poverty, war, disease, and pollution are destroying this planet and the human race. A radical new approach is needed. Transhumanism is the answer, I think we can all agree on that."

"Hear hear!" Mankiewitz shouted. All were in agreement.

Pietra 23 could see that from the perspective of the 21sters, Transhumanism made sense. These people couldn't possibly know what would happen 1,000 years in the future.

Then Karen Everard, the reporter, asked a question that Pietra 23 had never considered.

"What you're saying here *is* exciting. But your philosophy is totally materialist. What if the entire purpose of human life is to advance awareness past the biology? Past mere Transhuman enhancements and toward an overarching spirituality that unites consciousness everywhere across the universe? In this conception, Transhumanism – and even ASI – is inherently flawed and limited because it can never address, or even recognize, that possibility."

Harry Kaine exploded. "There is no such thing as spirituality, Karen! It is simply unproven nonsense."

"I wouldn't go quite that far but it certainly is unproven," Dr. Sauvage agreed. "The vast majority of those who believe in spirituality are scientifically uneducated."

"So you believe that when you die, you're dead," Everard said. "What happens to your consciousness after you die?"

Sauvage seemed uncomfortable with this. "It's precisely the reason we need to develop longevity technologies! Transhumanism is the only future for mankind."

"You seem very sure of yourself," the reporter replied. "The concept of God, or a higher power that exists beyond the material world, has been a part of every human civilization on earth. Surely Transhumanism cannot dismiss an idea that has taken root in every culture throughout human history."

"We'll risk it," Singh said. He spoke dismissively. "Transhumanism does go beyond biology, that's the point. Transhumanism enhances it."

"You don't know what you don't know," the reporter said.

Pietra 23 saw that Harry Kaine's face was red with anger. The reporter, clearly irritated now, shrugged with the air of someone who knows that the other person will never understand, no matter how much explaining is done. "I'm just playing devil's advocate. I had to interview the Interfaith metaphysical group last week, and I challenged them also. I got some interesting ideas from them. Unlike some I'm a true journalist; it's my job to question everything."

Kaine's mood switched suddenly from anger to charming tenderness as he smiled at the reporter. Pietra 23 saw Everard's face soften. Wow, she thought. Kaine's mood could shift quickly! She frowned, remembering from her study what 21sters called the anti-social personality. Kaine exhibited some of these tendencies, yet his enthusiasm was a positive attribute. He was an inspiration to others.

After the meeting ended Kaine approached her as they were walking out of the room. Again she felt the sexual component. "Would you like to get a drink? We can go to Densingers, where we first met." Slowly Kaine's hand reached for her cap. "Why don't you take this off? I want to see your full profile."

Pietra 23 knew she was in trouble, and blocked Harry's hand with her arm. She could not risk a sexual encounter but she felt an almost magnetic attraction to the man. She had to get out of here!

At that moment Kaine was approached by Mary Chen and Karen Everard, both trying to get his attention. Pietra 23 took the opportunity to slip out of the room. She noticed a look of irritation on Kaine's face, but she walked rapidly down the hall toward the front door. She knew wasn't going to make it to the front door before Harry got into the hallway. Pietra 23 saw an unused classroom with the door open and the lights off. She walked to the back of the classroom and saw a storage closet. She opened the door and got out her Shifter communicator, inserting the lead into her brain implant. Hopefully those guys would act quickly this time. As the minutes went by she waited impatiently for Alan 133 to get going. 3013 AD was getting less and less real to her and she wondered what it would be like to live permanently in 2035...

"C'mon Alan 133, dammit!" Pietra 23 swore, surprising herself with a 21st century Americanism. Just then she heard Harry Kaine's voice, talking to Mary Chen, as they walked down the hall. She heard Kaine say, "Wait a minute Mary. Excuse me Karen." Pietra

23 heard footsteps approaching the classroom door. Dammit, she swore again silently. Kaine must have heard her outburst!

As Harry Kaine stuck his head in the door he thought he saw someone in the storage closet. Then a flickering, then nothing...

Chapter 10

3013 AD

Pietra 23 appeared on the Shifter platform and looked around at the drab room with its gray permacrete walls and floor. She became flooded with disappointment and discouragement. She was back in 3013! She would have to do her Guardian shift tomorrow and then make her report to Christian 172. She was trapped in a dead-end future. Her goal from now on would be to spend as much time in 2035 as possible, even if it meant dealing with Harry Kaine.

Pietra 23 looked around the room at the Lis and the Alans and the one Sigmund. All were looking at her with what she recognized now as an apathetic gaze. The life force was beaten out of them. Until a short time ago she had been one of them.

"What's the matter, Pietra 23?" Alan 133 asked.

"This society is dead," she said with heat. "Dead and gone."

The Lis and the Alans looked at each other in amazement. "Pietra 23, your life force appears to be significantly enhanced," Li 355 said.

"Of course it is! You can't live in the 21st without being exposed to the vitality of a living society."

Sigmund 78 moved immediately to activate the booth that contained the Life Force reader. "Please step in, Pietra 23."

Irritated at Sigmund 78's apathetic tone of voice, she strode over to the booth. The men here were cardboard cutouts compared to the 21sters. Even the egomaniac Harry Kaine was an improvement!

After thirty seconds Sigmund 78 began to tremble. "I can't believe it. Alan 133, come over here."

Alan 133 scanned the readout. "167.4! Impossible."

"Is it that high?" Pietra 23 remembered the strong emotions she'd been feeling in the 21st. Maybe the reading was genuine.

"Could it be that life force energy is transferred if a clone is exposed to human beings with high life force readings?" Alan 133 asked. Hope again surged within him.

"How long will it last?" Li 308 asked.

Alan 133 recognized his duty. "First I must de-brief Pietra 23, and tomorrow she has to go on her shift. I think we should wait a week to determine how far her LRI falls." He looked over at Pietra 23. She was much more attractive and feminine now. He felt a faint stirring in his loins.

"I agree," said Li 355. "We have been using this room a lot lately and we are beginning to attract Guardian attention."

"The foot patrols have been increased in this area," Alan 49 said, checking his console.

"Very well," Alan 133 said. "Everyone stay away from here until..." he checked his chronometer. "Until September 30th. It's the 22nd today." He looked at Pietra 23. "After the debrief, report back here at 0600 on September 30th for a life reading check."

They all marched out of the building through the underground tunnel that led out into the ruins of an old subway station. From there, each took different routes through the safe zone back to their living quarters.

Pietra 23 lived her boring life for eight days, which consisted mainly of spying on her fellow citizens and reporting to dullards like Christian 172 at her local Guardian station. She felt her life force gradually decreasing as she dealt with the lifeless, purposeless clones that made up Guardian society.

When she reported to the Shifter room on the 30th, the day was like it always was: temperatures in the 60s or 70s, almost no wind. Even the weather was apathetic here in 3013. Alan 133 and Sigmund 78 were already there. Alan 133 noticed immediately that Pietra 23 looked less lively, and felt a stab of despair. "All right, please step into the booth."

After thirty seconds Sigmund 78 checked the reading. "131.1, still 18 points higher than last time!"

Pietra 23 stepped out of the booth. "I feel...depleted. Maybe we should all translate back to the 21st." She looked directly at Alan 133. "I saw on the news feed that the next generation of clones will be androgynous. That's the end."

A look of despair crossed Alan 133's face. "You're right. Desperate action is required. Any ideas?"

"Trying to divert Harry Kaine from his life extension agenda is going to be difficult," Pietra 23 said. "He's obsessed with the subject and he's already got a team of scientists around him. The only solution is to kill him."

Alan 133 and Sigmund 78 were shocked. Murder was unknown in Guardian society.

Just then Li 355 and Li 308 walked in. "I heard that, Pietra 23," Li 355 said. "It won't work. Our recent work with the Shifter shows that temporal energy is conserved. Time is like a rubber band. It can be stretched or changed temporarily, but it always wants to return to its former shape. Even if you could kill Harry Kaine, it's almost certain that someone similar will take his place."

Pietra 23 spoke sharply to Li 355. "Then it's hopeless. Those who want to live should translate back to the 21st, or some period before the Planetary Referendum in 2098."

"There was an old audiovid from the 20th about that," Li 308 said. "The sun was going to explode so the society sent their entire population back in time using what they called an atavachron."[8]

Li 355 smiled apathetically. "That wouldn't work. It would just create a time loop, because there would be no one living past this point in 3013. This entire timeline would gradually become detached from the main flow of temporal energy. Everything would die off."

Pietra 23 sighed. "Yes. That's what is happening to us now."

The group was silent for a few minutes as all thirteen Shifter Project clones arrived in Building 103.

"But what if we could permanently enhance a clone's Life Reading Index past 100?" Alan 133 said.

The clones in the room couldn't take in this proposal, except for Pietra 23. "Of course!"

She turned to Li 308. "Has anyone studied my last immersion/temporal path?"

Alan 133 shook his head. "We haven't had time. After I debriefed you we all went home. No one has been here for eight days."

Li 308 went to his Shifter console. "I can bring up anything that happened when you were in the 21st."

Pietra 23 was excited. "Bring up the meeting in the LIfe Sciences building. Can you measure the LRI of everyone in that room from here?"

Li 308 looked startled. "Never tried it before. But it should be possible."

They all watched as Li 308, with Li 355's help, programmed his console. In a few minutes a number appeared above the heads of each person in the room back in 2035, at the beginning of the meeting.

"Harry Kaine and the four scientists have readings between 170 and 180," Alan 133 commented. "Pietra 23, your reading is 147.3." As they watched, Pietra 23's reading changed to 147.7.

"Pietra 23 may be correct," Li 355 said. "A clone's exposure to viable human beings seems to increase its biological vitality."

Everyone began to chatter. Pietra 23 noticed a sense of hope pervading the room, as opposed to the usual dreary apathy.

"Without a broad increase in life force readings the human race will die," Alan 133 said. "Pietra 23 has proven that LRI's can be increased by association with 21sters. Therefore we must insert our clones into the past. If enough of us do this we may be able to create a viable community right here in 3013 that can challenge the Guardians."

Pietra 23 spoke. "That's right, but there's a danger. By the end of my visit to the past I didn't want to come back."

Li 355 was astonished. "You would rather stay in that barbaric time?"

"When I was there? Oh yes. When your LRI goes up it's such a wonderful feeling. You don't want to lose it." She looked around the room at the small group of rebels. "Your apathy is depressing."

The other clones looked blankly at Pietra 23.

"We're not apathetic," Li 308 said.

Pietra 23 laughed. "You don't know what you don't know, as that reporter told Harry Kaine."

Alan 133 became mildly assertive. "All right. The conservation law of temporal energy says that we can't divert Harry Kaine from his disastrous Transhumanist agenda, even if we go back in time and kill him." Alan 133 shuddered. "A thing impossible for us clones. So we must begin a program of enhancing our clones by inserting them back in time." Alan 133 looked at Pietra 23. "How long were you immersed before you had the desire to stay in the 21st?"

"About six hours."

"I don't think any of us can take more than that," Alan 133 said. He looked around the room. There were only thirteen clones here, the best and brightest in the city. "We need eleven clones to go back in time and two clones to manage the timeline and the Shifter. Who wants to go and who wants to stay?"

There was an uncomfortable silence. "Where do we go and what do we do?" one of the Lis asked.

"We'll all go to Midland in 2035," Pietra 23 said. "Stay on campus, go to public places where there are lots of people, soak up the life force energies of the 21sters." She could see that except for Li 355 and Alan 133, the other clones were terrified. Or as close to terrified as a clone could get.

"We either do this or we die a slow death," Alan 133 agreed. "We have no choice."

Everyone looked away, not wanting to be chosen.

Alan 133 looked at Pietra 23 for direction. She had more energy than anyone else in the room.

"Li 308 and Li 57, you two are the best time technicians we have," she said. "You two will stay and be in charge of insertions and extractions. No more than six hours for any person in the past. No clone can appear during the same time it has already been inserted."

Alan 133 was amazed at Pietra 23's decision-making ability.

The clones began complaining and asking questions. Pietra 23 raised her hand. "Stop. We're doing this and we start today."

"Today?" Li 308 asked in disbelief.

"Right now. Get something to eat. We'll print out some 21ster clothing for each of you. It's cold in 2035." She turned to Li 308 and Li 57. "We'll all insert into 2035 the morning after my meeting with Harry Kaine. We'll avoid him. Take off your Guardian unis and put on your 21ster clothes. We leave in an hour."

Guardian monitor station, Section Seven, Old New York

Christian 172 turned to Paul 26. "There has been unusual activity in Building 103 in the Old Sector. We must place monitoring devices within the building."

"That is expressly forbidden," Paul 26 said. "No monitoring of the Safe Zones. If the Overseers discover this we will be terminated."

"The Overseers themselves are incapable of monitoring the Zones. Therefore they will never see us install the devices."

"Unless a foot monitor observes the work and reports back."

Christian 172 spoke almost contemptuously. "This station controls the personnel assigned to foot traffic in that area. You will assign yourself to patrol the area during your day shift and report directly to me."

Paul 26 was unconvinced. "The Overseers may intercept communications from the foot monitors."

"You are a fool. Our technicians will tune their communication devices to send only to this station. You will begin your patrol on your next duty shift."

Christian 172 would himself patrol Building 103 during his off-shift hours, while his assistant Paul 26 monitored the area during his day shift. Several clones, including Pietra 23, had been late for their duty shifts during the past several months. He would compile a list of the disobedient ones and personally investigate.

CHAPTER **11**

Shifter building #103, Old New York, 3013 AD

One by one, nine reluctant clones and two motivated clones from 3013 AD stepped onto the Shifter platform and were translated to 2035 AD. Li 57 and Li 308 had discovered a basement in an abandoned warehouse in the industrial section of Midland, Illinois.

"This will suffice for now," Li 308 said. "We may have to discover more secluded places in the city if we plan to do more of this."

The eleven clones found themselves in a cold basement standing on a dirty concrete floor. Light from a pale sun streamed through a broken window above them.

"What are we supposed to do now?" Li 17 asked in a hopeless voice.

Pietra 23 felt energized. "We socialize quietly. To start out we'll go to the Midland Public Library, where quiet is expected and we don't have to speak with any of the 21sters."

The other ten clones, grumbling, were mollified by this. "Let's go then," Alan 133 said. "It's cold in here."

Pietra 23 led them out of the building onto a dirt road. After a quarter mile they turned left onto a concrete road. The clones could hear the movement of water on their right. Pietra 23 pointed. "Beyond that marsh is the Midland River. A branch of the library is about a mile and a half walk from here."

The clones began walking.

"Stop huddling together like frightened children!" Pietra 23 demanded. "Straighten your backs and walk tall. We are free men and women here in the 21st."

The clones were baffled by this statement, for they did not understand the concept of freedom. They spread out a little. After a minute a large 18-wheeler came barreling down the road behind them. Most of the clones panicked and began to run. One of them

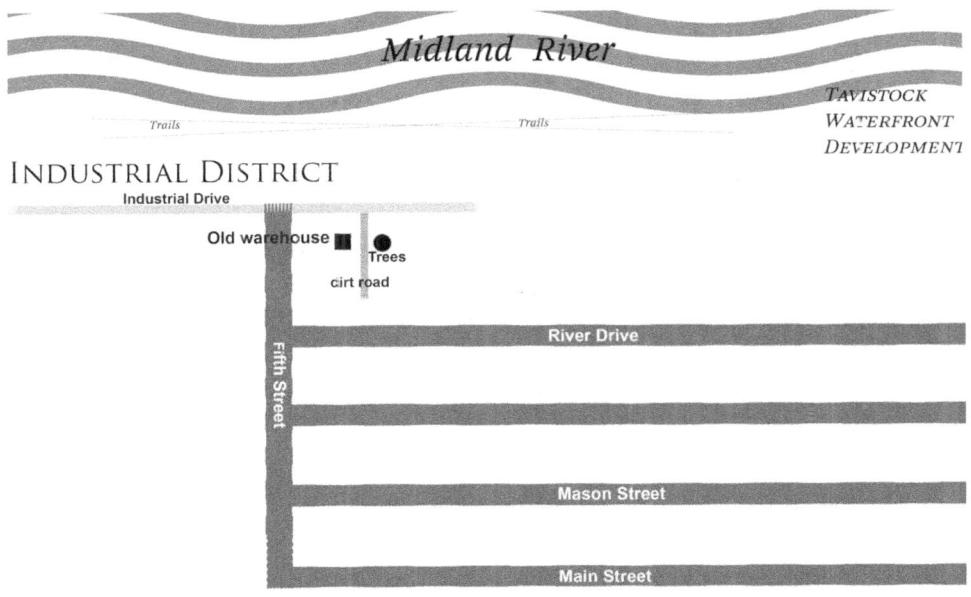

Midland River

TAVISTOCK
WATERFRONT
DEVELOPMENT

Trails   Trails

INDUSTRIAL DISTRICT

Industrial Drive

Old warehouse ■   ● Trees

cirt road

River Drive

Fifth Street

Mason Street

Main Street

The Abandoned Warehouse, Midland Industrial District, 2035 AD

crossed the road just in front of the truck. The driver slammed on his brakes. "Hey, what ya doin!!" he said angrily.

Pietra 23 stepped up to the truck on the passenger side. The driver rolled down the window. "Sorry mister, these are at-risk adults from the hospital. Just taking them out for a walk."

"Sure lady! But watch where you're goin!"

The truck slowly moved away in a haze of belching blue smoke and the crashing of gears. "That was good work Pietra 23," Alan 133 said, choking on the toxic smell of hydrocarbons.

"Go get Alan 49," she said. "He's running around like a chicken with its head cut off."

Alan 133 thought this was so funny he bent over laughing. Pietra 23 crossed the road and slapped a panicked Alan 49 across the face. "Get hold of yourself man!"

"You hit me," the clone accused.

Alan 133 crossed the street to join Pietra 23 and Alan 49. "Clones! Move to this side of the street so we can see approaching vehicles."

Everyone agreed to this and the group trudged along the grass on the side of the road. By habit everyone formed a straight line and walked slowly. Several vehicles approached but these were much smaller. "Passenger vehicles," Alan 133 remarked. He recognized them from the Guardian history programs.

They reached the end of the road, which had a sign that said "Industrial Drive." They came to another, wider road with several lanes of traffic, which had a sign that read "Fifth Street." Lights were suspended across the street from poles. "Street lights," one of the clones said.

Pietra pointed to a building about a half mile down the road to their left. "The library is there. We have to cross this road."

Suddenly cars began to whizz by, very close to where they were standing at the light. Some of the clones panicked again, drawing curious glances from two 21sters on the other side who were waiting to cross. Suddenly the vehicles stopped. "It's OK to cross now, but hurry," Pietra 23 said.

Two of the clones began to run across the road, frightened. "Walk don't run!" Pietra 23 shouted.

"Are you guys OK?" one of the 21sters asked Alan 133.

"Just at-risk adults out for some exercise," he said.

The two gave the clones a curious look, but passed on.

"What's an at-risk adult, Pietra 23?" Alan 133 asked.

"Someone who has trouble adjusting to society. I discovered that on one of the 21st history vids."

Alan 133 almost laughed again. "That describes us."

Pietra 23 was pleased that Alan 133 was adjusting so quickly. He seemed...stronger. "The rest of these clones are worthless here," she said to Alan 133. "You and I will have to carefully monitor the others."

The clones reached the library building and Pietra 23 led the way in. A blast of warm air welcomed them. Pietra 23 heard laughing and shouting at the back of the building. She boldly went up to the receptionist. "What's happening back there?"

"Children's play hour."

"Could we look in do you think?"

Alan 133 stepped up. "We're supervising some at-risk adults," he said. "They may get benefit from watching the children play."

The receptionist frowned. She was a smallish woman with jet black hair and red lipstick.

"We'll stay against the walls," Pietra 23 said.

The receptionist looked at the eight clones huddling behind like frightened mice. "Well...I guess it's all right, they look harmless. I'll notify the supervisor, but if she asks you to leave..."

"Then we'll go."

"Well done Pietra 23," Alan 133 said.

"That's a strange name," the receptionist said. "Are there more of you?"

Alan 133 became frightened, but Pietra 23 said, "That's what my father used to say."

The receptionist laughed and waved them on. Alan 133 and Pietra 23 herded their clones down the hallway, which had workstations and racks of data discs and printed books. "Books!" Li 355 cried. "Actual, real books!"

This elicited stares from the 21sters. "At-risk adults," Alan 133 said in explanation. Pietra 23 herded the clones into the play area and against the walls. Children were running around, yelling and playing games. The big playroom was crowded.

"How did you know to say that about your father?" Alan 133 hissed at Pietra 23. "Printed clones don't have mothers and fathers. Just parent 1 and parent 2. And why did the receptionist laugh?"

"It's called a joke, I got it from the history vids. The early 21sters were big on them until the Resistance tried to make everyone think alike."

Pietra 23 noticed how quickly she had re-adjusted to the high energy here in the 21st. She felt better even than during the meeting with Kaine and his scientists. The children in the big playroom were literally throwing off energy and she was soaking it in. She noticed that Alan 133 was doing the same. The other clones were pressed against the wall. All had their 21ster flexible head coverings on, to hide their brain implants.

"Would you get my ball?" one of the children asked. The ball had landed between Grace 5 and Sigmund 78. Pietra 23 gestured with her hands. "Pick it up and give it to the child."

Grace 5 tentatively reached for the ball. The child had come up to them and was staring with curiosity. The clone handed the ball to the child. When Grace 5's hand touched the little girl's, the clone started.

"Thank you lady!" the child said. Grace 5 and Sigmund 78 managed a tentative smile. The child ran off.

"I did it!" Grace 5 said, somewhat like Sheldon Cooper to the blue jay. Sigmund 78 was looking now at the children with awe. "Their life force readings must be off the chart!"

Suddenly the other clones began to loosen up. They watched the children playing for over thirty minutes, fascinated with their energy.

Alan 49 said to Li 355, "We have little clones but nothing like this!"

One of the children overheard and asked, "What's a clone?"

The supervisor frowned. "What are you teaching these children? I think you should leave."

Pietra 23 stepped in. "We'll leave now, thank you."

The clones filed out as the supervisor alerted the front desk. "I don't want those people in here anymore."

The receptionist acknowledged. "I'll tell them they have to leave the building."

The clones soon found themselves outside the library. A cold wind was blowing.

"Where to next, Pietra 23?" Alan 133 asked.

"The campus is less than a mile away. We can go into a public building there and get warm. Sit quietly on a bench. Hopefully there will be a lot of foot traffic."

Pietra 23 led the group, Alan 133 took up the rear. After a minute Pietra 23 heard an excited voice calling her name. She looked up to see Harry Kaine striding up to the group.

"What have we here?" he said, inspecting the little group from 3013 AD.

"Pietra 23, is this Harry Kaine?" Li 355 asked.

"Pietra 23?" Kaine's expression exploded with understanding. "You're a clone!" he exclaimed, almost out of his mind with excitement. "Number 23 of the Pietra line! That's how we number them in our experimental research!"

Pietra 23 was amazed at Kaine's perspicacity, for the Guardian cloning/printing program had evolved later from 21st century research. But Pietra 23's life force had been enormously enhanced by the day's activities. Three times she had been inserted back into the timeline where humans possessed enormous vitality. "Oh Harry, it's nice to see you! My friend here was just joking of course." She pointed to the group of clones. "These are at-risk adults I'm sponsoring for a day trip."

Kaine's face displayed skepticism. "I saw you disappear from that storage closet yesterday."

Pietra 23 laughed. She couldn't think of an appropriate response, so she signaled for the group to move along.

"I still want to have that drink with you," Kaine said. "Pietra 23."

"Of course, but some other time."

As the clones walked away Pietra 23 saw Harry's eyes on her and the group.

The group were getting hungry so Pietra 23 led the way back to the basement of the warehouse, where they would signal to Li 57 and Li 308 in the Shifter room via their brain implants. She hoped that Kaine wouldn't follow them. The blustery wind was at their backs but the group still had almost a mile to walk. She saw a public transportation bus move past them, and a bus stop. The next time they came they would have to bring a currency card.

The group arrived at the warehouse and descended the stairs to the basement. The clones placed their comm leads into their brain implants and waited.

Nothing happened.

CHAPTER **12**

Building 103, Old New York, 3013 AD

"Three persons are scouting the area," Li 308 announced.

"I see them," Li 57 responded from her console. "A Guardian foot patrol."

At that moment Li 308 received the comms from the inserted group in 2035. "How long before they can find their way in here?"

"Twenty minutes, if they use the front entrance."

"Get everybody back here."

Ten clones were successfully transferred back to the present, one at a time. The Lis had increased the Time Shifter's efficiency so they could now transport someone every 120 seconds or so. Just as Pietra 23 was about to be extracted, several Guardian police in their black and red unis entered the room. In the lead was Commissar Christian 172.

"Malcontents!" he said. "Traitors! You are using forbidden technology. All of you shall be terminated. But first we will extract your memories before your bodies are disassembled."

Christian 172's two assistants approached, intending to apprehend the miscreants. Their Submission modules were out. To their surprise, all but two of the clones resisted arrest. "I do not understand," Paul 26 said. "These other clones should exhibit submissive behavior."

Christian 172 stepped in front of Alan 133 and glared.

Alan 133 laughed. Christian 172 had one of the highest life force readings in all of New York. Alan 133 used to be terrified of the Guardian commissar, as were all who met him. But it was different now. Alan 133 slapped Christian 172 across the face, silently thanking Pietra 23 for her boldness and self-confidence. The Guardian commissar collapsed to the floor in shock.

"You have used violence," he mumbled, holding his jaw. "The Overseers shall hear of your behavior."

The other clones quickly surrounded Paul 26, Paul 11, and Christian 172.

Alan 133 stood over the commissar. "You will go back to your monitor station. You will report that the heightened traffic in this sector was a normal spike in foot traffic, and that all is normal."

Christian 172 and his assistants were appalled. Clones were submissive; had always been. That's what the brain implants were for! Yet the clones before them seemed almost superhuman; their aspect was terrifying.

The Commissar shrugged. "You would defy an order from the Overseers? If you refuse to come with us the Khan will seize all of you. You are all doomed."

Alan 133 nodded dismissively to Christian 172 and his assistants. "I think not." Alan 133 stood as straight and tall as he could. He felt full of energy after his trip to the past. "Go! You, and your Overseers, are the doomed ones. In a very short time you will discover this."

While Li 308 and Li 57 watched at their consoles in awe, the other time-traveling clones hustled Christian 172 and his two assistants out the front of the building, avoiding the secret back entrance.

"What have you done?" Li 57 asked the others when they came back. "You have endangered the project and all of us."

Alan 133 ignored this. He took a deep breath and smiled at his fellow time travelers. "You can feel it, can you not? The increase in life energy?"

The other nine clones nodded. It seemed that they were, as a group, now enhancing each other.

Li 355 spoke boldly to Li 57 at her Shifter console. "This society is too apathetic to curtail our activities. So what if the Overseers send their clones to apprehend us? Ten of us are more powerful now than a thousand of them. You have no idea how potent the enhancement of life force is."

"What did you feel when our Submission modules were activated?" Alan 133 asked the others.

"A headache," Sigmund 78 replied.

"Yes, a big headache, but it was bearable," LI 355 said.

The others confirmed this.

Alan 133 was almost ecstatic. "We have eleven clones with enhanced life force. With a few more trips into the past we may become permanently enhanced."

Sitting at his console and looking at the group of time travelers, Li 308 suddenly understood. Intimidation and coercion could only work if the Submission frequencies affected the brain implant! But these time travelers were now immune. "I see," he said. "Our society is almost dead; the life force and the vitality is almost gone. The Overseers have no real power against us if we become enhanced, for they are almost dead themselves."

Sigmund 78 saw it. "The Guardians are like zombies! We obey them simply because we always have. It's just a bad habit ingrained over the centuries."

Li 308 understood now what the ten enhanced clones were saying. He could feel his own vitality increasing just being in their presence. "We clones merely go through the motions of living. The next androgynous generation of clones will have even more decreased LRIs." A look of awe came over his features. "Why, we are barely in time to save humanity! In a few more generations the autonomic biological functions of our clones will be so low that our bodies won't work anymore. All of us will die!"

Li 355 agreed. "That is correct. The only answer is to re-enable our reproductive organs, if that is possible."

"There is no guarantee that time traveling will do this," Sigmund 78 said glumly. "Moreover, the Overseers can muster several thousand clones. Even those who are enhanced will be overwhelmed by sheer numbers."

Alan 133 shrugged. "The Shifter project is the only hope for humanity. For us and the Guardians, even if they are now too evil and stupid to understand." He turned to the two Shifter console operators. "Bring back Pietra 23 immediately."

Everyone had forgotten that Pietra 23 was still trapped in the past!

When Pietra 23 emerged on the Shifter platform the two console operators who had stayed in 3013 were astonished. Alan 133 seemed like a superman to them. But Pietra 23 had now gone three times into the past. She looked just like an authentic 21ster now; vital, alive, and powerful. She exuded a feminine energy. Alan 133 felt it.

"I feel like having sex," she said, shocking everyone. "But not with Harry Kaine!"

Alan 133 laughed. "Sigmund 78, measure the life force of everyone. We must all stay together as much as possible, resonating to each other's enhanced LRIs. We will all return to the past as many times as necessary. Then we can recruit others to join us."

Pietra 23 shouted for joy. "Freedom!"

The enhanced clones were overly optimistic. When Commissar Christian 172 reported glumly to his Overseer at Guardian HQ, on the top floor in the Tower of the Overseers,

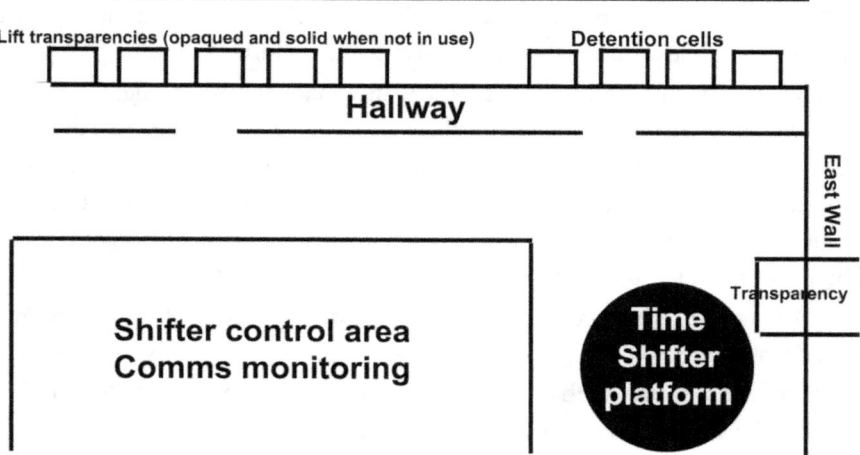

The Tower of the Overseers – Top Floor (Command Center)

Commander Patton 1 smiled. "Two can play at that game," he said, leading Christian 172 to a sealed room.

"A time travel device!" Christian 172 said, examining the machine. "It resembles the one the rebels have, except ours is bigger."

Patton 1 nodded. "It is the same technology."

Christian 172 had an idea. "We too can send our own teams back into the past, enhancing them."

"That is correct, Christian 172! Kill the traitors with our own enhanced clones."

Christian 172 agreed. "The infestation of enhanced rebel clones must be stopped immediately."

Paul 26 objected. "Our society will soon die if we do not enhance all of our clone bodies. We must cooperate to save the race. Fighting and killing each other is species suicide."

Both Christian 172 and Patton 1 regarded Paul 26 with contempt. "You are a fool," they both said in unison.

CHAPTER **13**

2035 AD

Winthrop Sauvage sat back on his sofa with a cup of hot coffee, feeling optimistic. He had just gotten home from his informal meeting with Harry Kaine and his scientists at the Life Sciences building on campus. Fortunately, Kaine had connections to several venture capitalists. Funding for his cryogenic research was hard to get. He had written several grant proposals but didn't expect any of them to bear fruit. It may be time to get out of academia altogether and into the private sector. Trying to live on an academic salary was ridiculous! He needed Kaine because he needed money.

Then there was his...addiction? Addiction, yes, to children. But perhaps not so unusual, for there were others he chatted with on his network who also used assumed names. He knew two people at Carleton who were part of the network...but merely viewing images is harmless. His need was real. For now it could be satisfied with pornography, but he admitted to himself that a day might come again when images would not be enough. His daughter Winnie had been his only mistake...he must call her.

"What do you want?" Winnie said. Her tone was brusque.

"Still haven't forgiven me, have you?"

"Still looking at child pornography, aren't you?"

He had no answer for that. "I'm getting out of academia."

"Harry Kaine."

"How could you possibly know that? I just thought of it two minutes ago."

"You were seen, Winthrop."

Winnie always called him Winthrop, not Dad.

"Room 235 at the Life Sciences building, 7 p.m., last night. Kaine was there."

"You're keeping tabs on me now?"

"You're dangerous, Winthrop. Our network keeps tabs on all child rapers and pedophiles in the city."

"I'm sorry for that Winnie. I...I lost control."

"You're sick, Winthrop. Get help."

"And become a registered sex offender? I'd lose my job and any potential funding for my research."

"It's all about you, isn't it? You fit the profile, Winthrop."

The connection was broken.

It isn't fair, Winthrop thought. Psychologists are now recognizing that pedophilia is a natural sexual impulse. Children are beautiful! Demonizing such things only hides them, and makes monitoring these networks more difficult. Something he was reluctantly thankful for.

Winnie was right of course. His cryogenic research was for himself, for he had a morbid fear of death. He was already 56! Only 30 more years or so left. He must either lengthen his lifespan, or cryogenic technology must be improved to await the day when the advances promised by Transhumanism and life extension were realized. That's why his research was so important!

Harry Kaine was his ticket. He'd talk to him this week about getting funding. But first, the new batch of images he had downloaded last night on the darknet...

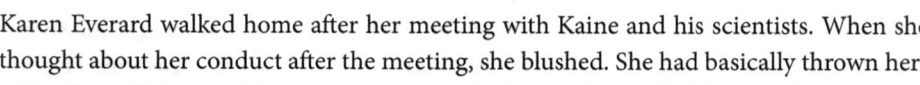

Karen Everard walked home after her meeting with Kaine and his scientists. When she thought about her conduct after the meeting, she blushed. She had basically thrown herself at Harry Kaine, and shown her contempt for Mary Chen. Why? Joe is right, Harry is a creep. But when in his physical presence she was enormously attracted to his maleness.

She was still mulling this over, walking with her head down in the cold, when she almost stumbled into Joe Courvall. She was in front of Joe's apartment building! How had that happened? She must have sex on the brain.

"Karen!" Joe said. "Nice to see you. What's up?"

Joe's tone was merely that of a friend. She felt vaguely disappointed. "Where are you going? Do you want to get a coffee?"

Joe blushed.

Now what was that about? She knew now he was interested in her, and they had been casual friends since their days at Carleton.

"Uh, thanks Karen. I'd really like to, but I'm meeting Kesha at the coffee shop."

Kesha? Her disappointment must have shown on her face, for Joe blushed again.

"Maybe tomorrow?" he said.

"Sure."

Joe rushed off as if to an important date.

Now she was curious. It was still only 8:30, she wanted an espresso anyway. Karen walked slowly toward the coffee shop on Mason. Fortunately the place was big and hopefully Joe wouldn't notice her coming in. She needn't have worried, for she saw Joe at a table for two against the wall, his back to her. Joe's eyes were on a woman with multicolored corn rows.

She's beautiful, Karen thought. She ordered her drink and shamelessly sat down at a little table behind Joe's, so that Joe's back was to her.

Karen put her espresso on the table and got out her comm device, which still had a punch-style keyboard. She needed to write down her audio info from the meeting with Kaine and his scientists. She got out her recorder and plugged it in, reviewing the conversation. She turned the volume of her ear buds low enough so that she could hear the conversation in front of her.

Joe was clearly infatuated with the woman, who looked to be in her late twenties. Trying to look like she was just another customer, Karen listened in on the conversation. She was, after all, a reporter.

"...so you can't tell me anything about your work?" Joe was asking. "I'm a boring engineer who tests circuit boards and optical processors for the communications industry."

Karen tried not to stare as the woman smiled. "Boring is good."

"Oh?"

Karen knew the look Joe was giving her. A cute, self-deprecating look that was very engaging.

"Oh all right. Winnie and I work for an organization that deals with abused and abandoned children."

"Right, COSA. My friend Ger is terribly interested in Winnie."

The woman's eyes sparkled. "Yes, I know."

"But I'm interested in you. Your name is Kesha."

"Spoken like a true engineer. What is the coefficient of your interest?"

Karen could feel Joe's pleased surprise at this riposte and they both laughed. For some reason she felt irritated. Joe was hers! He was part of their gang.

Karen could feel Joe's sexual interest, which she hadn't noticed when she was around him. As the conversation continued the woman told Joe about her work with children. "Lots of abandoned and abused children in Midland, and especially Chicago. It's horrible sometimes."

As the conversation went on, Karen could see that Joe was all in on Kesha. She'd better go. Her espresso was gone and she didn't want Joe to turn around or get up and leave. He'd surely see her. Karen quietly cleared the table, got up, and bused her cup. Just as she was walking out the door she heard Joe say, "If you ever want to get really bored stop by my office at 455 Fifth, number 203. Not too far from here."

Karen didn't feel like walking home in the cold. What was she doing here anyway? She activated an app and one of those electric self-drivers popped over. She pushed her currency card into the robot driver (which only had a torso and a head, to make humans feel better) and was mad at herself. What was wrong with her? She thought of herself as a calm, serene person. Her emotions had first been ruffled by Harry Kaine and that sycophant Mary Chen. Then she bumped into Joe and he had turned her down for Kesha.

When she got home she undressed and took a hot bath. As she soaked in the tub she wondered whether she was interested in Joe Courvall, or just irritated at Kesha and Mary Chen.

She was definitely going to find out more about the Two Bitches, Winnie Sauvage and Kesha What's-her-name. And that COSA organization they were working for.

———— • ✳ • ————

Winnie Sauvage and Ger were sitting in a fast food restaurant, drinking coffee and eating tacos. For the first time in his life, Ger was interested in a woman. Winnie knew she shouldn't get involved with a normie like Germaine, but she liked him. And he was a good listener.

"Do you know Kjirsten Chastaine?" Winnie asked.

"Yes, but how do you know that?"

Uh oh. She didn't want to say too much. "In my work with children we occasionally ask the police to investigate...uh...clients. We've worked with her father."

Ger could see her reticence. "Look Winnie, I'd appreciate a little honesty here. We've already talked about your work, I already told you about Ralph Zimring. He told me all about the Bosses." That was a lie, but he wanted to see her reaction.

Winnie blanched; her normally pale skin became even whiter, emphasizing her large, intense blue eyes. Ger thought she was absolutely gorgeous.

"Fuck. How much do you know?"

He didn't know anything except that the Bosses were very dangerous. "Enough to know that those people are dangerous and never to say their name in public."

"You just did."

Ger grinned. "I made a mistake. Never happen again."

It was hard to be mad at him, she thought; the goofy looking clown. "All right. Right now I'm in a private network of people who have been...abused. We understand child abuse because we've experienced it, and we're good at spotting offenders. We mostly work for COSA or Child Protective Services, or we're independent contractors within our network. But we're also paranoid, so we change our places of residence every six months or so."

"OK. Now tell me why you asked about Kjirsten Chastaine."

"Kjirsten knows this guy Liqao Chang who runs the IT Department for the Midland Police Department. Chang probably reads all of our communications." She spoke bitterly.

Ger didn't understand.

"Kjirsten's father, the police chief, is a notorious conservative who hates me personally and thinks our network exaggerates our cases and gives the police department and the city bad publicity. He says we are a 'public nuisance.' He ignores almost every request for investigation into abusive foster parents and other pedos. Chang is friends with Kjirsten, his daughter. The police chief thinks Liqao is God's gift to humanity, but he's really a smug little twit."

Ger laughed. "Yeah, Liqao is a little like that. I remember reading about you, come to think on it. You were COSA's spokesperson when they accused some guy in Child Protective Services of being a pedophile."

"That's right. Bob Butters. A real sociopath."

"You said some inflammatory things about the police chief, and CPS. Jack Chastaine threatened to investigate COSA if you didn't shut up."

"Yes, and we had to drop the case because Liqao Chang knew everything, to the smallest detail, about us. That's why I said he probably reads our communications to the Midland Police. Butters is still working for CPS."

"Wow."

"So now you know why I asked about Kjirsten."

"Got it."

Winnie batted her eyelashes. "If you have any influence with her, get her to ask her old man to reopen that case."

Ger smiled. "You aren't averse to flattery to get your way."

"Or sex."

"So if you slept with me, what would I have to do?"

Winnie laughed. Ger's expression was like a cat wanting to lap up some cream. "Oh hell. I'll sleep with you because I like you."

"You're not...nervous or afraid?"

"A little, but I've had so much counseling I'm almost over the abuse."

Ger smiled again. Winnie saw his whole face light up. It was like a ray of happy sunshine.

"Then I'm your man. When do we get started?"

Winnie thought this was so funny she broke down laughing. This guy was a typical man, looking at sex like it was an orgasm machine. She reached into her purse and got out a handkerchief and wiped her eyes. "Are you an engineer like your friend Joe?"

"Nope. I'm an inventor like that guy Tesla. But the question still stands."

"My place or yours?"

"Mine," Ger said. "We'll be totally alone."

Ger saw Winnie start at this and her smile vanished, so he smiled. "Don't worry, I'm harmless. I just meant that we could have privacy. You can leave anytime you want, I won't force anything."

"OK."

But when they got out of the popup taxi in front of Ger's house, Winnie was reluctant. "I'm sorry Germaine, I really like you, but—"

"I understand. Some bad memories?"

Winnie shook her head in acknowledgment. "I guess I still need more therapy."

"Can I at least have a kiss? Just a little one?"

Winnie laughed again. The little popup cab drove away.

"I like it when you laugh."

Winnie closed her eyes and lifted her head.

Ger placed his lips lightly on hers, but not for too long. "You taste good," he said.

Tears began to run down Winnie's cheeks.

"Sorry, I didn't mean—"

"No, Germaine. These are tears of happiness. My experiences with men haven't been...the best. I'm trying to convince myself there's such a thing as a good man."

Ger felt her intense emotions. "I'll wear a cape next time we meet. I'll be Good Man, to your rescue."

"Thanks Germaine. Someday I'll write my memoir and you'll know everything."

"Call me Ger, all my friends do. Can I see you again?"

Winnie impulsively threw her arms around Ger. It felt good. He didn't try to grope her, just gave her a little squeeze. "Yes. I'll call you. Get me another popup, would you?"

Ger used an app to call one of the little public transportation vehicles. He saw Winnie into the cab. He stood in the cold and watched until the car disappeared into traffic.

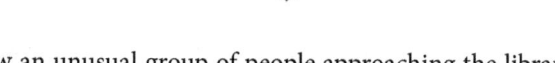

Kirra Bigbear saw an unusual group of people approaching the library, and groaned. It was late in her shift and her feet were tired. Her job today was to welcome visitors and direct them to the proper area. The spacious library was unusually busy; she had probably walked at least three miles already today.

There was something odd about this group. The woman in front looked capable, and the guy in back. The other nine were lost.

"What's happening back there?" the woman asked her. She had red hair sticking out of her pulldown cap, and a beautiful face with perfect symmetry of her features.

Kirra thought the question was rather abrupt, and a little rude. She spoke without warmth. "Children's play hour."

"Could we look in do you think?"

Kirra gave a disapproving look. These people looked weird in their mismatched clothing.

The man from the back stepped up. "We're supervising some at-risk adults," he said. "They may get benefit from watching the children."

Kirra saw that he, too, had perfect features. She looked at the others. There were two groups of three that looked remarkably similar. Male and female, all of them looked like they had just stepped out of a beauty parlor.

"We'll stay against the walls," the redhead said.

Kirra looked at the other nine visitors huddling behind like they were afraid of something. "I've never seen such perfectly formed faces, all of you! Why, it's amazing."

The redhead smiled; the man beside her looked frightened.

"You don't look like any at-risk adults I've ever seen," Kirra said.

"We're taking them around for exercise," the redhead explained. "We thought that watching the children play would...lighten them up a little."

Kirra shrugged. "Well...I guess it's all right, they look harmless. I'll notify the supervisor, but if she asks you to leave..."

"Then we'll go."

"Well done Pietra 23," the man said as they turned to go.

"That's a strange name," Kirra said. "Are there more of you?"

"That's what my father used to say," said the woman.

Kirra laughed and waved the group on. She came from a family of eight children and they had driven her father crazy. The man and the woman herded their charges across the main floor, which had computer workstations and racks of discs and printed books. "Books!" one of them cried. "Actual, real books!"

Kirra saw that the group was disturbing the browsers at the stacks. "At-risk adults," the man said in explanation, which seemed to satisfy everyone. Kirra relaxed.

Thirty minutes later she got a call from the children's supervisor and she made sure the strange group exited the building. Kirra heard "Pietra 23" mentioned, and the redhead answered. It was very strange.

A minute later a good looking man wearing a Carleton University name tag came in the side door and approached her at the front desk. It was Harry Kaine, the guy they had seen at Angelos. "Do you remember seeing a redheaded woman come into the library?" he asked.

"As a matter of fact, I did."

"Was there a group with her?"

It sounded like an interrogation but she was happy to help. Harry gave her his full attention and made her feel like the most important person in the world. "Indeed there was. A man and a woman and nine other at-risk adults." Kirra hesitated.

"Yes?"

"Well, they didn't look like any at-risk adults I've ever seen. More like people you'd see at a fashion show. You know, beautiful human beings. But kind of strange."

Harry's eyes it up. He regarded her silently, inviting her to speak.

"Two of them called the redhead Pietra 23, whatever that means."

Harry Kaine slammed his fist into his open palm in excitement. "Just the clo –, uh, woman, I'm looking for!"

Kirra was confused. "What did you say?"

"It's not important my dear. Which way did they go?"

Kirra could see how amped up Harry Kaine was. She pointed toward Fifth St. "They just left a few minutes ago, walking down Fifth."

Harry turned toward the front door and raced out of the building.

Kirra felt as if she had been interviewed by a VIP. Who is this Harry Kaine, she wondered.

Chapter **14**

Kirra Bigbear left work as usual at 5:30. She was to meet Liqao Chang at their favorite cafe on Mason. As she walked to her vehicle she noticed that the early May weather was warming a little. Finally!

Liqao was already there waiting for her when she arrived. She shook off a slight feeling of irritation. Her boyfriend was early for everything.

"Are you ready to order?" he asked just after she sat down.

Kirra looked at him, exasperated. "I've only been here for five seconds! Do you remember what we talked about last week?"

"Uh, yeah. Sorry. I'm too abrupt sometimes."

"Sometimes?" Kirra softened this with a smile. "Liqao, you need to relax. Meditate, play an instrument, go for walks. You're so hyped up all the time!"

Liqao spoke diffidently. "What you call hyped I call passion. I want to be even richer than Max Berglin and I have to work hard at it."

She had heard it before, and could see she had offended his male sensibilities. Maybe he just wasn't right for her...

Liqao picked up on this immediately. "You're not going to leave me are you?"

This was said in such a shocked and plaintive tone that it made Kirra laugh. But she noticed how sensitive he was to her moods. Kirra liked that. They had even talked about getting married, but Kirra was sure that Liqao wanted marriage more as a cultural imperative than his love for her.

"God Liqao, you need to lighten up."

Liqao laughed, the tension broken. "You are so...funny and cute and...observant."

"So are you." This is the side of Liqao she liked – the lighter and perceptive side, not the hard-driving businessman side. Kirra decided to change the subject.

Just then the waiter came and they ordered.

"I talked to Ger last night," Kirra said after the waiter left. "Ger said that you read the emails from a group called COSA and unfairly used them against the COSA spokesperson to quash a case. Is that true?"

Liqao blinked. He was shocked. Kirra could be quite blunt sometimes and wasn't hesitant about expressing her feelings or her opinions. He decided to laugh it off. "All's fair in love, war, and business."

Kirra smirked. "So you did."

It was no use trying to fool her, Liqao decided. "Yes I did. Police Chief Chastaine is being harassed by these people, who submit unsubstantiated claims with no proof. When they don't get their way they bitch and moan to the press. It's not fair because the police chief has to be measured and reasonable and can't defend himself."

"The job of the police is to investigate. You could be fired for what you did." Kirra could see that the implications were not lost on her boyfriend as Liqao squirmed in his chair.

"You won't tell on me will you?"

Kirra was irritated. Liqao was very sensitive and intelligent, and she knew he wanted her. But he was ethically challenged. Should she tell the police chief? Or the city's Ethics in Business office?

"If you're thinking about telling on me to Chastaine, don't bother. I already told him."

"You mean he asked you and you had to."

Liqao grinned like a little boy caught with his hands where they weren't supposed to be. "Well, yeah."

"You are a fiend, Liqao Chang. But you're cute and I can't resist you."

"Let's forget it," he said as the waiter brought their food.

As they ate Kirra tried to forget. But Liqao seemed to have a fundamental personality flaw. In matters of business or personal gain, he only thought of himself. Somewhere down the line he was going to get himself in trouble, and maybe take her with him.

Kjirsten Chastaine was nervous. Tonight was the night when she brought her boyfriend to the house. She wasn't looking forward to it at all. Dad...well, she loved him but he was, in his own words, "a hard ass." Things had changed a lot since he was born back in a more barbaric time in the late 1980s. Back then his kind of macho crap was still admired in some circles. Twenty years ago we even elected a macho president. The days of President

Conrad were over, but nobody ever told Dad. He still has a picture of the former president on the wall in his office.

Her comm device rang.

"Are we on for tonight?" M'basa asked.

She hesitated. "Yes...yes we are."

"You're worried about what your father might say to me. I understand. Don't let it bother you, I have a thick skin."

"I'm not worried about you, I'm worried about me! Dad can be pretty blunt. He drives me crazy."

M'basa Ogunfatidime thought to himself that the daughter was a lot like the father, but he didn't say it. "All right then. I'll be over at 7. If he kicks me out of the house we can go to the Blue Nile."

"That's not a bad idea! Why don't we skip the drama. I'll meet you over there." They both liked the restaurant, which served Ethiopian food.

"No you don't. You've been putting me off for months about this."

Kjirsten didn't deny it. "It's not like we're getting married."

"No, but I think your father will want to talk to the guy who's been sleeping with his daughter."

"None of my other boyfriends wanted to," Kjirsten grumbled.

"In my culture family harmony is very important. I at least want to know where I stand with your father." M'basa had no illusions. Nigeria was totally different from the United States. The culture here was much more adversarial than in his country.

"You're not telling me something."

M'basa laughed. "My father insisted!"

Kjirsten laughed with him. "I thought it might be something like that." She really liked M'basa because of his big heart and his warm smile. But sometimes he was a bit of a pushover where his family was concerned. Just like her. "All right, come over at 7. At least my mom will be there in case an argument breaks out."

M'basa arrived for dinner and began talking with Kjirsten and her mother. Ten minutes later the door slammed and heavy footsteps were heard from the hallway. Kjirsten and her mother exchanged anxious glances. A large man with a round, red face entered the kitchen and stopped abruptly.

"Who's this?"

"Jack, this is—"

"Thanks mom, I'll make my own introductions," Kjirsten said. "Father, this is M'basa Ogunfatidime, my boyfriend."

Jack Chastaine stared at the intruder in his kitchen. The first thing that came to his mind was that his daughter had used the word "father." That meant she was either mad at him, or about to be if he misbehaved. The second thought that occurred to him was that the boy had a good set of shoulders. The third thought was that if this asshole was sleeping with his daughter he'd...Then something occurred to him. "You're last name is Ogunfatidime? There was a player on the Michigan football team in the 1970s by that name."

The women were astonished at this speech, and for the thousandth time wondered about Jack Chastaine's thought processes.

"My grandfather," M'basa said.

"Holy shite! I always thought that Olatida Ogunfatidime was the most melodic name I ever heard. He didn't do much in his time with the program though," Jack said with a tone of disappointment.

This was too much for M'basa, who began to laugh. "No, Grandpa Ola wasn't a great football player. But he was a good man."

"That's important," Jack agreed. "And speaking of that, you'd better not be sleeping with my daughter."

M'basa had to suppress a laugh. This guy was right out of a comic book.

"Daaaaad!" Kjirsten cried. "I'm 24 years old! Do you think I'm still a virgin?"

"Jack! Kjirsten!" Lacy Chastaine said. "That's not a topic of conversation at the dinner table."

Jack seated himself at the big wooden table with his hands on his lap. "It sure the hell is, honey, when a stranger invites himself into my kitchen." He looked at M'basa. "Well, are you?"

There was nothing now but to tell the truth. "Yes. But my father thinks I should stay away from these Americans and marry a nice Nigerian girl."

Jack's big arms hit the table, startling the women. He stared at M'basa. "Well I'll be damned. That's just what I tell Kjirsten, but in reverse."

M'basa tried to keep a straight face but he couldn't. "Isn't that what all fathers say? And most mothers?"

Jack looked at his wife and began to smile. They began to laugh. Kjirsten, who had been looking anxiously at her father and mother, also started laughing. After several minutes everybody was calm again.

"OK," Jack said to M'basa. "You can stay. What's for dinner?"

"Leftovers," Lacy said.

It was like a scene from an old movie. The intruder was introduced to the parents, who were trying to get used to a stranger having intimate relations with their daughter. The ice was completely broken when M'basa asked Jack, "You're a fan of Michigan football?"

"Know everything about it from last season back to 1965." Jack spoke matter-of-factly. "Although the rule changes have made traditional football more like flag football these days. But I still enjoy it."

M'basa could see that Kjirsten was getting hungry and sensed her discomfort. "Why don't we forget leftovers and go out to eat? My treat."

Kjirsten's face brightened. Jack looked interested. Lacy looked relieved.

"You must have a good job young man," Jack commented as they all got up from the table.

"I do. I'm a systems analyst for the U.S. branch of Sundarum Electronics out of Chennai, India."

Jack's face relaxed and a sense of relief came over him. This M'basa was bedding his daughter but he wasn't a bum. "Yeah, I know the old man, a guy named Prasad Vetrithingal. He used to live in the city."

"Mr. Vetrithingal! Yes, great guy, getting on now. He started the whole thing. Grandpa Ola knows him, got me the job. I hope to be running the place in another ten years."

Better and better, Jack thought. Ambitious, halfway decent looking, his grandsons might not look too bad. "All right Lacy, where do you want to eat?"

"The Blue Nile," Kjirsten suggested with a smirk.

"I didn't ask you, I asked your mother."

"I've never been there," Lacy said. "What kind of food is it?"

M'basa answered. "Ethiopian. It's great tasting food and healthy."

Jack slapped a hand to M'basa's shoulder. "All right Gunga Din. Lead on!"

"I hope you don't beat me like in the poem," M'basa said.

"Huh?" Jack said.

M'basa grinned. "I'll tell you at the restaurant."

To Kjirsten's surprise, M'basa explained about the poem on the way. "The poem is by a British officer and set in India. Gunga Din is a native water carrier who gets beaten by the Brit soldiers because he doesn't get them water fast enough. He treats the wounded and eventually saves the life of the narrator. Gunga Din gets shot tending to the soldiers. As he's dying the soldier apologizes for his behavior and says, "You're a better man than I, Gunga Din."

Jack winked at Lacy. "An intellectual I see." The older man turned to M'basa. "Are you a better man than I?"

M'basa was nonplussed by this statement. But before he could think of a reply Jack said, "How do you think Michigan will do this year?"

M'basa mentally sighed with relief. "8-4, like always." He glanced at Kjirsten, who wiped her forehead with a gesture that said, "Whew! That was a close call."

After dinner was over they got up from the table. Joe Courvall was walking into the restaurant with a woman they'd never seen before.

"Hi Joe!" M'basa said. "Who's the lady?"

"This is Kesha." When Kesha saw the police chief she ridged up.

"It's that Kesha Montgomery!" Jack bellowed. "Are you the friend of that Sauvage woman?"

"I am. I want you to stop persecuting us."

"Me persecute you! It's the other way around!"

Joe moved Kesha along and turned back. "Nice to see you Kjirsten, M'basa. Hello Mr. Chastaine." Joe spoke this last in a sarcastic monotone.

The police chief looked suspiciously at Joe. The Courvall kid plastered an innocent expression on his face. Jack raised his finger and was about to say something, but thought better of it. "All right, Joe." His gaze followed Kesha as she found a table. "That girl is trouble."

Joe spoke blandly. "Certainly Mr. Chastaine."

Jack snorted and walked out, followed by the rest of the party.

M'basa stole a kiss from Kjirsten when they were walking to the car. "I think I'd better go. I'll call you tomorrow."

Kjirsten smiled at M'basa. "It wasn't *too* bad was it?"

"Not bad at all. Talk to you tomorrow."

After the Chastaine party left, Kesha regarded me with suspicion. "The police chief knows you?"

I shrugged. "I'm friends with M'basa and Kjirsten, his daughter. I've been over there a few times."

"He said I was trouble. He's the problem, not me."

I saw an underlying bitterness in this comment, but let it pass. As we talked and ate she told me about her work with abused children. I was getting a much better read on Kesha. She told me some of her life story, how she'd been raped in college by an athlete and sexually harassed as a grad student by her professor. Not anywhere near what Winnie Sauvage had gone through (according to Ger), but some terrible stuff. How could you get over that? The answer is you probably couldn't. My intuition told me that Kesha was now where she would be for the rest of her life: bitter at the treatment she'd received. I sensed that our relationship would be short-lived, but I didn't care. A fling with a slightly older woman; it was exciting.

As we left the restaurant Kesha said, "Do you want to come to my place?"

I didn't hesitate. "Yes. As long as you are comfortable having sex with me."

Kesha smiled. "I like sex. I just don't like getting raped."

I was glad that we were both on the same page about our relationship. I didn't want to hurt her.

Two days after the family went to the restaurant, Kjirsten was with M'basa on a Saturday night at the Mason Street coffee shop. They both had a good feeling about M'basa's meeting with her parents. Instead of the usual small talk and sex afterwards at M'basa's apartment, they talked about M'basa's home country of Nigeria, the history of the Igbo people, and about African culture. "The Igbo are my ethnic group, which has a history going back to the stone age. We are industrious people with a vibrant and colourful cultural display."

It was an unusual evening, for M'basa had never opened up so intimately to her before.

"I feel a deep bond with you, Kjirsten, that goes beyond just physical attraction."

"I feel it too, but remember that I'm the daughter of Jack Chastaine, police chief. The discussions in our home were about politics, police matters, and football. I have a practical mind."

"I understand. You're like Joe. He's an engineer, but I also feel a sort of...soul bond with him. And Karen. And even Liqao! Everyone in the gang I suppose."

"I'm sleepy," Kjirsten said finally, going to the bed and lying down. She was asleep in a few minutes.

M'basa felt energized after their talk. He went to the kitchen and made a cup of hot chocolate, something unknown to him before he came to America. He was pleased. These Americans were often shallow, interested only in personal gain, and often ignorant of anything outside their own country. Certainly Kjirsten's father was like that. Yet the Chastaine family had accepted him, a stranger and a foreigner, as boyfriend to their daughter. Tonight he had uncovered depths in his girlfriend he only suspected were there.

M'basa finished his chocolate and went up the stairs to the bedroom. He would not touch Kjirsten tonight; let her sleep.

Before he came to America he had studied American culture, which led him to the legends of the North American indigenous. He lay beside her thinking of the Navajo and their "message to the moon." The story went that when NASA was preparing for the Apollo project in the late 1960s, they did some astronaut training on a Navajo Indian reservation. One day, a Navajo elder and his son were herding sheep and came across the space crew. The old man, who spoke only Navajo, asked a question which his son translated. "What are these guys in the big suits doing?" A member of the crew said they were practicing for their trip to the moon. The old man got all excited and asked if he could send a message to the moon with the astronauts. Recognizing a promotional opportunity, the NASA folks found a tape recorder. After the old man recorded his message, they asked the son to translate it. He refused. So the NASA reps brought the tape to the reservation where the rest of the tribe listened and laughed but refused to translate the elder's message to the moon. Finally, the NASA crew called in an official government translator. He reported that the moon message said, "Watch out for these guys in the white suits. They have come to steal your land."[9]

M'basa almost laughed out loud. Yes, the white man is clever, but he is also acquisitive. He fell into a deep sleep.

————— • ✷ • —————

Lledren Cadwallader had been working in a soup kitchen for the past year, making minimum wage and wondering what he was doing with his life. His parents had sent him to Carleton but he hadn't been interested in studying anything on the curriculum. He had finally settled for a major in General Studies, a bullshit niche for those like him who were just there to get a piece of paper.

The last time he went home his father said, "I don't understand you Lledren." As usual, the old man was sitting in his recliner. "You had the opportunity for a good education and you wasted it. You're wasting your life working for nothing."

It was no good arguing with his father. Anybody born in the twentieth century was hopeless; their minds were set in an old energy, an old paradigm. Dad was a lot like Kjirsten's father, unable to get out of his narrow box of thinking. Dad's name was Trystan, which in Welsh means "noisy." A perfect name for a marketer. But Trystan was right this time.

"You're right dad," he had said. "I know I'm wasting my life but I don't know what to do about it."

"You can start by quitting that dead-end job and come to work with me."

Lledren groaned internally. "Dad, I'd be worthless to you. I have no interest in advertising or sales."

His father examined his son critically. Lledren was tall, with long hair and a disheveled look. The boy always looked out of place. "You don't seem to have an interest in anything, son."

Lledren brightened. That was exactly right. "You're right dad! I have no interest in anything this planet has to offer me."

"That's a strange thing to say, Lledren."

"But it isn't! The world" — he waved his arms around — "is all backwards. We think backwards, we do everything the wrong way."

Trystan didn't understand any of it. "The world is a wonderful place, son. You just have to find your niche."

Lledren shrugged. Sure, but how do you do that? "Can I do my laundry here?"

"So that's why you came."

"Well, yeah. I'm tired of my apartment."

"Stay overnight then and talk to your mother. Your old room is still available."

"Thanks dad. I will."

He got his laundry done. That evening after talking with his mom, Lledren felt a little better. Trystan didn't even bug him about his job. He went upstairs to his old room. His old posters were still on the walls. The room had been cleaned but nothing had been touched. Lledren felt this was a sign of respect. He went to bed in his room, feeling comfortable for the first time in weeks.

That had been a year ago; he hadn't been home since.

Lledren worked five ten-hour shifts every week. Working at the homeless shelter was rewarding, but it could also be stressful. Today a mentally ill man had attacked another man and stolen his plate. It had taken him and Karel Friedman, the wealthy progressive and activist, over twenty minutes to calm the man down and restore order.

Lledren didn't understand why Karel worked at the kitchen, or his motivations. Everyone knew that Karel had been a socialite and a womanizer as a younger man. Maybe he got religion, as his father would say. After they cleaned up and prepared the canteen for the morning shift, he and the activist sat on a couple of old chairs in the storage room.

Lledren was blunt. "Why do you work here, Karel?"

Friedman laughed. "My father left me a lot of money and I have a guilty conscience. I try to do something constructive every day."

"I have never been able to understand what I'm doing here," Lledren said.

To his surprise Friedman said, "Yeah. Life on earth. Seems pointless sometimes, doesn't it?"

"You get it, Karel. I look around, I see nothing I'm interested in. Except maybe try to figure out my purpose in life."

Friedman, who Lledren always thought a lightweight, said, "That's why we're here Lledren. Not to accumulate stuff, win awards, or get adulation. I tried all that when I was younger. That's all part of the Matrix. You gotta find your higher purpose."

Lledren felt his eyes tearing up. Friedman had nailed it. "That's it. So I'm not crazy because I don't want to work in my father's marketing firm. I'm not a weirdo because I couldn't care less about finding a job."

Karel nodded. "That's right. Finding yourself is far more important. It's one of the reasons I volunteer here."

"What's your life purpose, Karel? Most of my friends already know theirs. It's frustrating."

Friedman shook his head. "No, Lledren. Most people just do what their parents groove them in to do, or what society programs them to do. Almost everyone you think is centered and wise, isn't. They just look like they are, or they are faking it. Reaching the point where you question yourself is the first step toward wisdom."

Lledren's jaw almost dropped. That seemed right. Maybe he wasn't a loser after all. Maybe he was working through some deep life issues. He felt a lot better. "I think we should have more of these conversations."

Friedman smiled. "Any time." He looked at his timepiece. "C'mon, let's get some sleep. I'll be back for another shift on Friday. We'll talk more then."

When Lledren got home he felt that something inside him had awakened, but he didn't know what it was. He felt...different, in a good way.

Lledren woke up the next morning and saw the same dingy apartment bedroom with the open closet door, worn curtains, and a dresser with old socks on top. He looked at his timepiece; it was almost time to get up. Yet today the normal apathy he felt about his life had lifted a little. At the soup kitchen he was in a good mood all day.

"What's gotten into you Led?" LaQuanda asked as they were cleaning up.

"I had a good talk with Karel last night. Been feeling good since then."

"What was it?"

"I..." He looked at LaQuanda's cheerful, chubby face. The feeling inside him said that it was OK to tell her, that she would understand. "You see Quan, I always felt I was just drifting through life with no purpose. Everyone else seems to have a purpose."

LaQuanda's face expressed humorous disbelief. "Oh dear. Do you think I have a purpose?"

"You look motivated every day you come in here."

LaQuanda laughed. "Lled, I work all the hours God sends because I have to pay the rent. I don't have any skills or a trade." She looked around the dumpy soup kitchen. "I'm making the best of it. I don't think you can call that a purpose, but I've grown to like this place. I like helping people even when they are surly and nasty."

"I feel the same way, but you're always cheerful."

"Maybe that's what life is about. Finding satisfaction in whatever you do."

Lledren's eyes widened. "That's right Quan! I mean, what you said is just an old platitude, but I understand it now."

"Your face is a lot brighter, Lled. You look better today."

Lledren looked into her face and saw a childlike pleasure and happiness for him. An unusual feeling of joy went through his body.

"You know I'm gay, right?" he said finally.

"Silly boy! Nobody gives a shit about that."

"Try telling that to my old man. He wants grandkids. My parents are so inhibited, especially my father. You know, born in the twentieth century and all that."

"I know it. My mother is the same."

They both sat silently in the back room on benches, their work done for the night. Lledren said, "I want to be friends with you forever, Quan."

She reached over and gave him a hug. A big smile came over her face. They sat staring at each other for a second, then Lledren burst out laughing. LaQuanda joined him.

After closing up they walked together silently. "I gotta get a popup and go home," she said. "I'm tired."

Lledren gave her another hug. "Thanks Quan. I don't know what you did, but I feel like a flower that just bloomed."

One of little electric vehicles appeared around the corner and LaQuanda got in. She waved. "See ya!"

Lledren was excited and didn't feel like going back to his dingy apartment. He called a popup and had it drive him to his parents' house. His parents were late risers and liked to stay up past midnight.

He let himself in, intending to go up to his old room without making a fuss. He had a standing invitation to show up anytime. He took a peep into the living room and saw his parents watching a vid. Dad was on the recliner and mom was in her wing chair.

His mother was excited to see him. "Lledren! Come on in!"

"To what do we owe this honor?" his father said a little sarcastically.

"Thought I'd sleep here tonight. My apartment sucks."

Trystan sighed. "For the hundredth time, quit that goddam soup kitchen job and work with me."

Instead of getting angry or depressed he felt cheerful. "Ain't never going to happen, pops."

"You're aimless, son. Your life has no direction. If you can't find your purpose then at least make some money while you're figuring it out."

Lledren faced his father squarely. "You just nailed it dad. I think I know what I want to do with my life."

"I hope it involves quitting that crummy job."

Lledren grinned. "Nope. I can use that job to my advantage. I'm going to start a VR blog. Don't know why I never thought of it before." A VR blog, that was it! VR software and equipment allowed you to present everything in 3D. The Visualize company would pay you for great content. He'd start recording at the soup kitchen. It was the perfect place to show the issues facing society. He would not shove anything down the public's throat. He would tell poignant stories about people that demanded solutions to human problems.

Trystan Cadwallader noticed a liveliness coming from his son. He'd never seen that before. "You mean you want to be a writer?"

"Sort of...a virtual reality blogger. I know how to get some monetary support. If I'm good, it'll work out."

"I'll pay you a lot better than the government."

"Thanks dad, I mean private support. Working in that kitchen is the perfect environment for me. It reminds me of what I want to write about. It's good copy."

Trystan shrugged. His son had weird ideas, always jumping from one thing to the next. "All right. Not what I'd do, but at least you're showing interest in something."

Lledren came over to his father. "Get up. I want to give you a hug."

Trystan was startled but got himself out of the recliner. Lledren put his long arms around his father and gave him a big squeeze.

"What's all this, son?" Lledren had never been demonstrative. He was a loner; he rarely even hung out with that crowd from Carleton.

"You just showed me where my life should be going."

Trystan was baffled. "I didn't do anything!"

Lledren slapped his father on the shoulder, something he knew the older man would understand. "Well, you did."

Lledren sat with his parents for an hour watching their vid. It was the most comfortable he'd felt here since he was a child. He was excited about his new project. Ideas were percolating in his head.

When he went up to his old room he called Joe Courvall. "I know what I'm going to do with my life," he told Joe.

"Really? What?"

"I'm going to write/produce a virtual reality blog about current events and social injustice, and how it affects people who come into the soup kitchen."

"Good call. You've always been into social issues."

"Yeah. There's a lot in here —" Lledren pointed to his heart and his head — "that no one knows about." Lledren paused for a moment. "I've kind of...discovered myself Joe. Whatever I have inside is going to come out."

"I always wondered why you didn't get into journalism. The old media giants are dead."

"I never had any confidence I could do it. Now I do."

"Good luck."

"Luck has no part in it. If you hear of anything, let me know. I need to develop a network of sources."

Lledren could see Joe debating with himself for a few moments. "Look into COSA, Children of Sexual Abusers organization. You might find something interesting there."

"Thanks Joe. I've also got a few people I can talk to down at the kitchen. A progressive lawyer works there. He ought to be good for a few leads."

I could see how excited Lledren was. "I'm happy for you."

———— • ✳ • ————

A week later Kesha found out that I told Lledren to look into COSA. We had a big fight. She said she didn't want to see me anymore and that I was just another dickhead, and that men were all stupid. Oh well. I told her I thought Lledren would give them some good publicity, but Kesha thought differently.

I asked Karen if I was wrong about it but she just shrugged. I don't think she's too keen on Kesha.

Working at the *Chronicle* was boring sometimes, Karen Everard thought as she sat in the Mason Street Cafe having a cappuccino. Two weeks ago she had been on a roll. Her feature article about the Interfaith metaphysical group had been published. The week after that, the one on the Transhumanists, led by the magnetic Harry Kaine. That one had generated a lot of reader interest for the paper. Since then it had been a corny article titled, "Coming of Age at Mason and Third – In 1985, it had everything a young man needed." Then a really boring piece on the Midland Kiwanis Club. She could hardly believe that such an outdated organization was still around in 2035. She had just finished an article on a bond bill that had narrowly passed a City Council vote. Now she was looking for something with teeth. She was still writing up her piece on the disappearing woman on the Diag, but its importance had faded in her mind.

Frustrated, she walked out of the cafe and went down State St., wandering in and out of shops. She saw the sign at the end of the block overhanging a seedy-looking building: "East Side Kitchen." She knew this was a hangout for people down on their luck, where they could get a meal, keep warm for a while, and hopefully get a billet at the homeless shelter for the night.

She had never been in the place, but she knew Lledren worked there. For some reason Joe told Lledren last week to investigate COSA. What was that about? Her reporter instincts kicked in. It was a cool late May evening and she was getting cold, so she walked in. The first person she saw was Lledren, handing out food and looking excited to be doing it. This was curious, for she had never known Lledren to get excited about anything. He looked up to see her staring at him.

"Karen! What are you doing here?"

At that moment a man in rumpled clothing swung open a metal door from the back and approached Lledren at the front counter. "Do we have any more wheat flour?"

It was Karel Friedman, the famous public defender and progressive irritant to the conservative Midland City Council.

Karel turned and stared at her. He seemed mesmerized. Karen was uncomfortable under his gaze. "This guy is staring at me, Lledren."

Lledren smiled at her like a brother. "Sorry. Karel thinks you look...nice. You do."

Karen noticed that many of the clientele, mostly men, were also looking at her. She felt Karel was a little creepy.

"A sight for sore eyes," Karel said.

She ignored the older man. "Are you free to have a coffee?" she asked Lledren.

"I'd love to, but my shift doesn't end until 6. Then I've got to interview Karel here and edit my COSA material."

Karen was surprised. She noticed some equipment sitting on a small table at the back of the room. "Are you a journo now?"

"Yes! I'm filming for that VR company, Visualize. Check out my VR blog, posted the first story today."

Karen was impressed. This was a motivated Lledren she had never seen before.

Lledren waved his hand around the facility. "People are still going hungry even after the budget re-allocations. It's a crime. I'm showing this in living 3D so that people won't be able to ignore it anymore."

"That's great Lledren."

"C'mon Lled, the people are waiting," Karel said. Karen realized that a couple of customers were waiting in line, and getting impatient.

Time to go. Karen called a popup and went home. Why would Lledren of all people be investigating COSA, the child abuse organization? That might be a good story if she could get something for the *Chronicle*. She could also question Lledren about Kesha and this Winnie Sauvage woman Germaine Robinson liked. Word on the street was that Winnie's father, Winthrop, was into child pornography, and that Kesha and Winnie Sauvage were tight. Were Winnie and Kesha pervs hiding as child protectors? It had been done before. That was another possible story. She would to talk to Lledren and Joe tomorrow.

CHAPTER 15

3013 AD

The day after their trip to the past, Pietra 23 and Alan 133 addressed the enhanced clone group in the Shifter room.

"Your brain implant interface is missing," Li 355 observed, indicating Pietra 23. The slit at the back of the head normally contained a black neurointerface chip, allowing it to pick up the signals from the Tower of the Overseers. These signals gave life-giving support to the clone bodies, but also allowed clones to be monitored and controlled by the Submission module within the brain implant.

"That is correct. My LRI has stabilized at 187. I no longer need it here."

The clones looked at each other in astonishment. 187! Such a reading was unheard of.

"We can survive here without the brain implants," Pietra 23 said. "To make sure my LRI is permanent I will stay here for the next two days, alone. Then we will measure my LRI again."

The clones were excited but also worried. "You will miss your Guardian shift," Li 57 said.

Pietra 23 spoke contemptuously. "My Guardian shift! What can these impotent Overseers do to me now?"

Alan 133 was impressed. Pietra 23's vitality was superior to everyone in the group.

"From now on, address me as Pietra. I no longer consider myself a clone."

The others gasped, including Alan 133.

"Yes Pietra 23," Li 57 said. She was at her usual place at the Shifter console. She and Li 308 still had not gone into the past because they had to monitor the timeline and run the Shifter.

"Did I not say Pietra?" she barked.

Li 57 cringed. "I'm sorry...Pietra." She was ashamed to realize that she had spoken in the Submissive, as if Pietra 23...Pietra...was an Overseer or a Commissar.

Pietra nodded to Li 57. "That is well, Li 57. We do not bow to these Guardians, and certainly not to our own people."

Li 57 wanted to tell Pietra 23 that such boldness was a thing impossible for her, but she was afraid.

Pietra noticed this. "If you have something to say, Li 57, say it."

Li 57 summoned all the courage she had. "I am far more afraid of you than I am of Christian 172. Perhaps the Guardians are right. Clones need direction and control."

Alan 133 was shocked. "Are you not inspired by Pietra, Li 57? Look! She is magnificent! And so can we all be."

Li 57's clone body began to shake. "I...I am frightened. No, no, this cannot be right." Li 57 got up and began to back away from her console.

"What are your plans, Li 57?" Pietra asked.

"I will return to my Guardian duties...I can no longer do this." Li 57 ran out of the Shifter room, and exited via the exposed front entrance.

Pietra frowned. "That clone is endangering us all. Perhaps it is better that she leave the group."

"Does anyone else wish to leave us?" Alan 133 asked, looking at Li 57's friend, Li 308, who sat at his console. "We are now short one of our best temporal technicians." He glanced at Pietra; Pietra nodded. "We want no one here against their will."

No one spoke.

"That is well," Pietra said. She bowed to Li 308. "Can you handle the Shifter by yourself? Alan 133 and the other Lis can help you if you tell them what to do."

Li 308 became inspired by Pietra 23's trust. "Yes, Pietra 23."

Pietra laughed and wiggled her finger.

Li 308 corrected himself. "Pietra, not Pietra 23."

Pietra smiled encouragingly at Li 308. She must remember to treat these clones like 21ster at-risk adults. She wondered whether at-risk adults in the 21st had low LRIs. She would ask Li 308 to research this.

"Will we all feel like you after more trips into the past?" Alan 133 wondered.

"I certainly hope so."

"Hovercraft above," Li 308 announced suddenly.

Soon footsteps were heard at the front entrance. Two dozen clones came into the building, led by an Overseer.

"Patton 1!" cried Li 308. "We are lost!"

"You are to come with us," Patton 1 ordered. "All of you are subversives."

Pietra smiled. "So. Your running dog Christian 172 informed on us."

Patton 1 nodded grimly. "What did you expect?" He looked into the room and saw the Shifter. "It is true. You have prohibited technology."

If Patton 1 expected Submissive behavior he was to be disappointed. Pietra stood her tallest and raised an arm, her finger pointed at the Overseer. (She had learned this trick from a 21ster movie.) "Remove yourself and your clones from our presence. We are free men and women!"

Patton 1 took a step backward, feeling the force of this statement and this clone's energy. Then he stepped forward. "You are clones! Paul 26, activate their brain implants at the highest Submission setting."

Pietra was amused. The other clones began to feel a headache, but nothing more. Li 308 cried out and slumped in his chair at the Shifter console.

"Your controllers are useless here," Alan 133 said with delight.

A trooper pointed to Pietra. "That clone has no brain implant interface."

Astonished and then angry, Patton 1 ordered a dozen more of his clones into the room. "Seize them!" he cried.

As two of Patton 1's clones approached her, Pietra placed one hand on each of their chests and shoved. The two fell into the others, knocking several of them down like 21ster bowling pins. Patton 1 watched in silence. Pietra realized that her strength was enormous here, and did not wish to inflict pain on the helpless clones.

From the back, one of the Overseer's troopers rushed Alan 133. Although not as strong as Pietra, Alan 133 easily pushed the clone back into the minions of Patton 1, who were gazing fearfully now at the enhanced clones.

"I had to see it for myself," Patton 1 said to Pietra. "Christian 172 was correct. You are dangerous lunatics, we can do nothing to stop you for now. But we *shall* stop you. On the timeline, and here, after we send our own troops to the past."

Alan 133 felt fear as the Overseer marched his clones out of the facility. He went over to Li 308, who was moaning in pain. "The Overseers are gone now, brother. Soon you will feel better."

"It's on now, as the 21sters say," Pietra said to the enhanced clones. "There may even be violence. If you wish to leave us, do so at once. In a few days it may be too late."

Patton 1 returned to the Tower of the Overseers in his personal hovercraft, along with the Overseer troops. He summoned Christian 172 to the Tower. The two were taken to the top floor of the building.

As they exited the lift, Commander Patton 1 and Commissar Christian 172 saw a huge work area with a Time Shifter, surrounded by banks of consoles in a circle, all manned by temporal technicians. Patton 1 noted with satisfaction that their Shifter was three times as large as the rebel one in Building 103.

"Report!" Patton 1 ordered the head technician. This clone sat on a raised platform with a console, upon which rested a huge 3D tank that was now filled with a roiling white mist.

"No further temporal activity by the Resistance as yet."

"Good! I was afraid they would immediately send their team back to the past for further enhancement. Monitor the timeline. If you discover any disturbance, communicate immediately with me or Commissar Christian 172."

Caesar 13, the head technician, bowed his head in submission.

Patton 1 thought furiously. They would have to send a team to the past immediately for enhancement, to counter the Resistance. How many clones? He had only counted twelve in Building 103. To ensure success he would need at least twenty enhanced clones. More would be too hard to control.

"Recruit a team of twenty clones," he said to Caesar 13. "We know now that exposure to a more vital timeline enhances LRIs far too high. Insert your clones to the same temporal location as the Resistance. Disrupt their operations. Kill them all."

Caesar 13 was shocked. "*Kill* them? How is this to be done?"

"You fool! Recruit the most antisocial clones, enhance them, send them down the timeline. When the Resistance appears, hurl your troops against them. Must I tell you everything?"

Patton 1 walked out of the Shifter room into the hallway, mumbling to himself. If only Khan 1 had not built the Shifter ten years ago, in an attempt to "protect Guardian society from temporal interference." Patton 1 had thought it unnecessary, a paranoid delusion. Now, the situation with the rebel clones was spiraling out of control. They had undoubtedly built their own Time Shifter using the designs stolen from the one in the Tower.

Khan 1 was a fool! All would have been well had the Shifter not been constructed. He and Khan 1 would have lived the rest of their lives unchallenged at the top of society. Why meddle with a good thing? His sensible advice had been ignored.

Patton 1 walked to an opaque lift door in the wall. On his signal, the transparency cleared and he stepped in, returning to his quarters in the Tower, only one level below the Khan himself. Khan 1 had private apartments across the hall from the Time Shifter complex, which also contained the control and communications center for all of New

York. The Supreme Overseer received feeds from every clone, monitor, and hovercraft in the city in his private quarters. Khan 1's LRI was only 95, but he was the most cunning clone in all of New York. Cunning and ruthless, but stupid.

Patton 1 considered Caesar 13, the head time technician, who had built the Khan's time device. Could that diabolically clever clone have given the rebels the knowledge to build their Shifter? That clone was known to associate with the so-called "reformer" group. He would keep a much more careful eye on Caesar 13's activities from now on.

Patton 1 set his brain implant for stimulation and activated the special console in the wall (available only to the highest Overseer cadre). As the provocative frequencies traveled through his body like a drug, an idea occurred to him. Why not depose Khan 1 and rule himself, the unquestioned king of New York? With twenty enhanced clones he could destroy the entire Guardian force! And kill the imbecile Khan 1. Patton 1 licked his lips, overwhelmed with emotion, as he thought of hurling the Khan's lifeless body from the top of the Tower. After that he would move to the top floor and destroy the Shifter, ensuring his rule for the rest of eternity.

———— • ✹ • ————

Alone in Building 103, Pietra was drawing up a plan to salvage the human race in 3013. Even though the resistance clones had now been identified by Patton 1, it was agreed that they should continue to act normally for the next two days and avoid Overseer attention. The enhanced clones would behave as if they had been cowed by Patton 1's attack and report for their Guardian shifts while Pietra stayed alone in the Shifter room. In this way she could test to see whether her enhanced LRI was permanent. Meanwhile the other clones would gather valuable intelligence.

At the end of the day Li 57 walked into the Shifter room through the hidden entrance. Pietra regarded her silently. What was the bitch doing here? Pietra smiled to herself; she was thinking and acting like a 21ster.

Li 57 approached her as she sat at the small cafeteria table. Pietra could tell that the clone was steeling herself to say something.

"I request permission to return to the group. I can no longer live comfortably in Guardian society."

Pietra was unsympathetic. "That's what you said about us. You abandoned the group when we needed you the most."

Li 57 stood taller. Pietra was impressed; this clone was overcoming her fear.

"I understand that and will accept any punishment that is meted out."

"Why the change?"

"Because...because I can feel my life force energy receding. I didn't realize that my LRI was being enhanced just by contact with our group. In a few weeks I will just be a clone again." Li 57's face showed anguish. She pointed toward the front door. "Out there it's...intolerable. I understand that now."

Pietra smiled, for this statement resonated with her. "Very well, Li 57. You are reinstated. Do you report to Station Seven and Commissar Christian 172 as usual."

"What should I do if he uses the submission module? Like Li 308, my LRI is not high enough to resist the submission frequencies."

"Avoid this by acting stupid. Tell Christian 172 that the others were frightened by the assault of Patton 1 and his troops, and have reported for their shifts. Do not display emotion."

"Is this true? Have the others reported?"

"Yes, but we are merely allaying suspicion until we can all go back to the past again. You and Li 308 must also become enhanced. You are the only two clones who have not yet traveled to 2035."

At this Li 57 became excited.

"Calm down Li 57!" Pietra said. "If Christian 172 sees that face he will be very unhappy. But there is little worry. These Overseers and Commissars are weak and stupid. Christian 172 will already know that the others have reported and are on Guardian duty. Look placid and dull and you will be all right."

At this Li 57 smiled, for being in Pietra's presence was raising her spirits. "I understand now, Pietra." The clone assumed the stolid aspect of a typical Guardian clone.

Pietra laughed. "Yes Li 57, that will do nicely."

The reinstated clone walked out.

Well, Pietra thought, that was a lucky break. They had regained their best temporal technician!

That night Li 308 arrived, asking about the status of Li 57. Pietra told the clone that she had rejoined the group, and that both of them would have to go back to the past to be enhanced. Li 308 was not excited about this, but he was thrilled (as thrilled as a Guardian clone could be) about the return of Li 57.

Li 308 was about to return to his domicile when Pietra asked him about how time worked. "Are we on a separate timeline here in 3013? If we go back to the 21st century, is it likely that we can change the future?"

"We are in a temporal thread that is gradually detaching itself from the temporal manifold," Li 308 replied. "A dead end."

"How did that happen?"

Li 308 scratched his head. "Li 57 and I think that when life force readings decline, the temporal manifold detaches the thread from the main flow of time. Somewhat like a cyst that contains dead material that is no longer usable by the body. Eventually the cyst falls away."

Pietra was shocked by this. "So there's no hope then."

"I did not say that. The temporal manifold is not a line where time flows only one way, from past to future. It contains many potential threads, or timelines. Past affects future, but future also affects past."

"So there's hope that our intervention in the past can change our timeline."

Li 308 spoke fatalistically. "We don't know yet. Temporal energy is conserved, Pietra. That means time tends to bounce back to what it was, like a rubber band. The temporal flow resists change."

Pietra was disappointed, but her enhanced life force reading did not allow her to mope.

Li 308 interrupted her thoughts. "One thing we do know: There aren't any humans after 3209 AD."

Pietra was appalled. "That's when humanity dies?"

"You might say that. Timewise, death happens when the temporal energy of a time thread goes to zero. When we look through the Shifter to the future we don't see anything past 3209: no planets, no stars, no universe even."

Li 308 saw how shocked Pietra was. "But that also has a positive aspect. We know that there aren't any humans past 3209 AD on this timeline to interfere with what we're doing. It's us against the Guardians for the future of the human race."

Pietra gulped. "Patton 1 will surely make good on his threat to enhance his clones He will probably act quickly. We must expect and prepare for attacks from those fiends."

Li 308 shrugged. "Pietra, we only have one choice for survival: to revitalize this time thread and merge it with the main flow of temporal energy and its life-giving support. We have almost 200 years to do this. There is hope. Past affects future."

Pietra nodded her agreement. A positive attitude was necessary to make changes. She had learned this in 2035.

"I'm sorry, Li 308, that I frightened Li 57 away yesterday. She knew all of this?"

Li 308 nodded.

"If you see her before I do, apologize to her for me."

"OK." Li 308 left.

After her talk with Li 308 Pietra stumbled back to a pull-down bed in the wall of the cafeteria. She woke up the next morning with a start, dreaming that her LRI had decreased and that she was just a clone again. But when she got up she felt no loss of vitality or energy. She avoided her Guardian shift, eating one of the manufactured SynthPak meals for breakfast. This food was beginning to taste like cardboard to her enhanced biology, but it still sustained her.

Pietra worked all day, planning their next move. She was sure that the Overseers were going to send their clones back to the past, enhancing them. Then it would be many dozens against their 13 when Patton 1's troops returned to 3013. The Resistance needed more clones! Pietra thought about this for a while and dismissed it. Recruiting others to resist in this dead, brainwashed society was pointless. Clones had been taught to be submissive for almost a thousand years. So it was 13 against the entire Guardian society.

The best solution was to prevent the Overseers from sending their clones back to the past in the first place, so they could not become enhanced. The only certain way to do that was to assault the Tower and destroy their Shifter. However, even she trembled at the thought of a direct attack on the Tower of the Overseers. It was the most feared and imposing building in New York, the center of Guardian power for almost ten centuries. How could she get her clones motivated? No, it wouldn't work unless all in their rebel group could be further enhanced to 21ster levels. That meant more trips back to 2035.

Suddenly she felt a stab of fear. What if Patton 1 attacked them when they were in 2035? The Overseers could eliminate them before they even got back to 3013! Then Patton 1 and his sociopaths would be able to rule unopposed until the end of time in 3209.

Pietra was panicking now. She got up from her chair and walked around, trying to calm herself.

She sat down again and spent the entire day thinking it through. An attack had not happened on any of their previous trips to the past. Therefore it would not happen or she would have a memory of it. If Patton 1 decided to attack them in 2035, it would have to be during a new trip the rebels would take to the past. The group would have to be very careful when they reinserted into 2035.

What if Patton 1 decided to attack the rebels in 3013? Well, the Overseers wouldn't do that until they had their own troops at 21st normal LRIs. They had already been defeated twice using their normal security forces. Yet the Time Shifters allowed anyone to spend lots of time in the past and come back to the future an instant after leaving 3013. The Overseers could have already enhanced dozens of their troops. Pietra began to panic again. The attack could come right now. At this very moment Patton 1 could be on his way to destroy their Shifter with dozens of enhanced clones!

Pietra bolted from her chair, intending to go outside to see if that demon Patton 1 was outside. As she reached the front door she calmed herself. Patton 1 would want to eliminate them all at once, when they were all together. Everyone was scattered now.

Pietra walked back to her chair, starting to feel excited. There were 7 males and 6 females in their group. Enough to re-start human society here in 3013 if their reproductive organs began to function again! If they acted quickly they could insert into the past and return to 3013 with a powerful force, before Patton 1 could get his troops enhanced. Then the Tower could be attacked and the Guardian Shifter destroyed.

After that, the rehabilitation of Guardian society.

Pietra went over the plan again. She was beginning to dimly understand the complexities of time travel. She remembered something Li 308 told her. The time techs, after many insertions and extractions, had gained in knowledge. They knew now that it was impossible to go back and forth in time to any point you had already lived through. There was some kind of temporal barrier that wouldn't allow it. So the rebels couldn't insert into 2035 during any time they had already been there, or they would meet themselves. And they couldn't insert back to 3013 from the past to any time before they left 3013, for the same reason. Otherwise you could have many copies of yourself all over the timeline and make a big mess. The same would hold true for Patton 1 and his sociopaths.

This thought calmed her. Then she had another thought. What if Patton 1 attacked them in 2035 from a point in the future here in 3013? The Overseer troops had not been to the past yet, so they could attack the rebel group on any of their previous trips to the past! There was no temporal barrier for them. But no. She had no memory of this, so it had not happened. But could it?

Pietra got up from her chair again, frustrated. Time machines made planning almost impossible! She sat back down and spent a lot of time going over everything again.

Eventually she understood that there was only one option: the rebels had to insert back into 2035 and emerge as quickly as possible, before Patton 1 could create an enhanced force of his own. They had to insert first thing tomorrow morning for a long excursion to the past, getting all of their clones to 2035 LRI readings, and return right away. Then, an immediate attack on the Tower before Patton 1 could get set. She faced the reality that it was her job to lead the group. She spent the rest of the evening steeling herself for this mission and planning how to attack the Tower.

Just before she went to bed Pietra used her chronometer (which doubled as a communications device) to send encoded messages to all the resistance clones: "Report 0600." Of course the Overseers could decode this, but she hoped it wouldn't matter.

It was now almost midnight. She had been alone for two days and still felt no loss of life force or vitality. She was cheered by this. She ate another synthpak in the small cafeteria at the back of the Shifter complex in Building 103, and then walked back and lay down on the cot that came down from inside the wall. In five seconds she was asleep.

The other clones reported for duty in Building 103 exactly at 0600 the next morning. Among them was Li 57. "I am very glad to see you," Pietra said.

"Why have you called us in so early?" Li 308 asked.

Pietra told them of her plan, and the ultimate goal of attacking the Tower and disabling the Guardian Shifter. She was met by looks of shock and disbelief.

"Why are you so afraid?"

Even as she asked the question she knew the answer: the LRI's of the other clones were still not high enough. Only when the others had reached 21ster levels would it be possible.

This would be a good time to test her LRI. Pietra went into the testing chamber and was tested by Sigmund 78. In a minute the reading was displayed. "187," Sigmund 78 said in an awed voice. "Same as before. That is even better than 21st human normal! Can it be true that you are no longer a clone?"

Pietra was ecstatic. "I feel strong and powerful," she said. "I feel like a woman. My brain implant interface has been out for over a week now. The next step is to see if my reproductive organs have been activated. But that will have to wait."

She made all of the clones step into the chamber.

"Alan 133, you're next." The reading was 156. The others were between 138 and 144. Li 57 and Li 308, who had not yet returned to the past, had readings just over 110.

"These readings are all at least 40 points higher than your baselines," Pietra said. "Even Li 57 and Li 308 have increased their baseline reading by over ten points, just by association with us!"

The clones were amazed.

"Another long trip into the past and we shall all be in the 170s," Pietra said. "Please get used to the idea that we shall have to disable the Guardian Shifter. We need to do this as soon as we get back."

Pietra observed the look of hesitation and doubt on their faces. "We can talk about that later," she said. "We need to go back now. We might as well return to the warehouse just after we left. The goal is to spend an entire day in the 21st and circulate among the 21sters, soaking up their life force energy just as we did the last time."

Sigmund 78 grumbled. "Let's return in summer when it's warm."

"No," Alan 133 said. "Pietra is right. The more clothing we have on, the less obvious it is that we are clones. Our printed bodies are too perfectly formed and will cause suspicion."

"Very well," Sigmund 78 replied, seeing the sense in this. "But I hate those 21ster clothes."

They retrieved their 21st clothing from the previous trip to the past. One by one, they were transferred by Li 308 and Li 57 to the warehouse in the industrial section of Midland, one hour after they had been extracted from the previous trip.

Patton 1 returned to the Shifter area after his session with the brain stimulator. He saw Caesar 13 sitting quietly at his station, monitoring the device.

"What are you doing here, Caesar 13? I ordered you to assemble twenty clones for immediate insertion. The rebels must be killed."

"This has already been done. I merely await their arrival."

"You have identified and summoned twenty clones in…90 minutes?"

The chief technician stifled his irritation. "Yes sire. To identify twenty antisocial clones is a trivial matter. Assembling them merely requires a summons to their implants."

Patton 1 was annoyed at the head technician's dismissive attitude, but he was pleased that Caesar 13 had referred to him as "sire." Was this an omen that supported his newly-made plans to overthrow Khan 1? He regarded Caesar 13 carefully as the technician went about his tasks. This one was highly intelligent and clever. He would use him to ascend to Supreme Overseer. Then, of course, he would destroy him.

Patton 1 walked around the facility, observing Khan 1's setup.

Caesar 13 watched Patton 1 carefully, noting a change in his demeanor. Clones hardly ever exhibited more than mild emotion, but this Overseer was excited about something. From his raised platform Caesar 13 saw Patton 1 circling the monitoring stations, checking the displays, observing his time technicians, and sometimes staring balefully at the Shifter with a look of determination. An hour before, he noticed that the rebels had recently inserted again down the timeline to…2035. He was obligated to tell Patton 1 of this development.

Patton 1 interrupted his thoughts. "Is all in readiness for the insertion into time?"

Caesar 13 bowed his head. "All is ready. The clones should be here shortly."

Patton 1 checked his chronometer. "What is taking so long?"

At that moment a hovercraft appeared through a transparency at the docking station attached to the outside wall of the building. Twenty clones, led by Christian 172, walked into the Shifter room.

A technician spoke loudly. Tesla 8, at her console, confirmed that the shipment on the hovercraft was correct. "These are the clones we're looking for."

Patton 1 smiled with satisfaction. He was still all abuzz from the implant stimulation. As the twenty clones trooped into the Shifter area, he studied them. How well had Caesar 13 chosen? He saw that these clones were rough and undisciplined. Patton 1 saw one of them roughly squeeze out a technician as he sat down on her console chair. When the technician tried to regain her seat, the clone shoved her to the floor. Satisfactory!

Caesar 13 reluctantly resorted to the Submission modules at the lowest possible level. This was necessary to suppress the anti-social activities of the twenty clones as his team of time technicians prepared them for insertion. All he wanted was to get them out of here and into the past. He spoke to Christian 172, whom Patton 1 had appointed to lead the group.

"See the images in the display tank. By the order of Patton 1, these are the 13 Resistance clones that must be neutralized."

The twenty clones gathered around the display. They eagerly memorized the faces, for it was a chance to revenge a lifetime of slavery on someone. These clones didn't care who they hurt, as long as there was a target for their hatred.

"You must attack the right targets," Christian 172 said firmly. "Only these and no others or an unwanted disturbance will be created. If you disobey you will be terminated."

Caesar 13 knew from the history tranches that in 3013 or in the 21st, a certain percentage of the population enjoyed destruction. Christian 172 was of this ilk. And Patton 1, and Khan 1, and all the "leaders" of Guardian society. For the millionth time Caesar 13 wished that things were different. Their small group of reformers had no effect on Guardian society, and must meet in secret. Yet he could rebel in his own small way: He decided not to tell Patton 1 about the rebel insertion.

"Are you ready, Commissar?" he asked Christian 172. "If so, gather your clones and stand on the Shifter platform."

"Scum!" Christian 172 said. "Come here."

The twenty clones were afraid and refused to approach the platform. "Animals, you have ten seconds to come over here before I fry your brains."

Caesar 13 almost laughed out loud and felt better about his disobedience. Patton 1 and Christian 172 were the best and brightest of Guardian society! Sociopaths, both of them. Suddenly he smiled. This Overseer project was doomed to failure.

"What are you smiling at, technician?" Christian 172 said. He spoke menacingly.

Caesar 13 bowed his head in submission, but continued to smile to himself. "I await your orders, commissar." He had risen to his high position through intelligence and ability. He was an old hand at the game of going along to get along. Besides, these incompetents needed his expertise.

Christian 172 thought he saw a smirk on the face of this arrogant clone, but he was too eager to visit the past and destroy the rebels to bother with punishment.

"The Shifter has been programmed for 21 clones," Caesar 13 announced portentously. "Please step onto the platform."

Christian 172 herded his reluctant clones using his Submission module. When all were in position Caesar 13 began the insertion protocol. When the gigantic metal rings slowly began to turn above them, several of the clones ran off the platform, a look of abject terror on their faces.

"Fools!" Christian 172 cried, turning up the Submission frequencies. Heads bowed, the fleeing clones returned to the platform like zombies in a 20th horror movie.

Christian 172 laughed, but Caesar 13 felt a pang of sympathy as he slowed the machine. These clones were sociopaths but deserved some little consideration.

"Are you ready commissar?"

"Ready."

With the Submission modules of the twenty clones activated, Caesar 13 reactivated the Shifter. The huge metal rings began to slowly rotate faster and faster. A blue haze, shaped like a sphere, began to form around the group as the rings became a blur of motion. Ten seconds later Caesar 13 heard a big pop! The platform was empty; the blue sphere was gone and the rings instantly stopped their movement.

Caesar 13 permitted himself a smile. He would carefully follow the adventures of these imbeciles. It ought to be good for a laugh, even though he knew that his implant would suppress it. A few smiles anyway. For certain, Christian 172's party would be in for a surprise when they arrived in the 21st.

2035 AD

The Guardian group found themselves in a small forest just outside the abandoned warehouse that the resistance clones had used just a short time earlier. The rebel clones were nowhere in sight. A 21ster walked by on a dirt path about 100 feet away, dressed in 21st winter clothing.

"It's cold!" one of the clones said, stamping his feet.

Christian 172 realized that Caesar 13 had failed to provide them with 21ster garb They were in standard blue and black Guardian unis, but their extremities were exposed to the freezing air. "The fool!"

Christian 172 signaled from his brain implant to return to 3013.

3013 AD

Caesar 13 received the signal, but decided to let the party stand in the cold for several minutes. Only their heads and extremities would suffer. It would be good for them. He noticed Patton 1 was still in the Shifter room, monitoring the display.

"What is the matter?" Patton 1 said. "All is well. Why does the party stand in the woods and not track down the rebels?"

These clones are really stupid, Caesar 13 thought. He would play it like the obedient servant, just following orders. Caesar 13 fiddled with his console, pretending to reset the device. In actual fact this Shifter, once enabled, was continuously operable until it was shut down, somewhat like a cookie on an old 21ster browser session.

After 15 minutes Caesar 13 began the transfer. Ten seconds later the party materialized on the platform, rubbing their hands and feet together.

"Fool!" Christian 172 said accusingly to Caesar 13. "You failed to provide us with proper 21st clothing. We stood in the forest for 15 minutes in the freezing cold."

Caesar 13 decided to test the boundaries of their stupidity. "I asked you if you were ready and you said yes. It takes 15 minutes for the Shifter to reactivate."

Patton 1 saw the outraged expression (for a clone) on Christian 172's face and laughed. Patton 1 suspected that the chief technician was being disingenuous, but he delighted in his commissar's frustration. "This clone is correct, Christian 172. It happened exactly as he said."

No mention was made of the 15 minute delay. Caesar 13 concluded that these "elite" clones knew nothing about the Shifter or its operation. He and his twelve technicians had built the Shifter on the orders of the Supreme Overseer Khan 1, but apparently none of the Overseers had bothered to learn how it worked!

The twenty attack clones on the platform hurried off and headed for the lifts just past the Shifter room entrance.

"Come back here you fools!" Christian 172 cried.

His command was ignored. The clones huddled around the opaqued transparencies against the hallway wall, trying to summon a lift. Christian 172's face got redder and redder. Patton 1 was almost doubled over with laughter.

Caesar 13 tried very hard not to laugh out loud. How can these dimwits run an entire society? Well, they didn't do it very well. Everything here in the Tower – the monitoring station for all of New York – was automated, requiring no thought or intelligence.

Christian 172 finally calmed down enough to reactivate the clones' Submission modules. They stumbled back into the Shifter room, grumbling and complaining like 21st century drunks with a bad hangover.

Caesar 13 could see his time technicians staring in disbelief at the unfolding drama. Their looks said, "*These* are the clones who rule us?" One of them apparently felt sympathy for the group, for he activated a printer and programmed it to produce 21st century clothing. "At the back you will find appropriate garb. It's all next to the printer."

Christian 172 headed for the printer and found clothing. "What are you waiting for?" the commissar said, speaking to the other clones.

"I'm not wearing that stuff," one of them said. "It looks stupid."

"I'll give you stupid," Christian 172 said, increasing the power of the Submission frequencies. Meekly, the clone went back and put on the clothing.

After another half hour, everyone was dressed properly.

"Where are these renegade clones now?" Christian 172 asked.

An intelligent question! Caesar 13 thought. "The renegades are proceeding down Fifth Street in Midland, Illinois. They are now at Fifth and Mason. Destination unknown."

A look of confusion crossed the commissar's face. "Have we forgotten anything?" he snapped at Caesar 13.

"Unknown. I am but a Shifter technician, unable to comprehend the intricacies of a complex Overseer mission."

Again Patton 1 regarded the chief technician, suspecting insubordination. He formed the intention to blast Caesar 13's Submission module. Then he realized that he needed the chief technician to oversee the Shifter operation.

Caesar 13 saw Patton 1 frown and a look of frustration cross his face. It was all he could do to keep his expression as bland as possible.

Patton 1 glanced over at Christian 172 and found a more interesting target. His commissar was fumbling with his controller, trying to herd his milling, grumbling, antisocial clones back onto the platform. This was so funny that Patton 1 began to laugh again. "Commissar Christian 172," he said. "Why do you make me wait? Assemble your clones and execute your mission."

The look on Christian 172's face was priceless, Caesar 13 thought. Excited about tracking down the renegade clones but daunted by his collection of unruly antisocials, his face was marked by a combination of anger, frustration, and hopelessness. He looked like a 20th century clown! Caesar 13 wondered whether the commissar had forgotten anything else. When he glanced at Patton 1's face, he could see that the Overseer was wondering the same thing.

Finally Christian 172 got the group reassembled on the platform. Before they were ready, Caesar 13 activated the Shifter. Again some of the clones tried to exit the platform in a panic as the rings began to turn. The last thing Patton 1 and Caesar 13 saw was a frantic Christian 172 desperately pressing his Submission controller as some of the clones fell to the platform floor. Then they disappeared.

Patton 1 was clearly enjoying himself. "You started the Shifter too soon," he said to Caesar 13, speaking in an accusing voice.

"I am sorry, sire. The optimum temporal moment appeared and I took advantage." Caesar 13 wondered if he had gone too far, but he hadn't enjoyed himself this much in his entire life.

Patton 1, pleased by another use of "sire," was in a forgiving mood. "That was enjoyable, but the commissar had better complete his mission successfully."

Caesar 13 bowed his head in acknowledgment.

Satisfied, Patton 1 returned to his quarters for another session with the brain implant stimulator and his plan to overthrow Khan 1. Caesar 13, although insolent, would make an able assistant. For a while.

Caesar 13's LRI was 133, the highest recorded Life Reading Index in all of New York. He was the only clone in the city with the intelligence to fully understand how the Shifter worked. The paranoid Khan 1 watched him like a hawk, but Caesar 13 also secretly observed the Supreme Overseer. This was easy because the Shifter installation also included the city's central communications hub, which he was also responsible for. He knew as much about Khan 1 as Khan 1 knew about him.

Just before Patton 1 activated his brain implant stimulator he sent an order to Christian 172's private quarters, to be seen on his return from the past. "We have no use for these clones after they have destroyed the Resistance. Terminate them."

Caesar 13 saw the communication and was revolted.

This order to kill the antisocial clones (with the approval of Khan 1, no doubt) was the last straw for Caesar 13. He decided he could no longer live in this decaying, degraded culture. His only option was to use the Shifter to escape into the past. Fortunately, the rebel clones had already marked the timeline with their ventures into the year 2035. The City of Midland was as good as any other place, he supposed. Much colder back then, but manageable. While performing his duties, he devoted a part of his mind to plan how he would escape back to 2035, and how he would live in the past.

Caesar 13 turned his attention to his display tank to see how Christian 172 was making out in 2035.

CHAPTER **16**

2035 AD

Christian 172 and his goons from 3013 found themselves in the forest again, by the abandoned warehouse where the renegades had inserted. Christian 172 knew that the renegades were proceeding down Fifth and were at Mason. Caesar 13 had shown him a map of Midland before they left. He thought he remembered where Main was but had no idea about Mason. The only option was to find Main and look for the rebels.

It was cold! They would have to walk, for they had no access to public transportation. He had forgotten to ask for currency cards. Furious (for a clone) the commissar vowed to make Caesar 13 pay for his incompetence.

His job was to terminate or apprehend the rebels, but these rebel clones were enhanced. How long would it take in this primitive century to accelerate the life force readings of himself and his attackers?

Christian 172 decided to find a public building with many 21sters milling about, and get warm. As they trudged down a dirt path out of the forest he had to turn up the Suppression frequencies to keep his group in line.

The group approached a paved road. Several hundred feet beyond that he saw an intersection with traffic lights. Vehicles sped by on the road, seemingly at random. Christian 172 was able to herd the clones satisfactorily, but they moved slowly. As they approached the light he realized that his group of twenty clones was attracting unwanted attention. They crowded the intersection, bumping into each other and complaining.

"Where is Main Street?" he asked one of the pedestrians about to cross the road. This person had pressed a button that seemed to stop the traffic. The woman seemed not to understand him. These primitives spoke with a strange accent. "Where is Main Street?"

he repeated. The woman's face showed strong displeasure as she pointed to the left. "About a mile that way down Fifth Street." She strode away hurriedly, obviously irritated.

She was irritated! What about him? These 21sters were stupid! The commissar decided that he would locate the renegades as quickly as he could, the sensibilities of these primitives be damned. He rushed his clones across the street and then turned the corner onto a road with paved sidewalks. These people are crazy, he thought, with their insane motor vehicles spewing foul-smelling hydrocarbons and speeding along with no supervision! Everyone on the sidewalks gave way, giving them alarming looks. Christian 172 rather enjoyed this. Perhaps this century was not so bad after all.

His plan was simple. Find the rebels, surround them with his clones, and send a signal to Caesar 13 to activate the Shifter. Then herd the rebels into the Tower detention cells that were next to the lifts.

After walking down the street for a while he saw what looked like Pietra 23 enter one of the buildings. He rushed his clones over there. His goons were shoving each other, making raucous noises, and generally making a nuisance of themselves to all the 21sters they passed on the street. Christian 172 enjoyed this, but he had to turn up the Supression frequencies again to control them.

At the building entrance, Christian 172 sent a blast to his clones through their brain implants. "Make no sound, malignants! I will send you the signal to attack." As they entered the building, the commissar saw the red head of Pietra 23 with her cap on. All of the renegades were with her. As silently as possible he approached the group, who were standing in the lobby listening to a lecture with several dozen 21sters.

Christian 172 immediately recognized the speaker as Harry Kaine. The Founder! A sign said, "1,000-year Space Mission to Save Humanity. Featured Speaker: Harry Kaine. Guest speaker: Winthrop Sauvage, cryogenics." A blonde woman was recording the speech, and the crowd was listening intently, some commenting to each other. He must act now before these 21sters and the rebels became aware of his clones. Fortunately the voice of the speaker was amplified very loudly, and the rebels were all grouped together near the speaker.

Through his implant he sent instructions to his goons. 'Surround the renegade clones so they cannot escape. I will activate the Shifter through my brain implant.' But Christian 172 had not counted on the enthusiasm of his clones, who ran toward the rebels, shouting insults. People began to scream and run out of the building. Christian 172 frantically signaled Caesar 13 through his implant to extract his clones and the rebels before his clones killed the rebels. That pleasure would be his when they got back to 3013!

By this time two uniformed security guards had appeared and rushed toward them. Harry Kaine ran into the group, shouting, "The clones are back!"

The Shifter activated and the two building security guards heard a popping sound.

"Fran, did you see what I just saw?"

His companion was staring at the empty lobby. "There were about 40 people here, Julius, and they just vanished into thin air."

"We have to find them!" Julius said.

But there were no hiding places in the lobby. It was empty. "Either those people disappeared or I'm going crazy," Fran said.

"How are we going to report this?" Julius asked.

The two security guards looked at each other.

"My mama didn't raise no fool," Fran said. "There was a disturbance, people ran out the building. That's it."

Julius wiped his brow. He was shaking. "That sounds right to me. I need a drink."

"You go ahead, I'll make the report."

CHAPTER **17**

3013 AD

Christian 172 found himself standing on the Shifter platform in the Tower, surrounded by his goons. "Seize the rebels!" he shouted. His clones, already roused to a fever pitch of excitement after their attack in 2035, began to assault everyone on the crowded platform, including himself. "Get the rebels you fools!" he screamed, reaching for his Suppression module. The commissar was able to get his clones subdued, but he had to increase the setting to its highest in order to control them. Soon his gang of attackers was sitting on the Shifter platform, squirming uncomfortably under the Submission frequencies from their implants.

Christian 172 was confused. Without his enhanced clones he could not secure the rebels, but if he released them he himself would be in danger, for those fools might attack him again. That thought prompted him to look around the platform. He noticed that five of the 21sters had also been transported! Including the great Harry Kaine, a legendary figure worshipped by the Overseers. Christian 172 was stunned to see the Founder, the father of Guardian society, not ten feet away from him.

A detachment of Patton 1's red-and-black clad security forces suddenly entered the room and approached the platform. "Fools!" Christian 172 cried. "What took you so long?"

At that moment the rebel clones ran off the platform, shoving the security forces out of the way. Christian 172 pressed his Submission control and realized it would not work on the rebels.

"Caesar 13!" Christian 172 shouted. "Seal off all the exits from this floor!"

Caesar 13, from his raised station at the main control console, reluctantly complied.

Between the rebel clones, who were manhandling the Tower security forces, his twenty attacking clones, who were still squirming in pain on the Shifter platform, and the Great Man himself standing close to him, Christian 172 became totally confused. "Stand down until I make sense of this situation!" he shouted to Patton 1's guards. The rebels situated themselves along the front wall, just ten feet from the lifts in the hallway. Patton 1's security forces stood nervously watching them. All had heard of the superhuman strength of these rebel demons from the Guardian news broadcasts.

Pietra, standing against the wall, looked longingly at the Shifter. She wanted to destroy it, but Christian 172's twenty enhanced clones were on the platform. If the Overseer turned off his Suppression module, those lunatics would attack them. They had to get out of here!

Harry Kaine slowly walked off the Shifter platform to one of the transparencies on the east wall of the Tower and looked out. "Why, this looks like New York! But a New York I've never seen before." Hovercraft filled the skies; the complex New York road system was reduced to mere foot paths. He turned to the commissar. "Where are we?"

Patton 1 entered the room and saw Harry Kaine. He turned to Christian 172. "I asked you to capture the renegades and who do you come back with? Five of the 21sters, including our Founder! You are a fool."

Christian 172 had no patience for this and spoke angrily. "Sire, these misfits" – he turned to his gang of clones – "began to attack the 21sters and myself as well as the renegades. I had to subdue them or they might have killed the Founder. As you see, I have them all."

The commissar pointed to the rebels, who were awed to be on the top floor of the terrifying and formidable Tower of the Overseers, the control center for all of Guardian society. Even Pietra was staring, eyes wide, at the huge time shifter and the time technicians sitting at their consoles.

Harry Kaine regarded the clones with interest and excitement. "You are all beautifully made. Are we in the future?"

"According to your reckoning, this is the year 3013," Caesar 13 replied from his raised console that overlooked the entire Shifter area.

Harry accepted this without a qualm. "But this is wonderful. Our Transhumanist goals have been realized!" He looked around the room and saw Pietra. "Pietra 23! Nice to see you again. I knew you were a clone!"

"Not anymore," she replied. "I no longer need my brain implant interface."

"Brain implant?" Karen Everard said. Her head had cleared and her reporting instincts were kicking in. "Are we really in the year 3013?" She pointed to the Shifter. "Is that a time machine?"

I couldn't believe what I was seeing. Everyone from the future looked perfect; their bodies were beautiful. I noticed that there were five types of identical looking humans in the room.

Kjirsten and M'basa stood with their mouths open. This place...it felt so different! Kjirsten thought. M'basa felt within his heart that something was terribly wrong here. "What are those people doing on the platform? They are under constraint against their will."

Harry Kaine looking out the wall transparency at the hovercraft floating by. He noticed the monitors on the side of the building. He couldn't see any of the familiar buildings of the New York he knew. "The city is strangely empty," he remarked.

Patton 1 had had enough. "Caesar 13, send these 21sters back to the past where they belong. Take care with our Founder! Nothing must go awry in the past." He motioned for Harry to get back on the platform.

"Yes sire." Caesar 13 programmed the Shifter for another run.

"Oh no you don't," Harry said, striding toward the hallway. "I want to see this brave new world."

Christian 172 managed to herd his struggling clones off the platform using his Suppression module. They were mad to attack the rebels standing against the far wall, but the commissar's instinct for self-preservation stopped him from releasing them. What if these malignants attacked him again?

Myself, Kjirsten, M'basa, and Karen were still on the platform. The rings began to turn, very slowly. We were being sent back to 2035, but Harry Kaine would be stranded in the future! M'basa and I glanced at each other for a split second. If Harry stayed in 3013, that could radically alter the timeline.

"Harry, come back here!" M'basa shouted, running off the platform to fetch him. "We have to get back!"

M'basa began struggling with Harry, dragging the smaller man toward the platform.

"You fool!" Patton 1 cried. "Don't hurt the Founder!" Patton 1's security guards rushed forward to help Harry. M'basa was able to brush them off but he had to let go of Harry in the process, who raced out of the Shifter room and into the hallway.

"Oh crap," I said. I grabbed Karen's hand and motioned to the others as I ran off the platform. "Come! We must all stay together!" Patton 1 turned quickly to Christian 172. "Unleash your clones! Seize the 21sters and the rebels!" At that moment one of the transparencies opened along the hallway wall and a messenger clone, on routine Tower business, stepped out of the lift. The rebel clones ran into the elevator. We all followed them, shoving the security clones out of the way. All of us managed to crowd into the lift before it opaqued.

Patton 1 was shouting insults at Christian 172 while Christian 172 frantically turned off the Submission frequencies of his twenty clones. "Get the rebels!" he cried, sending a signal to open one of the lift transparencies. As one, the clones ran as fast as they could, smashing into each other in their eagerness to get to a lift and track down the rebel clones.

'Stupidity is its own reward,' Caesar 13 thought to himself.

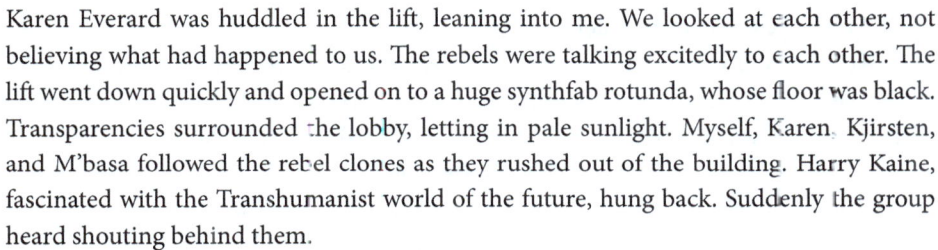

Karen Everard was huddled in the lift, leaning into me. We looked at each other, not believing what had happened to us. The rebels were talking excitedly to each other. The lift went down quickly and opened on to a huge synthfab rotunda, whose floor was black. Transparencies surrounded the lobby, letting in pale sunlight. Myself, Karen, Kjirsten, and M'basa followed the rebel clones as they rushed out of the building. Harry Kaine, fascinated with the Transhumanist world of the future, hung back. Suddenly the group heard shouting behind them.

"It's those crazed thugs of Patton 1!" Li 355 shouted.

"Keep together!" Pietra shouted. "No stragglers!"

Harry Kaine saw twenty screaming clones rushing toward them and took off. "Wait for me!"

We turned right out of the rotunda and ran across a manicured lawn toward a crumbling building that looked like an old amphitheater. "Where are we going?" Karen asked me.

"I have no idea!"

We huffed and puffed across the lawn, the gigantic black Tower of the Overseers behind us. Pietra led the group as we hastily scrambled down a permacrete walkway past the crumbling amphitheater, and then onto a platform that was above a stairwell that led underground. The yelling clones were almost upon us now. I could see the people walking on the streets staring at us. Hovercraft stopped their movements just above the confrontation. "Cease and desist immediately," a voice cried. It was coming from a hovercraft.

"Come!" Pietra shouted to the group. "There's an old subway entrance here on 42nd Street!" Pietra led everyone down the stairs and into a filthy, dusty subway tunnel.

The attackers were just above us now on the permacrete platform above the stairs, screaming insults and shouting wildly.

"Entering the tunnels is forbidden except for the tunnel monitors," Pietra said. "We may be safe here."

One of the attacking clones suddenly rushed down the stairs and the others quickly followed. "Run!" Pietra cried. We all took off as fast as we could into the tunnel. Our pounding feet made a huge cloud of centuries-old dust, obscuring us from the view of our pursuers. We were choking and coughing but Pietra 23 urged us forward. "Don't look back!" she cried. "These are Patton 1's enhanced clones!"

For several harrowing moments we heard pounding footsteps behind us, then a babble of voices. "I can't breathe! Let's get out of here!" a frightened voice shouted behind us. We heard thudding feet going back the other way, raising more choking dust.

"They are leaving," Pietra shouted as she gagged on the huge cloud of dust. "We must get fresh air!"

We had no choice but to go back. Our lungs were filling up with fine particles from the thousand-year-old subway floor.

We heard our pursuers run up the stairs as we followed behind. We stood heaving on the platform, trying to clear our lungs. The twenty clones ahead of us were now on the permacrete walkway that led to the carefully manicured lawn and the Tower. "We can't go back in there!" one of them shouted. "To freedom!"

As one, the twenty enhanced clones took off down the walkway. A road led toward what was once lower Manhattan (the old Sixth Ave, now just a footpath) and the clones followed it, choking and coughing.

Several hovercraft diverted from their flight paths and began to follow the clones while we coughed on the small platform above the stairs, still trying to fill our lungs with clean air.

"Cease and desist immediately!" voices from several of the hovercraft shouted.

Pietra saw the attacking clones suddenly lose their enthusiasm. They were milling about now, still gasping for air. The hovercraft operators must have activated their Supression frequencies. Just then one of the hovercraft landed and troops rushed out, marching the clones into it. The hovercraft lifted off.

"Did they see us?" Sigmund 78 asked, gasping for air.

Pietra sneezed. "Who knows? Stay here and don't move. Clear your lungs." She stood just outside the stairwell and monitored the sky. After several minutes the overflights of Overseer hovercraft resumed their standard grid pattern: a slow, steady flight in straight lines. She carefully exited the stairwell and looked around. Pedestrian traffic had resumed its normal pattern. People were walking slowly in single file, in the approved manner. No gawkers were present. All looked normal.

"I'm going out there," Harry Kaine said. "I want to see what the future looks like."

Pietra knew they couldn't go back using the secluded tunnel system, where they would be safe. For some reason the dust in this tunnel was much thicker than in the Old Sector. Several hours would pass before it settled in the unventilated subway. They would have to proceed on the footpaths and risk discovery by the hovercraft. Pietra consulted with her rebels. "Our goal must be Building 103 and our makeshift Time Shifter. The 21sters must be returned to the past."

Kjirsten and M'basa, standing on one of the stairs that led to the subway tunnel, were speechless. They were finally able to breathe normally.

Karen Everard was staying close to Joe Courvall. He had remained calm and had adjusted quickly to the impossible situation they were in. She could see that Harry Kaine was anxious to get outside and explore.

Pietra motioned to the group. "Do you see how the others are walking? Do as they do. Walk slowly, single file, keep your head down. Do not display curiosity. Don't hurry or rush about, stay on the foot paths." She looked at Harry Kaine, who was babbling on about the glory of cloning and Transhumanist advances. "You saw what happened to the others," she said to him. "Don't create a scene." She remembered an old saying from the 20th history tranche. "When in Rome, do as the Romans do."

Karen could see that Harry wasn't listening.

Kaine spoke impatiently. "Let's go!"

"Follow me," Pietra said. She walked slowly off the platform and turned right down Sixth Ave. The foot paths followed the old permacrete roads, now overgrown with weeds and brush on either side. Here, the paths north and south were separated, just as the vehicle traffic in 21st New York had been.

The rebel clones began walking slowly toward the old 24th Street and Building 103, followed by myself, Kjirsten, M'basa, Karen, and Harry. Pietra told me she wanted to send us back to the past as quickly as possible, but that we couldn't rush it. Harry Kaine began running around in excitement, stopping clones on the foot paths, asking them questions about how they liked living in the New York of 3013.

Karen stared at the man in disbelief. Did he not know he was endangering everyone? She noticed that one of the hovercraft had stopped its slow-moving flight and was turning toward him. Pietra walked over and spoke to Harry. "Get back in line! We have fifteen blocks to go and we must not attract the attention of the Overseers."

"Pietra! I didn't think you were such a stick-in-the mud."

"Have it your way, but get back in line."

The group began walking slowly again.

Harry walked off the path again to explore a unique looking building about 100 feet from the road.

"Harry!" Pietra said. "Stay on the path. That's a Guardian monitoring station."

Harry didn't hear her and continued walking toward the structure. He noticed that in the New York of 1,000 years into the future, most of the structures of the New York he knew were gone. Large buildings such as the Empire State building and Madison Square Garden were now just empty shells. Everyone wore the same blue and black form-fitting suits (except the troops that had come out of the hovercraft. These had worn red and black). He and the others from the 21st century looked out of place.

Karen noticed that the hovercraft now began to point at Harry Kaine, gradually losing altitude. "Cease and desist!" a voice said loudly. "By the order of Patton 1, return to the foot path or your Submission implant will be activated."

The rebel clones continued to walk, but the 21sters had stopped and were gawking. Pietra walked up to them. "Resume your walking. Do not attract attention. We must leave Harry Kaine or the entire fleet of hovercraft will be upon us."

Karen and myself, Kjirsten and M'basa followed Pietra back to the group and resumed our slow pace. Karen looked back and saw that three of the hovercraft had landed. Ten red-suited troops emerged, but looked confused when Harry Kaine did not submit. "The Overseers don't know that Harry doesn't have a brain implant," Pietra explained. "Keep walking."

Karen saw the troops rush Harry, who pushed them all away easily. They fell like pick-up sticks in a heap. Another hovercraft landed, and more troops came out. These attacked Kaine but met with the same fate. The red-suited troops, confused, went back to the hovercraft, which took off with speed.

"Fortunately for Harry, weapons have been outlawed for six centuries. Continue to walk as normal," Pietra said.

"Where are those planes going?" Karen asked.

"Back to the Tower of the Overseers. I have no idea what they'll do next. We have to get back to our Shifter without attracting any more attention."

Harry ran back to the group. "Did you see that? I'm a superman here!"

"Get back in line Harry," I said. "And shut the fuck up."

M'basa put an arm around Harry Kaine's shoulder. "If you get out of line again, asshole, I'll end you."

I stood on the other side of Kaine, holding him by the waist. "I'll second that."

The air went out of Harry Kaine and he looked depressed. For the first time he gave serious study to this New York of the future. The shops, restaurants, and businesses of his day were gone and cleared away, replaced by weeds and sickly looking shrubs. Those structures that remained were made of some building material he'd never seen before, and resembled a depressing Moscow street of apartment blocks. Housing, probably. But the people were beautiful! Surely that was an advancement. He noticed that there were no birds flying, or insects, although the temperature was about 75 degrees. The air here was stale, as if the life energy of the entire city had atrophied. He threw off his winter coat, as the others had done.

M'basa looked at Harry. "If we release you, will you be a good boy?"

"Fuck you, whoever you are."

M'basa and I increased our grip.

"You're hurting me."

"We'll hurt you more if you don't listen to this Pietra," I said. "She seems to know what she's doing here."

"All right. Let me go. I want to speak with her."

"Don't do anything stupid."

M'basa and I followed Kaine as he went up to Pietra at the front of the line.

Karen saw Joe moving away from her. As he moved beside Pietra she felt a pang of jealousy. Who was this strange redhead with the slit in the back of her head? She looked a lot like the redhead at her interview with Harry Kaine at the Physics and Astronomy building. Was she the same redhead who flickered on the Diag? Of course! Pietra must be a clone, just as Harry Kaine said. She looked at the heads of the other blue and black clad persons as they passed. Each of them had something black stuck into the slits. Brain implants. This is a cloned society!

"Tell me about your life," Harry said to Pietra as Karen caught up to Joe.

Pietra glanced witheringly at him. "We are clones living in a dying society. Look around you. Our time device tells us that by the year 3209 this timeline, and all the life in it, will be dead."

Harry Kaine reacted as if he had been slapped in the face. "I...I don't believe you."

Pietra adopted a saying of Patton 1. "Then you are a fool."

The walk back after that was uneventful.

Harry Kaine ignored everyone and kept to himself, deep in thought. He didn't like Pietra and he sure as hell didn't like Karen Everard's friends. Karen herself had shown her true colors on this trip to the future. She was myopic and couldn't understand the importance of human advancement. Harry looked around him at the barren landscape, and the automatons walking single file down the street. No, this couldn't be the future; it was a trick. The sooner he got back to Midland the sooner he could advance humanity forward along the correct path.

When the team approached Building 103, they saw a hovercraft on the street in front of the building.

"The Shifter!" Pietra cried, and ran through the front entrance. Three Overseer clones, dressed in red and black unis, were crawling over the device. Li 57 and Li 308 were cowering in fear, guarded by two men in black and red unis.

Pietra didn't waste time in pointless talk but yanked the first Overseer clone off one of the rings, which he had just begun to disassemble. Two others drew metal batons and were about to smash the control consoles when the rebel clones rushed them. The five security clones were unceremoniously thrown into the hovercraft and ordered to lift off. These minions, frightened by the enormous strength of the rebels, were only too happy to get out of there.

"We shall have to guard the Shifter day and night," Pietra said.

Li 308 and Li 57 rushed to examine the damage.

"Superficial," Li 308 said with a sigh of relief as he finished his inspection. "They activated our Submission modules and we were helpless."

"You two will have to be enhanced," Alan 133 said. "You're no good to us as weak as you are." Li 308 and Li 57 reluctantly agreed.

"First we have to send these 21sters back to their century," Pietra said.

"Where shall we send them?" Alan 133 asked Pietra.

"The empty warehouse. They will not be observed there."

Alan 133 identified the proper spacetime coordinates. He showed me a 3D map of Midland in his display tank. "The warehouse is here, in the industrial section of the city by the Midland River."

I nodded. "It's hard for me to believe that you are a clone. You look perfectly human."

"Looks are deceiving," Alan 133 replied. He turned around to show me his brain implant. "Those of us who have traveled into the past have enhanced our life force readings. If you look carefully at the other clones in our society, they look good on the surface but their bodies have little vitality."

I remembered how easily we had defeated the attackers. Harry Kaine had taken care of a dozen of them all by himself. Pietra called them "Overseer clones."

Alan 133 spoke to Li 57 and Li 308. "Please program the Shifter to these coordinates."

Before I could ask another question, Alan 133 went to his console. "Here, this will show you the history of earth on this timeline after 2035." Curious, M'basa, myself, Karen, and Kjirsten (who had not uttered a word since we had gone to the future), and Harry Kaine stood in front of the display tank, which was filled with a fine white computational mist. It was a larger version of Ger's display tank! Alan 133 brought up one of the recordings Li 308 had made of the timeline. I found out later that this was the one he and Pietra 23 had looked at before they sent her to the past for the first time.

Alan 133 glanced to his left. The legendary Harry Kaine, the most important human being in history, was standing right next to him! As he fiddled with the console Alan 133 wondered how such a self-centered person, who did not seem overly intelligent, could have been the historical focal point for the demise of the human race. Perhaps it is because such persons are laser focused on only one thing, and do not consider the consequences of their actions on others. Patton 1 and Christian 172 fit that mold, he thought.

The 21sters watched as the white mist solidified into an image. "Hey that's me!" Kaine said, pleased. "Where am I?"

"This is the famous speech you made at the United Nations in Geneva, on June 4, 2048. It propelled worldwide interest in Transhumanism and was the beginning of the end for humanity."

Kaine frowned.

"That's only 13 years from now!" Kjirsten exclaimed. "I mean, from then. No, that's not right."

M'basa laughed and held her hand. "It's confusing, isn't it?"

Alan 133 moved the timeline forward. Fifty years later in 2098, Harry Kaine used the latest advances in human life extension to convince the world's governments to build the Transformation Centers. A Planetary Referendum was held, in which world citizens voted overwhelmingly to authorize the Centers. The display showed enhanced Trans-humans exiting the Centers, glowing with health, and the worldwide excitement it created. Harry Kaine had persuaded governments to offer the treatments for free.

Kaine smiled smugly. "You see? I was right! In 2098 I am 93 years old but I look 40 years younger."

"Not so fast Harry," Alan 133 said with as much disgust as he could. The display now showed those for whom the transhumanist treatment hadn't worked. Alan 133 narrated while Li 308 ran the temporal recording. "At first their bodies were simply cremated, with lavish and publicized funeral ceremonies for 'the brave souls who are advancing human progress.' However, as the number of applicants increased, the number of failures grew into the tens of millions. These were simply thrown into mass graves; their relatives told that there wasn't time or money for a funeral. A selected few still received the public ceremonies.

"A worldwide backlash quickly began, as citizens demanded to know why their loved ones had disappeared. Gradually, the World Security Force was created to bring order to the chaos. All citizens were required to undergo the treatments. This force eventually developed into the planetary enforcement arm of the Overseers, who called themselves the Guardians."

The display showed the imposing Tower of the Overseers in the old Bryant Park, by far the tallest building in New York.

Alan 133 turned to the 21sters. "This is the building we all escaped from.'

"By the year 2120 all humans had received the treatments. Everything went well for a while until the enhanced bodies began to lose the ability to reproduce. This came as a total shock to everyone, even the authorities. The world population plummeted. Eventually, to save the race, cloning was instituted in the 2300s. After that, for a century or so, a small subset of the cloned population was able to have children and everyone was encouraged. Even so, the world population continued to experience a demographic collapse." Alan 133's face expressed anguish.

Li 355 picked up the narrative. "During the 25th century, for reasons no scientist could explain, all of the cloned bodies became sterile, and so for every generation after that. We still don't understand why this happened."

Alan 133 recovered his composure. "LI 308, show these 21sters what the future looks like after 3013."

"The display will now show the most probable future after 3013, if nothing changes."

Everyone, including the rebel clones, were now gathered around the display tank.

"The next generations of clones after 3035 will be androgynous, for each clone generation for the past 600 years has been slightly less vital than the previous one. It will no longer be possible to maintain the official fiction of male and female, for no one believes the Guardian propaganda anymore. Our clones have been below the minimum life force readings for reproduction for a long time."

Sigmund 78 looked at Alan 133. "The next generation of clones will be able to reproduce!" he said sarcastically, mimicking the Guardian newscasts.

"Our glorious scientists have made new medical breakthroughs!" Li 57 mocked.

"The Supreme Leader, Khan 1, assures us that we are moving forward!" Pietra cried.

The display showed Guardian society gradually decaying after the 25th century, as clones were being born slightly weaker and weaker as the generations passed. The worldwide population was reduced from 10 billion to less than 100 million.

"Li 308, advance the temporal record to the year 3209," Alan 133 said.

The display showed a dead earth, with no life on it. "The night sky shows no stars, no light."

Everyone except Kaine was sickened. "My God," Karen said. "Are you saying that the entire universe disappears?"

"Impossible!" Kaine scoffed. "This is a trick. A pseudo-scientific simulation."

Alan 133 ignored Harry and answered Karen. "No, of course the universe doesn't disappear. We think that the earth, in this timeline, simply disappears from the universe."

Everyone stood in stunned silence. "Some of us have never seen this before," Sigmund 78 said. "Is this really our future?"

Alan 133 nodded. "IF nothing changes."

"So that is why you were so insistent that we go to the past and become enhanced," Li 57 said.

"That's right," Alan 133 replied. "We rebels are the only ones who can save humanity now. It is our job to reunite this temporal thread with the main flow of time. Right now we are in a little cul-de-sac. The umbilical cord that ties our timeline to the main flow of temporal energy is gradually being cut. We must re-establish it."

All looked accusingly at Harry Kaine. "Don't blame me!" he cried. "I haven't done anything yet!"

"You will, if these people are right," Kjirsten said.

Harry had recovered his composure. "Quite the opposite, beautiful Kjirsten. Transhumanism will merely have to go down a different path. It is my job to see that it does."

Li 308 looked to Alan 133. "Conservation of Temporal Momentum applies."

Alan 133 nodded.

The 21sters were confused.

"What are you talking about?" I asked.

"Time is flexible," Alan 133 replied. "However, if you attempt to change it, let's say by going back in time and trying to kill a historical figure, events will unfold so that the outcome that has already occurred will stay the same. Minor events might change around the main event, but the primary result doesn't change."

"So what you're saying is that if I go back in time and try to kill Adolf Hitler before he was supposed to die, something will happen to prevent it."

"That's right," Alan 133 said. "Major events like the birth of a person, once it has already occurred, have lots of temporal energy associated with it. Those primary events can't be changed by going back in time. Something 'coincidental' will happen to make sure you can't do it."

"The gun won't fire, or someone will come into the house just before the murder was to happen so that the murderer won't succeed," Kjirsten suggested.

"Precisely," Li 308 said. "The possibilities are endless."

"That's what we call the time conservation law," Li 57 added.

Harry Kaine was cheerful. "So you can't stop me no matter what you do!"

M'basa stepped grimly toward Harry Kaine. "But you said that this timeline was out of the main flow of time. So I could snap this little shit's neck here and do humanity a favor."

Kjirsten was alarmed. This was a side of her boyfriend she had never seen before.

"No!" Alan 133 said, grabbing M'basa and holding him off of Kaine He was amazed at his own strength. "If you kill Kaine here, time will adjust by severing the temporal cord of this timeline to the main flow of time. You will be responsible for killing everyone in our society."

"That is right, Alan 133," Li 355 said. "Killing Kaine could have disastrous results for our timeline, and yours too."

"So the time conservation law also works for time travel to the future as well as the past," I said.

Li 308 nodded. "That is correct. Past and future are connected...to put it simply, in a circle. Time flows both ways, and it doesn't care which way around the circle it goes."

My understanding was exploding. "If Harry Kaine doesn't come back to our timeline because he's killed here, the past in 2035 will have a gigantic hole in it. That means events have to radically change along the timeline from past to future. Which means everyone here might not exist."

"Yes," Karen said. "3013 might turn into a cyst on the flow of time"

"Well said," Li 308 acknowledged. "You've hit on the crux of the problem. Time, like life itself, is interrelated into a seamless whole. But if a pimple appears on the skin of time, it can safely be excised without harming the whole."

"And our timeline is the pimple," Sigmund 78 agreed.

M'basa stepped back from Harry, appalled at what he had almost done. "It would literally be genocide," he said.

Harry Kaine brightened. "Well then! I'm free to pursue my dreams." He looked at M'basa. "You can't touch me!"

M'basa wagged his finger at the smaller man and spoke deliberately. "We'll be monitoring you back in 2035, Harry. You had better watch your step."

Kjirsten knew that when M'basa spoke slowly he was holding back strong emotions.

M'basa was remembering the bloody conflicts during Africa's genocidal civil wars that had killed millions of his people. He understood how normal people, under the right circumstances, could turn into killers. He had almost become one! Yet Harry Kaine was dangerous.

M'basa and I stepped toward Harry. Kaine held up his hands. "OK, OK! I'll watch my step."

I saw the little smirk on Kaine's face. "Duping delight, Harry," I said. "You say that now but you have no intention of changing."

Kaine stammered, an acknowledgment that I was right. Under the scrutiny of the group, his smile disappeared. "How did you know?"

"I'm an old soul," I replied. "I know things."

Karen, myself, M'basa, and Kjirsten stared at each other. "We're all old souls," Kjirsten said. "We've been together before, in previous lives. That's what binds us together."

"That's the kind of new age crap we have to stamp out of society," Kaine complained. "Old souls my ass! There is no such thing as a soul." He pointed to his body. "*This* is what we are, and you can't prove that I'm wrong."

Karen Everard shook her head sadly. "You don't get it Harry. The test of life is to see whether you can get past your biology to the higher self. All you see is the obvious."

Kaine spoke angrily. "What you call the *obvious, is* reality!" He looked around at his 21st companions in disbelief. "You are all delusional."

Karen shook her head. "The Interfaith people are right. You, Harry Kaine, are a deadhead. You have no spiritual awareness."

"*You* miss the point, my lovely reporter. There *is* no spirituality! There are no old souls, whatever the hell that is. There is no 'higher self.' It's a bunch of bunk."

Li 308 looked at his console. "It's time to get you 21sters out of here. Foot patrol traffic is increasing in this area again."

Harry Kaine looked relieved.

"Make sure we don't come back before we left," M'basa said.

"Not to worry," Li 57 said at her console, figuring out the correct spacetime coordinates for the Shifter. "There is a wall of temporal energy blocking each segment of time you've already occupied."

I was amazed. Apparently time regulated itself! "I feel better about whoever designed the universe. That guy must be pretty smart!"

Karen laughed. 'You're funny, Joe."

I looked at Karen. I was amazed that I had chosen Kesha over her.

Harry wanted to tell these empty-headed morons that there is no designer of the universe, but they were too stupid to understand.

"Are you ready, Li 57 and Li 308?" Pietra asked the two Shifter operators.

"Ready."

"You must step one at a time onto the platform. We can't do mass insertions and extractions like at the Tower."

"I'll go first," Harry Kaine said.

"No you won't," M'basa said. "Last for you, boyo."

One by one Karen, myself, Kjirsten, and M'basa stepped onto the platform. The rotating rings and the blue energy surrounding the device were fantastic; but the pop! at the end, and the disappearance into thin air of Karen, was frightening.

After Karen had gone I panicked. "What have you done with her?" I shouted to Li 308.

The time technician pointed to the display tank. Karen was staring in disbelief in the old warehouse, completely disoriented. "She is back in 2035, safe and sound. In an old warehouse in the industrial district of Midland."

The others stared, fascinated, as Karen adjusted to her surroundings. "It's cold in here," she said, wrapping her arms about her torso and stamping her feet on the dirty concrete floor.

"Li 308, print winter garb for the 21sters," Pietra said.

"Make extras for Karen!" Joe said. It was a lot warmer in 3013 than in 2035. We had all unthinkingly shed our coats when we ran out of the Tower of the Overseers.

I went next, taking Karen a coat, a pullover hat, and a pair of gloves. The clothing wasn't fashionable, but functional. Kjirsten and M'basa watched Joe hold Karen's coat for her.

"Something developing there?" Kjirsten asked M'basa, who smiled.

As Harry Kaine stepped last onto the Shifter platform, he took a last look around the Shifter room and at the clones. There was a lot of good here but it could never happen if nutcases like Karen Everard and her ilk had their way. There were plenty of them around in

2035, babbling metaphysical absurdities and embracing religions that were nothing more than superstitious nonsense. How could rational people believe such drivel? The answer is, no reasonable person could. Karen and her group were irrational and crazy. They were far more dangerous than whatever he would do in the future. He must continue his work and ensure that Transhumanism went down the correct path. A path that would greatly benefit humanity!

Harry was astonished when the rings began to move faster. A soft blue sphere of energy developed around the platform as the metal rings whizzed, incredibly fast, above him. His body felt a kind of pleasant tugging, as if he were being magnetized. Then he felt another magnet...at his destination? The blue energy intensified and penetrated his body. Suddenly a pop and then he was standing on a filthy concrete floor in what looked like an old warehouse. The others were waiting for him. They didn't look happy to see him.

CHAPTER **18**

Patton 1 was furious with Christian 172. He confronted the commissar in the Shifter area after the rebels escaped. "Why did you not pursue the rebels?"

"The reprobates we recruited were too afraid of the ancient subway tunnel," Christian 172 explained. "Then, when our clones came out, instead of chasing the rebels down they made a break for freedom. I am holding them in the Tower basement, awaiting your orders."

"The Khan wants them terminated for treason against the state. He insists that we send another team back to the past for enhancement. He wants the rebel Shifter destroyed." Patton 1 turned on Caesar 13. "Fool! Recruit twenty more clones for insertion into the past. This time do not use the dregs of society."

Caesar 13 bowed his head in submission, sickened by the Khan's order to murder the twenty clones in the Tower basement. "This may take two or three days."

"I don't care how long it takes. Obey my orders!" Patton 1 stomped out of the room with his security guards, followed by Christian 172.

At the main Shifter console station, Caesar 13 considered his position. The twenty clones in the secure and heavily guarded Tower basement were beyond his reach. They could not be saved. He was certain he would also be terminated once his usefulness was over.

He must escape 3013, and do it as soon as possible. To do that he must buy time for himself and his techs to prepare for their journey to the past. He had given himself three days to recruit the next batch of clones, but he hoped to be gone before then. If his time techs came back to 2035 with him, the Khan's Shifter would effectively be destroyed because no one else in the city knew how to operate it. Therefore the Overseers would not be able to enhance their clones, and he and his techs would be safe from attack in 2035. The rebel clones would also be safe in 3013.

Caesar 13's worst fears were confirmed that evening when he tuned into the Guardian news broadcast. "Many citizens saw a disturbance outside the Tower today by a group of malcontents." Vid was shown. "Twenty unauthorized clones emerged from the 42nd Street Tunnel and attacked our security forces. These lunatics were rounded up and sent to the Tower basement."

No one who watched the broadcast could fail to understand the fate of those poor clones, the ones he himself had recruited. Caesar 13 felt that he was an accessory to murder. And now he was supposed to identify twenty more clones! This he must do to protect himself and his time techs, but was there was a way to save those poor clones?

Suddenly he felt a sense of impending doom for himself and his techs. As the chief time technician it was up to him to act quickly.

The next day Caesar 13 approached Tesla 8, his most experienced temporal technician. This clone, like himself, was frightened by Patton 1's order to terminate the twenty clones, and fed up with the stupidity of the Overseers. Caesar 13 walked over to Tesla 8's station and leaned over, pretending to check a console setting. "Are you interested in escaping down the timeline?" he whispered.

Tesla 8 almost jumped out of her seat. "Yes!"

"Remain calm. Discreetly survey the other techs." Caesar 13 indicated an oval-shaped depression on Tesla 8's console. "Press this to communicate secretly, beyond the prying eyes of the Overseers."

"What if we are overheard?" Tesla 8 whispered.

"I have set up a backchannel, completely invisible. These cretins are too stupid to know what it is, even if they suspected."

Tesla 8 almost laughed out loud, but stifled herself. The expression of any strong emotion in this society could get you sent to a Guardian Monitor station, which often resulted in a painful interrogation.

Caesar 13 walked back to his raised console and sent a backchannel comm to Tesla 8. "I will need your help. To live in 2035 we must procure identification documents, currency/credit cards, places to live, and what the 21sters call 'bank accounts' so that we may survive successfully in 2035. To do that I must myself travel back to the past."

"Why not print the required documents here in 3013?" Tesla 8 sent. "We know exactly what they look like."

"Because in 2035, society was already regimented, although not so much as it is now. I must physically apply for bank accounts, licenses, and rental domiciles, so as to be registered with the appropriate government/banking entities."

"I understand." Tesla 8 was herself an aficionado of the 21st history tranches. "To fill the bank accounts we can print what they called 'cash.' Our printers can easily duplicate the currency paper printed by the U.S. Treasury. Even the serial numbers on the bills won't be a problem."

"Excellent! I knew you would be a valuable addition to the team. See to it."

"When will you leave for your trip to the past?"

"Tomorrow night during the next scheduled Shifter maintenance run. I will program the device and store the spacetime coordinates in my secreted data depository. Then I will step onto the platform and you can activate the device."

"That might be difficult," Tesla 8 replied. "Maintenance runs do not typically transport anything, and so do not fully complete the Shifter cycle."

"I have already 'greased the wheels,' as the 21sters would say. Last night I messaged Khan 1, Patton 1, and his sycophant Christian 172 that a complete maintenance cycle is necessary if they want to enhance the next batch of clones. To avoid suspicion you must bring me back as soon as possible in real time after I insert. Can you do that?"

"Yes," Tesla 8 sent. "I will survey the other techs tomorrow. Then we will know how many are going back with us to 2035."

Caesar 13 was satisfied. He was certain now that the Overseers didn't even know how to operate the Shifter, or understand anything about how it worked. Khan 1 had created a small, controllable group to operate the installation and was totally dependent on himself and his time techs. The Supreme Overseer only cared that the entire installation functioned properly as the central communications/surveillance hub for all Tower monitoring devices and substations throughout the city. All of these comm feeds were sent to Khan 1 in his private observation area on the other side of the lifts.

Caesar 13 was pretty sure that Khan 1 intended to destroy the Tower Shifter at the soonest possible moment after the resistance had been crushed, and the rebel Shifter as well. And he and his techs too. Once the Shifter was destroyed he and his techs would no longer be needed. By ensuring that no more enhanced clones could be created, the Overseers could maintain their control of society.

It was insane. These walking zombies would preside over their own deaths, ruling over a society they knew was doomed. It was the reason he and his time techs had to get out.

Caesar 13 faced the fact that the Khan would have to terminate the second batch of enhanced clones after their work was done. He had wracked his brains all day but could not come up with a plan to save them. Caesar 13 understood that even if he was able to escape to the past he would be a murderer. It was something he'd have to live with for the rest of his life.

The next day Tesla 8 reported on the backchannel. Only she and Einstein 3 were going to the past with him. The rest were too afraid.

Caesar 13 was disappointed. It would make his job of preparation in 2035 easier, but he had a fondness for all of his technicians. They were the smartest group of clones in Guardian society, and had worked well together for a long time.

"I leave tonight during the first scheduled maintenance run," Caesar 13 sent to Tesla 8 on the backchannel. "You are scheduled to act as my assistant. I have informed the Overseers that I myself will be the test subject. Khan 1 has assured me that if I do not come back, he will terminate you and all the other techs. Be careful."

Tesla 8 shuddered when she received this communication. "I'll be there," she sent back.

That night at 4 a.m. Caesar 13 and Tesla 8 entered the Shifter area. Two other techs were there, monitoring the comm stations and the Shifter on a skeleton shift. However, Khan 1 and a dozen security clones were also present. This was unexpected! Caesar 13 and Tesla 8 suppressed their surprise and fear, keeping their expressions as bland as possible.

Caesar 13 tried to act as if all was normal business. Yesterday evening he had packed his satchel with Tesla 8's 'cash' and 21ster winter clothing. He had stashed it in his personal storage area at the main console. He had also identified 20 clones and sent the list to Patton 1 for his approval.

He walked to his raised console, Khan 1's eyes on him all the way.

"Tesla 8, prepare the Shifter."

Tesla 8 checked Caesar 13's spacetime coordinates and the insertion point, an old warehouse in Midland, Illinois. The abandoned building was empty and there was no foot traffic at five in the morning, 2035 time. She nodded to Caesar 13, who grabbed his carry bag and walked slowly to the platform.

"Halt!" Khan 1 shouted. "What do you carry in that satchel?"

Caesar 13's heart jumped as one of the Khan's security guards hurried forward. He tried to remain calm and keep a look of bored apathy on his face. "Merely primitive 21ster century clothing. It is cold at the insertion site." He opened the satchel, where he had placed his winter garb on top of the cash. If the guard poked around and saw the currency, his treachery would be exposed.

The guard looked into the satchel. "Just clothing, sire," he said.

Caesar 13 didn't wait to close the satchel. He stepped onto the platform and casually dropped the carry bag on the platform, trying to look bored despite his heart jumping around in his chest. The rings began to move, the blue energy gathered itself around him as the rings moved faster and faster. As he waited for the insertion to 2035 he began to calm down. He was safe, at least for the time being.

Caesar 13 heard a pop! and then he was standing on a cold concrete floor. He was immediately aware of two things: the temperature was very cold, and the air, even though frozen, smelled alive. He noticed a tingling in the cells of his body. Was this the "enhancement" the rebels experienced? As he looked down at his satchel he saw something crawling on the dirty floor. An insect! Fascinated, he knelt down to inspect the little creature, busily making its way on some unknown errand.

Caesar 13 felt his eyes water. Life! There was life here in 2035, even in the frozen cold. He knew that 2035 had plenty of surveillance cameras in the cities, but none were here. The time traveler felt an enormous burden lift. The constant watchfulness necessary to live in 3013 was absent here. He was free!

He stripped off his Guardian thermofilm and placed it in the satchel. Immediately he felt the cold everywhere on his body. He quickly put on his 21ster clothing. He buttoned up his winter coat, put on his pulldown hat and his mittens and his fur-lined boots, and walked out of the decrepit building into the darkness of 2035.

Caesar 13 knew exactly what he was going to do. Enter at 5 a.m. local time and go to an all-night diner to taste some 21ster food until the banks opened. Then visit three banks and deposit $9,500 cash in each one. That would be enough to get them started. Each account would be issued a currency card by the bank so they could use public transportation, buy food, and pay what 21sters called "rent." He had carefully prepared his identity data packets for thirteen time travelers, observing the timeline to see how it was done. Now it was only necessary to submit the forms in 2035. Most of that could be done via the WorldNet, except for the vehicle license and renting the domiciles, which he would have to apply for in person. From the timeline he had already identified three "apartments" for rent. He would go to the rental office and pay a "down payment" for all three. Then he would find a "computer" store and buy a 21ster computing device and a mobile communicator. After that, in a cozy cafe out of the cold, he would submit the identity forms. If all went well he would be through before the sun set. Then he would go back to the warehouse and signal Tesla 8 through the Shifter interface in his brain implant. There was only one problem: If Tesla 8 was not able to return him back to 3013 immediately after he left, the paranoid Khan 1 would terminate all of his techs. A delay of even ten seconds might be fatal. He pushed that out of his mind for the time being and got on with his mission.

As he went about his business, Caesar 13 noticed that no one bothered him. There were no Guardian monitoring stations, he felt no tugs on his brain implant, and people mostly ignored him. He must talk with a funny accent, for people noticed it and he had to repeat himself several times at the bank, the rental office, and at the computing store. His currency card, which he had to bring up on his mobile, worked fine.

There was a problem obtaining the vehicle license for himself, for a holo had to be taken. The place was packed with people.

"Please take off your hat," a woman said as he stood in front of the imaging device.

Caesar 13 didn't know what to do; he couldn't expose his brain implant! Fortunately the machine operated automatically and the image was taken. The woman looked at the holo suspiciously, sending Caesar 13's heart racing. "Well, that will be OK," she said. "Your forehead is exposed. We're too busy anyway to do it again."

When he got to the cafe after his day's work it was crowded with early evening clientele. He managed to find a single seat at the very back, squeezed next to a disposal container. That's right! These 21sters threw much of what they used in the "garbage."

After he completed the sending of his identity forms for himself, Einstein 3, and Tesla 8, Caesar 13 felt energized. It must be because he was in a very vital timeline, and was soaking up life force energy from everyone around him. In fact, he felt more alive after twelve hours in 2035 than he ever had since his cloned "birth."

The three temporal technicians were now called Caesar Smith, Tesla Jones, and Alfredo Einstein – this was a little private joke between himself and that clone. Caesar decided that he would stay here in the cafe until the forms were approved, which shouldn't take long. Then he would print them with his 21ster "computer."

Two hours later, after drinking several glasses of hot ginger honey tea (there was nothing like this in 3013) he was ready to walk back to the warehouse. It was dark now. Then he thought of the popups of this period; programmable vehicles with robotic heads. Why walk? It was over three miles to the warehouse. He was used to walking everywhere in 3013, for no ground transportation was allowed. But he was tired. He would test the mobile communications device and called one from a transportation app. The app worked fine; Caesar 13 was elated. This done, he went to the front and paid for his food. This time he used a currency app on his mobile, which worked flawlessly. He walked out of the cafe and entered the popup, which turned its robot head backwards. He held up his mobile, which had an information graphic that showed his identity and destination. That worked as well! The vehicle took off.

Caesar 13 rode in the popup to his rented domicile. He deposited his 21ster computer, his mobile communicator, the 2035 identity documents, and the currency cards for all three time travelers, inside a desk drawer. He felt enormously pleased with himself. The sense of personal freedom here was exhilarating. He called another popup to take him back to the warehouse and prepaid, considering himself now a real 21ster.

The popup arrived and scanned him when he got in. The little vehicle took off and parked in front of the warehouse on a dirt road. When Caesar got out, the little car turned around and he heard a honk! from it. Startled, he turned around quickly, expecting to see an official ready to question him. But the vehicle moved off cheerfully and Caesar 13 relaxed. The honk was probably just a form of acknowledgment.

He entered the warehouse and stuck his Shifter interface into his brain implant, which would send a signal to Tesla 8. He carried his satchel, which now contained only his Guardian thermofilm. Just before the Shifter extracted him Caesar 13 panicked, for he had forgotten that he had several thousand dollars of Tesla Jones' currency bulging in his coat pocket, for he had brought extra that he hadn't deposited in the bank. Frantically fumbling at the satchel's zipper, he stuffed the currency into it.

The warehouse, illuminated by pale moonlight coming through the windows, began to fade. He heard a small popping noise. Suddenly he was back on the Shifter platform in 3013. He was aware of going from a living atmosphere to a dead one. He stifled his fear and tried to look bored. The scene was as he left it. Khan 1 stood by with a dozen of his security clones who surrounded the Shifter platform. Caesar 13 glanced over at Tesla 8, who looked like she had just swallowed a cow.

"The maintenance run was successful," Caesar 13 said, trying to look bored and stifle the elation he felt of his success in 2035.

"Why are you wearing those 21ster clothes?" Khan 1 demanded.

Caesar 13 began to panic. He had forgotten to change back into his Guardian uni before he signaled Tesla 8 in the warehouse! Not only that, but his extra cash was lying on top of his Guardian uni in his satchel.

Caesar 13 tried to school his features to the bored apathy of the common clone. "It's much colder in the past," he said weakly.

Khan 1 turned on Caesar 13. "You were only gone for a second. Your mission was only to test the success or failure of the time device!"

Caesar 13 was now in a full panic. In his excitement to explore 2035 he had never thought of that! To even hint that he had spent an entire day in the past would be fatal. He wracked his brain for a good excuse.

Khan 1 pointed to one of the guards. "Inspect that satchel!"

The guard walked over as Caesar 13 unzipped the satchel, his hands fumbling at the zipper. Several of the currency bills were exposed, sitting on top of his uniform.

"What is this?" the guard demanded, holding up several one-hundred dollar bills.

Caesar 13 fought down his panic. He shrugged, forcing himself to appear calm. "I have no idea. The imbecile who printed this satchel must have left it in there."

It was lame and Caesar 13 knew it. But he had to play it like all was normal.

Khan 1 stared balefully at him, unsatisfied. Caesar 13 thought quickly, damping down his panic.

"I didn't want to be cold, sire, as Commissar Christian 172 was when he went to the past. Sometimes the time device...doesn't operate as it supposed to. That is why we need to test it."

Caesar 13 didn't wait for the Khan's reaction. He walked slowly to his Shifter station under the suspicious eyes of the Khan. The guard grumbled and threw the currency at him. Caesar 13 picked it up and threw the colored paper on his console, as if it were pieces of scrap paper. He began to casually take off the winter coat, hat, mittens, boots, and the various undergarments, all the while being scrutinized by the Khan and his guards in their red and black security unis. A lifetime of pretense, hiding his emotions and intelligence from the Overseers, served him well now. Even though he was shaking inside, he kept his countenance and his posture bland and apathetic. Caesar 13 got his yellow and black technician's thermofilm from the satchel and put it on, avoiding the rest of the paper currency which was spread haphazardly inside the bag. He tried to look casual as he stuffed the 21st clothing into the satchel, which he threw into a little storage space underneath his console at his Shifter control station. "I will need these again for the final maintenance check tomorrow."

The paper currency was still on his console.

Caesar 13 saw Khan 1's frown of suspicion as he walked over to examine the currency bill. The Khan, wearing an all-black uni indicating that he was the Supreme Leader, motioned to his guards. He looked directly into Caesar 13's eyes. "Take this clone to the basement and terminate him."

Caesar 13 froze in fear. Later, he thought that his split-second hesitation saved his life, for he wanted to scream in terror as any 21ster would have done.

As the security guards approached Caesar 13, Khan 1's face showed bitter anger, hatred, and then frustration. The Supreme Leader realized that without Caesar 13's assistance, the rebel clones could not be defeated. The time-travel device and its chief technician must be kept fully operational until then.

"Halt!" Khan 1 commanded his guards as they were about to seize him. "This clone must be kept alive for a little while longer."

As the guards retreated, Caesar 13 went to his knees and groveled silently before the Khan.

Khan 1 stomped out of the room with his guards. Caesar 13, trembling now with shock, tried to get up and fell to the floor in a heap.

When Khan 1 saw this a few minutes later at his comm station, he felt much better. He suddenly realized that the other twelve time techs might be able to run the Shifter without Caesar 13. After that malignant completed his tests tomorrow, the Shifter would be ready and Caesar 13 would be unnecessary! It would be a pleasure indeed to terminate that arrogant clone.

CHAPTER 19

2035 AD

One by one, the time travelers arrived in the warehouse in a state of shock after their trip to 3013. For me, it was like returning to life from a near-death experience. The people in 3013 were either antisocial, or walking dead. Even the air and the plants were hanging on to life, like a grizzled tree struggling for survival on a rocky mountain slope.

Kjirsten looked terribly shaken by the experience. M'basa had his arm around her and was talking to her softly. Karen and I looked at each other in disbelief.

It took about two minutes between each insertion. I got to see Kjirsten and M'basa "pop" into reality. One moment there was nothing. Then, for a split second, a turbulence in the space around where the person was coming in. Then a living, breathing human being manifested on the cold concrete floor of the warehouse.

To my engineering mind it was both magical and scary, but Karen seemed to accept it as just another advance in technology. When M'basa appeared he was really angry, and vowed to roll up Harry Kaine when he appeared.

"Please don't," Kjirsten pleaded. "Just put your arms around me."

When Harry popped into reality Karen confronted him. "After seeing the dystopian future you will help to create, have you changed your plans?"

Harry smiled smugly. He seemed energized by our trip to the future. "Not at all, my dear. Our little adventure has inspired me to do better and avoid the obvious pitfalls." Harry looked her over appreciatively. "Would you like to go for a coffee?"

This was said with so much charm and attentiveness that Karen almost said yes. "Uh, I'll pass for now."

"Some other time then!" Harry said cheerfully. Whistling, he walked out of the warehouse.

"Do you *like* that guy?" I asked Karen.

Karen liked Joe but this was going too far. "And what if I do?"

I stepped back. My fondness for Karen was being overwhelmed by jealousy. "A woman who could like Harry Kaine isn't my type. He's a slime."

"He's charming," Karen shot back. Then she smiled. "And he's also a slime."

The tension was broken and my anger evaporated. I smiled back and we laughed together. I imitated Harry. "Would you like to go for a coffee? ... My dear?"

Karen held out her arm. "Yes please."

I took it and we walked out.

Kjirsten looked up at M'basa. "Did you mean it back there when you threatened to end Harry Kaine?"

M'basa spoke passionately. "I did. You see, millions have died in our African civil wars. In Darfur, in Rwanda and Burundi, the Congo, Somalia, the Central African Republic...humans killing humans is sickening and irrational. Self-obsessed guys like Harry Kaine are always behind destructive movements, but his threatens the entire human race."

Kjirsten had never thought of it in this way.

"Americans traditionally don't consider other places in the world, except when something pops up on their screens. Somebody once said that foreign wars are the way Americans have learned geography."

Kjirsten felt better now that she was back in 2035. "True, but we now know about the global military industrial complex. America was just their headquarters for a while."

M'basa was enjoying this conversation but he was getting uncomfortable. "It's cold in here. Let's go someplace warm and get something to eat."

Two days after their return from 3013, Karen Everard got home after a pleasant evening with Joe Courvall at his apartment. They had simply talked as friends, and the time had passed quickly.

"Joe, did we really go into the future?" she had asked him.

"Yes. But it's been difficult for me to process everything. I'll never forget it."

"I have to do a story about this, but I can't tell the truth because people won't believe me. Just like they wouldn't believe what happened out on the Diag."

"We know what happened now. That woman was Pietra 23, and she was inserted back into the timeline from the future."

"I just had an idea! I'll write it as a story and publish it under my name. It's not something the *Chronicle* would want to print."

That was when I first got the idea of writing this memoir. "We've seen things no one in the world has seen. Should we tell the Gang about our travels in time?"

Karen laughed. "Sure, but if Liqao believes it I'll buy him two plates of focaccia rolls!"

"You're funny, Karen."

Karen noticed that Joe spoke with admiration for her. She liked it. "I have to find a way to tell this story in a believable way."

"Just preface what you say with, 'Once upon a time.' Make it seem like fiction."

Karen shook her head. "It happened; people have to know the truth about human augmentation. Joe, Harry Kaine is dangerous and we have to make it hard for him. If we do nothing, in 13 years he's going to make that speech at the UN and eventually wreck the human race."

"I learned something in my brief talk with Li 308 just before we got back. Harry is just a catalyst, a focal point around which historical events – or should I say temporal energy – come together. Focus less on stopping Harry than on educating everyone else. Expand your audience to all of humanity."

Karen was impressed with this reasoning. "I've got a problem though. If I tell the truth and play it straight-up I'll be criticized as a nutjob. If I make it fictional, the story loses its impact."

"Write it up as a series of articles on Transhumanism and what might happen in the future. Since we lived it, the articles will be totally authentic. You can present the good side of Transhumanism as well, if there is such a thing, and satisfy your journalistic integrity."

Karen was looking at me with frank admiration.

"God Joe, you should be a diplomat or a counselor."

"I'll open a circuit board design shop with a free counseling session for orders over $250."

"What will you call your shop?"

"Uncle Joe's Circuit and Psyche Fixit Emporium. Free donuts on Tuesdays."

Karen laughed.

"You have a great smile Karen. You should do it more often."

"Do you think I'm too serious?"

"Yes, but so am I. Maybe we can lighten each other up."

We left it at that.

I was at my desk at Phoenix, three days after I got back from 3013. I was having trouble concentrating on my work after my experiences in the future. It was just after 5 and I was ready to walk home when I got an urgent call from Ger. "Get down to the Taubman lab, Joe. I just discovered something I think you'll be interested in."

I was ready for anything to take my attention off of 3013, so I walked over to the Taubman Building without going home to eat dinner. When I got there Ger was standing in the lab. The rings were turning fast. There was a distinct blue energy surrounding the platform and the space above it was shimmering.

"Oh my God, Ger."

"You look like a ghost, Joe. What is it?"

Ger was my best friend but I was hesitant to say anything.

"C'mon dark side, spill it."

So I told him about our little adventure to the future. "The...people...from the future look exactly like the ones in your recording, Ger." I pointed to Ger's makeshift time viewer. "The spherical blue energy field and the disturbance in the space above the platform is exactly what I observed in 3013 just before we transported in time."

Ger was blown away. "I thought something had happened to you guys. Karen, M'basa, and Kjirsten look like they fell off a ski lift."

"Yeah."

We both looked at the turning rings. "I don't have enough power to get the rings turning any faster, Joe. I've gone as far as I can with this thing."

"For fuck's sake Ger!" I don't usually swear but I couldn't help myself. "In 3013 they call this thing a Time Shifter. When the rings turn fast enough the blue energy intensifies and then..." I described the popping sound and how the person on the platform disappeared. "Li 308...er, the people in 3013, say that a spacetime bubble forms around the platform and that the rings become inertialess. I saw it with my own eyes. They move so fast you can't even see them. After the popping sound they stop instantly."

Ger gulped nervously. "Uh, you know, I think I'm going to stop my experiments with this thing. I've learned all I can from it."

"I think that's a good idea." I described what it felt like to be transported in time while we watched the rings turning.

"Turn that damn thing off," I said. "What did you want to show me?"

Ger slowly walked over to the console and the rings stopped turning. "Here's something a lot more cheerful. Watch this. Yesterday I was experimenting with the geometry and the arrangement of the magnetic arrays. When I did I discovered something interesting."

Ger went to the device and fiddled with the magnetic arrays that surrounded the rings. The rings began to move, very slowly.

"Wait a minute. You turned the power off, right?"

Ger grinned. "That's right. So how are these rings moving?"

Ger asked this question as an instructor would to a student. Suddenly a light went off in my head and I forgot about 3013. "Of course! Magnets in phase pull on each other.

When they are out of phase, they push against each other." I remembered playing with small permanent magnets as a child. The force they generated was powerful. I remember trying to push them together as hard as I could, without success.

"That's correct. The energy of magnetism is permanent; not like a battery that runs down. My theory is that arrays of magnets, with the correct geometry, can perform a powerful push and pull that can move the rings without being plugged into the electrical grid."

"Or turn a generator shaft."

"Or induce an alternating electric current through a wire."

My jaw dropped. "But...this thing is an overunity device! Get us off the electrical grid. Free energy!"

Ger's face broke into a smile. "You get it, Joe. You have insight, you're not a normie."

I was very pleased to hear this because I'm just an engineer, not an autist or a genius like Ger. I gazed at the device. "But...why didn't anybody think of that before?"

"I don't know."

"Instead of clean magnetic power we went with inefficient fossil fuels, coal-fired electrical plants, and nuclear energy."

"That's right."

"Ger – you have a Nobel Prize here!" My mind was filled with ideas.

"We have to develop this idea after I've wrapped up my time research."

A feeling of excitement zapped through my body. "I'm in. You're going to need a good engineer."

"My thoughts exactly."

We talked for several hours until I got so hungry I left to get something to eat.

Winthrop Sauvage spent the entire evening self-indulgently looking at child porn and masturbating. When his addiction had been satisfied, he engaged in his usual post-orgy bout of guilt, self-deprecation, and rationalization.

He felt the impulse for more but stifled it.

"I'm sick," he said to himself.

"No you're not," his alter-ego replied firmly. "Millions of people are like you. Stop beating yourself up."

"I sexually abused my own daughter."

"You just succumbed to a moment of madness. You're a good person."

This dialogue, as always, continued until he felt he was going crazy. After that he got drunk to quiet his inner demons. He fell asleep on the couch.

When Winthrop woke up he had a bad hangover. He made a pot of strong black coffee and drank until he felt a little better.

This cannot continue, he thought. I have to come clean and get help. Winnie is right.

Today was Sunday. He had written a funding proposal for his cryogenic research and sent it to the NIH, but it was a long shot. The university wasn't interested. Tomorrow he would call Harry Kaine. Then he would contact that reporter, what's-her-name, from the *Chronicle*, who had written an article on COSA and child abuse. He would unburden himself to her as much as possible without getting himself thrown in jail.

At work, Karen Everard began to sketch out a series of articles on Transhumanism, which would use what she had seen in the future to make her stories sound realistic. Bob Guza, her editor at the paper, had already called the university and made appointments for her to talk one-on-one to the four scientists she had met during her earlier meeting with Harry Kaine. These would be more in-depth interviews. It would be interesting to ask these scientists about a Transhumanist future, having actually seen it herself. When she got home she began writing her memories of 3013 AD. She worked into the night and finally went to bed at two in the morning.

The next day she had to research a piece about a claimed increase in pollution in the Midland River. Just as she was about to break for lunch she got a call.

"Hello, I'm Winthrop Sauvage. Is this Karen Everard?"

"Yes, this is Karen Everard. Weren't you a speaker at the '1,000-year space mission to save humanity' event?"

"Uh, yes. But that's not what I want to talk to you about."

The man sounded afraid. "Did you see anything...unusual at the event?" she asked. That was the one where forty people (including herself) got translated to 3013 AD.

"Not really," Winthrop said. "While Harry was talking I was in a conference room going over my speech. I heard that there was some sort of disturbance, but thought nothing of it. By the time I was ready to give my speech there was no one left to hear it! I suppose that is unusual."

Karen felt relieved. "What can I do for you Mr. Sauvage? Are you the father of Winnie Sauvage, the one who works for COSA, the child abuse organization?"

There was a hesitation on the other end. "Yes, I am Winnie's father. It is around that very subject I would like to have a...private conversation with you."

Karen sensed a story. "Yes? Where would you like to meet, and when?"

"As soon as possible, somewhere we can be private."

145

Karen thought quickly. "I can meet you right now, Mr. Sauvage. How about one of those private conference rooms at the public library? The East branch."

Karen heard a sigh of relief at the other end.

"That would be perfect, Ms. Everard. What I have to say is difficult. If I don't get it off my chest soon..."

"All right then. I can be there in half an hour."

Karen walked into the City Desk office and saw her boss. "I think I might have a lead, Bob. Going to meet someone. Can you wait on the Midland River story?"

Bob Guza checked his spreadsheet, which contained each day's edition and the stories slotted for them. "There's no rush on that one, although the Save the River environmental group isn't going to like it." He looked up from his monitor. "What do you have?"

"I don't know yet, Bob. It may be nothing, but I'm trusting my instincts on this one."

Bob Guza smiled. "Go for it. But be back if you can before five for the staff meeting."

"I'll try."

Karen walked out of the Chronicle building on Third Street and Blake. It was a nice day and she wanted a walk, but the library was too far away. She called a popup and got there just as a nervous-looking man in a long coat was walking up to the front door.

"Are you Karen Everard?"

"Yes. Winthrop Sauvage?"

The man looked nervously around. They were standing at the front entrance to the building. Suddenly he pulled out a data pod and thrust it into her hand. His eyes were bulging. Here was a man almost ready to break, Karen thought.

"Please, take this and do what you will with it. I...I have to tell someone about this."

After another frantic glance around the entrance, the man ran off.

'That was anticlimactic,' she thought. But when she got back to the office and played the recording, she couldn't believe what she was seeing. An hour later she finished and walked into Bob Guza's office. "You have to see this."

Bob stuffed the pod into his player and displayed it on the large screen on the wall in front of his desk.

"Earphones," Karen said. "Mute the speakers and close the door."

The recording began. "I am Winthrop Sauvage," a voice said off camera. "I am making this recording because I can no longer live with myself. You won't see me but everyone knows who I am."

A graphic appeared on the screen. "This is the name of a private network I belong to. It exists for those who enjoy looking at child pornography. I don't know who runs it, but they are sick people. I am sick, and am going to get treatment."

The recording began with a series of images that first showed young boys and girls in cute dresses and uniforms. Then, half-naked and naked children in sexually suggestive poses. Then a series of vids that turned Karen's stomach.

"I think you've seen enough," the voice said. "There's worse. I don't know much, but I do know that these networks are nationwide. They certainly are not isolated to Midland. I suspect they are worldwide."

The recording shifted to a data table that contained information on the network's IP address, its possible location, the number of users, and the various subs on the network.

"He hasn't provided any names," Bob said.

"No, but we know his daughter Winnie is involved with COSA. I've already talked to her briefly, but maybe it's time for an expose."

"If she'll talk."

Karen shrugged. "We won't know until we ask."

After Winthrop Sauvage got home he felt sick. He thought he had done the right thing but now he wasn't sure. If those people at the *Chronicle* weren't discreet, he was ruined. And he may have implicated Winnie as well, although she already told him that the people who run the trafficking networks already knew about her. So maybe it won't be too bad for her.

He had cast his fate to the wind, as the saying went. It was out of his hands now; he no longer felt the crushing responsibility to tell someone.

He knew he wasn't going in for treatment. A new batch of vids was going to be on the network tomorrow...

Karen Everard called Winnie Sauvage.

"You don't know me Winnie—"

"I know you. You're that reporter for the *Chronicle*. If you're calling, Winthrop must have told you something."

"He did. He gave me a recording. Would you like to see it?"

"No, but I'd better. The people in COSA are in danger every day from the creeps who run the child trafficking networks. We're only left alone if we keep quiet about them, and your vid might expose something that wouldn't be good for us."

"Oh my God."

"Yeah. Come over tonight at 8 p.m. 2477 Warrington, a rental house next to a city housing duplex on the north side."

"I'll find it. See you at 8."

When Karen arrived (M'basa and I came along) the driveway was filled with cars and they had to park two blocks away. When Karen walked in someone said, "Is this the bitch with the pod?"

M'basa, who was behind her, said, "Yeah this is the bitch with the pod."

Karen recognized Kesha, Winnie's friend, as the speaker. Winnie was standing next to her. "Sonny, you might think you're tough. You're just a little mewling kitten compared to the psychopaths who run these networks."

To Karen's surprise, M'basa bowed.

M'basa was thinking of the murderers who had committed genocide in Africa. One of his uncles had died in the Hutu-Tutsi conflict a half century ago. These child traffickers were undoubtedly the same sort of monsters. "You're right. I apologize for my anger."

Kesha was mollified. She waved her hand around the living room. There were at least thirty people here, some standing in the small kitchen, some on the stairs. "We're here because whatever Winthrop put on that vid might affect all of us. Some of us drove here from Chicago to see this. Is everyone ready?"

Karen handed Winnie the pod and the recording began.

Karen noted the reactions of the COSA people watching the vid. Most of them didn't even blink. Apparently this was old hat for them.

"Winnie, what is your father doing?" Kesha asked after it was over.

"In typical Winthrop fashion he's throwing his problems onto someone else. He's been doing it his entire life."

Winnie turned to Karen. "What do you intend to do with this? If you publish it and identify the network, some of us here will be threatened. The Bosses only tolerate us if we keep our mouths shut."

"And if I don't publish this what will happen? Children will keep getting abused. It's just what they want."

"You will be a target too!" Kesha said. "Think about that."

Tempers were flaring now as the others were shouting and arguing. After a half hour police chief Chastaine walked in with Liqao Chang. "What's going on here?" Chastaine said. "You people are making way too much noise."

In response Kesha flipped the pod back to the beginning and it began to play again.

Everyone sat through it once more. "What the fuck is this?" Chastaine said. Karen saw that Liqao was shocked to his core.

Winnie lost her temper. "You stupid, ignorant moron! What do you think COSA has been trying to tell you for the past three years?"

The police chief was subdued. "I didn't know," he said softly.

"Now you do," Kesha replied. "What are you going to do about it? That perv at Child Protective Services is still there."

"We want a copy of this recording," Liqao said.

"I'll have it for you tomorrow morning," Karen replied.

Many in the group groaned. "I was afraid of this," someone said. "The Bosses can be hard on whistleblowers."

"To put it mildly," Winnie replied.

Kesha spoke contemptuously to Chastaine. "This fathead won't do anything about the child abuse at Child Protective Services. He's an enabler."

"That's not fair," Liqao said in defense of his boss.

"Who is this lightweight?" Kesha asked.

"He's the guy who read our emails and squashed the case against Bob Butters, that pedo in CPS," Winnie replied.

"You!" Kesha cried. "You're worse than the chief of police here!"

Liqao and Chastaine looked at each other. "Keep it down," the police chief said. He turned to Karen. "I want a copy of that vid on my desk first thing tomorrow."

The police chief and his IT assistant left.

Winnie, Kesha, Karen, M'basa, and all the others talked until almost 3 a.m. I hung back because Kesha was still mad at me. I was glad Ger wasn't here to upset Winnie Sauvage. All of the COSA people agreed that the cat was out of the bag now. "Lay low, keep your mouths shut, just do your job," Winnie said to her fellow COSA employees.

"I have a feeling someone is going to get erased over this," someone said finally.

Everyone left, hoping it wouldn't be them. But they all recognized that the information couldn't be suppressed. Too many people had seen it.

Winnie and Kesha wondered what the reporter and that idiot police chief would do. Certainly the Bosses knew about the meeting by now, it had been widely promoted that day among the COSA group. There had probably been a plant at the meeting. If so, they now knew what was on that pod, and that the authorities also knew.

The next morning before work Karen Everard dropped off a copy of the pod in a sealed package addressed to police chief Chastaine, marked "For Your Eyes Only." Then she dropped off a recording she had made of the entire meeting at the COSA house to Bob Guza's office. Then she had to get busy writing up that stupid story about pollution in the Midland River for tomorrow's edition.

At 3 o'clock the news broke on Channel Five.

"Winthrop Sauvage, the noted cryogenics researcher, was found dead in his home this afternoon, after an apparent overdose of amphetamines last night. Dr. Sauvage did not report for work this morning, and friends said he had been feeling depressed. Drugs, and several whiskey bottles, were found around the couch in his living room, and a suicide note."

Winnie saw the report on a screen at CPS, where she had gone to check on a child just assigned to new foster parents. When she got to her car she turned her mobile to Channel Five. She began to cry, great racking sobs; a feeling of abandonment and loss. She didn't think she'd feel anything. Her father had raped her, and done nothing after she'd been trafficked. But he was still her father.

She saw the handwritten suicide note and did not recognize her father's writing.

It was clearly a warning.

———— • ✱ • ————

I called Ger after I got home from work.

"You heard about Winnie's father's death?" I asked.

"Yeah, she told me."

"Everyone in COSA could be a target if her father's vid comes out."

"It all depends on what Karen does with it. And Liqao, and the police chief. Winnie says that if they keep it to themselves it might be OK."

"I don't think Karen will do that. You heard what she said at the meeting."

"No, Joe. Tell her not to do it. I got clued in a little by Winnie after the meeting. Whoever is running the human trafficking networks are not fucking around. If you don't believe me, talk to Ralph Zimring at Berglin Enterprises."

I was afraid for Karen. It would be just like her to publish a story. "It's off with me and Kesha."

"Winnie doesn't want to see me anymore either. Some of the shit she was telling me about those trafficking networks is scary."

"Yeah. At least you have Tanya. If you want her." Tanya was the receptionist at the Taubman Research Building who had a crush on Ger.

"And you have Karen."

"Maybe not if she publishes what's in that vid."

"Talk to her, Joe." Ger broke the connection.

I was feeling overwhelmed. Our trip to 3013 was indelibly etched into my memory, but it had been pushed into the background by the disgusting and frightening images on the trafficking recording. And now the death of Winnie's father.

I was worried about Karen. We were supposed to go out tonight but she'd been working late ever since the Winthrop Sauvage story broke. Maybe it wouldn't work out with her, like with Kesha. She had a temper, and our careers were completely separate, but I decided I wanted to develop the relationship. We were both busy writing up our experiences from 3013. I had already called Kjirsten and M'basa about that, and was trying to talk to Harry Kaine, but he was putting me off, the bastard. I had been dreaming every night about the Shifter, the Tower, Khan 1, Patton 1, Christian 172, and their assassin clones. Then there was Caesar 13, the clone who ran the Time Shifter. He was a sort of rebel but had to answer to the Overseers. It seemed like a movie, but I knew it was real. People – or clones – our descendants – were actually living there, up the timeline, in a society that was dying. I could hardly believe that it was the future of the human race. Then there was the trafficking story. I had to tell Karen to back off; it was too risky. I'd talk to her about it after the Gang of Eight meeting on Saturday.

I had my phone out and dialed up Lledren. I saw that he was still at the soup kitchen. "You're working late."

"Yeah, just finished V-filming and editing my latest V-report."

"I haven't checked that out yet. Did you hear about the death of Winthrop Sauvage?"

"Oh yeah, but that's out of my line. Check out my V-blog. Just finished a series tonight on the homeless here in Midland."

I was momentarily diverted. "There are still homeless in Midland? I thought the city allocated resources for that."

"They did, but most of the funding got spent on more public transportation. Progressives are screaming. We had somebody die last month of exposure."

"I'm sorry to hear that."

"Did you want something Joe?"

"Yeah. I've called a meeting of the Gang. Something happened to me, M'basa, Kjirsten, and Karen that we need to talk about. Would you be up for another trip to Angelos?"

"Sorry Joe, right now my finances won't allow it. But hopefully that will change."

I knew better than to offer to pay for Lledren's meal. "OK buddy. I'll set it up at Ger's place this Saturday at 7."

I called everyone in the gang. Only M'basa objected. "Saturday night Kjirsten and I have sex and I don't want to miss that."

"Sorry man. That's the only time I could get everyone together."

I could see M'basa sigh.

"Besides, Kjirsten already agreed."

"Traitor! You're on my shit list now, Joe."

Jack Chastaine went home the night of the COSA meeting, disturbed after seeing Winthrop Sauvage's vid. He didn't believe in organized pedophile networks. The occasional perv here and there, sure. A few images floating around, probably taken by Sauvage himself. The vids of sexual abuse, though, made him angry. He couldn't dismiss them, and it made him want to put a bullet in that pervert Sauvage. He'd arrest him tomorrow and get a search warrant for his house. Then he'd send him in for psychological evaluation. If he found evidence of child abuse he'd lock him up and give the evidence to the D.A. He went to bed that night and had bad dreams. He didn't tell his wife a thing about it. Lacy would get too upset.

Jack woke up the next day wondering whether those two troublemakers at COSA had been right all along. Then the news of Sauvage's death was reported later in the day. He personally inspected the crime scene at Sauvage's house that afternoon. It was not a suicide. The handwriting on the letter didn't match the dead man's. Plus, the arrangement of the crime scene told him that the bottles had been placed at the scene to make it look like the man had been drinking. A couple of MDMA tabs had been conveniently left on the side table by the couch. Sauvage was a known drinker but each of the bottles had only one Sauvage handprint on them, as if they had been bought, emptied, and someone wearing gloves had placed the man's hand on the bottles after he was already dead. A drinking man would have handled the bottles more; there would be multiple prints on each one. That got him thinking that perhaps this was the work of amateurs. The local Armenian mafia? Maybe, but the Katelians and the Nalbandians had gone mostly legit after that ruckus with Max Berglin several years ago. Besides, the local mafia had no beef with, or connection to, Sauvage. But if the killers were local, the child porn network might be local as well. No. Liqao told him that the IP address of the server in Sauvage's vid was near Langley, Virginia. The goddam CIA! Or what was left of it. Those boys were way out of his league. The storage pods on Sauvage's computing device had been erased. No evidence of child porn existed anywhere in the house. He had interviewed the man's irritating daughter Winnie, who told him that her father definitely had child porn on his computing devices and watched them. *Someone* had been in the house and covered up their tracks.

What really riled him was the message he'd received on his mobile after coming back from Sauvage's residence. A single sentence: *Look the other way and all will be well.*

Winnie and Kesha both received messages. *Get that fat police chief off the Sauvage case. And keep your fuckin' mouth shut.*

Both women were terrified, especially Winnie.

She knew how the Bosses operated. They had no regard for human life whatsoever; they were stone-cold killers who would take out anyone to protect themselves and their networks. She was in far more danger than anyone in COSA, for she had seen the trafficking networks at the top levels. She would need to be very, very careful. She was glad she had broken it off with Germaine Robinson. She had no time for normies now; she would only associate with her own people.

———— • ✲ • ————

Liqao Chang went home after work the day Winthrop Sauvage was killed. He felt terrible, and partly responsible for the man's death. His own work on the Butters case had possibly allowed a pedophile to continue to work at Child Protective Services. After seeing that child porn vid last night he was convinced that the two COSA women had been right all along about Bob Butters. Why did he have to rush into things? For the recognition of course. And that bonus last year didn't hurt either. It's what Kirra had been telling him ever since they got together: he was too impulsive and self-centered.

Karen Everard also received a message. *Let the Sauvage story die if you know what's good for you.*

She showed it to Bob Guza in his office. "Do we run with this? Or is it too dangerous?"

"I'll have to talk to the owner. If he says we don't publish, we can't."

"Then he's in it too."

"No, Karen. I've talked to that Winnie Sauvage. She wouldn't tell me much, but she is literally scared to death. She got a message too. Apparently the people who run those networks —" Bob made a throat slashing gesture.

"Wow. You see that kind of thing in entertainment vids."

Bob shrugged. "The world is still a pretty dark place."

"Who are these people, boss?"

Guza smiled. "If I knew that I'd get a Pulitzer."

"Call the owner now, would you Bob?"

Bob sighed and grabbed his communicator. He had the private line of the owner but was to use it only in a dire emergency.

"Hello Mr. Vorheis? Bob Guza here at the *Chronicle* in Midland Illinois. We're sitting on something explosive that needs your immediate attention." Bob put the call on speaker

"Send it to me. I'm real busy now."

"Sorry sir but that isn't possible. This information is way too sensitive."

Karen saw the owner's eyebrows raise. She'd have to learn that trick to dismiss the overly curious. "I have to make a special trip up there if I want to see this, is that what you're saying?"

"Correct. It involves pedophilia and child trafficking."

"Oh my God. No, sorry Bob, can't get into that. It's a no-go."

"Why, if I may ask? This story has national implications."

Karen saw Vorheis' eyes widen in fear.

"The subject cannot be discussed. The *Chronicle* is a local publication, Bob. Keep to local subjects." The comm was cut.

Bob Guza looked at Karen. "I wonder if he got a message too."

"He was afraid, boss. Maybe he did."

"If you want to publish this story, Karen, it cannot be under the *Chronicle* label. You'll have to do it independently."

"We could do a dual byline."

Bob shook his head. "I've got a wife and kids. I can't put them in danger."

Karen left Bob's office disappointed. She now had two blockbusters: the child porn story and the time-traveling clones. She couldn't write about one and had to pretend about the other. Bob wanted another story on Transhumanism, which had generated a huge response from the public. So she had to interview Harry Kaine's scientists in more depth. Bob had set up the meetings. She was to interview Dr. Mary Chen and Dr. Armand Singh, two of Kaine's Transhumanist scientists, on Friday.

When she got home that afternoon she sadly drew a line through Winthrop Sauvage's name on her interview list. He had been in the Kaine group. She was understanding more and more that Harry Kaine was the focal point on the timeline, just as Li 308 had said up the timeline in 3013.

That night Karen Everard called Kirra Bigbear. Whenever there was an intractable problem for anyone in the gang, Kirra or Joe always had insight and wisdom. Karen explained the child porn vid and the story that needed to be exposed. She explained about the threatening messages.

"So what should I do? Publish, or eat the story? If I publish I endanger myself and the COSA people. If I don't I'm not doing my job as a journalist. And more children get abused because I didn't do anything."

"You aren't saying the problem the right way," Kirra said. "Publishing or not publishing, that's not your major concern here."

"What do you mean?"

"I think this problem revolves more around your ego. You're really more concerned with your journalistic status than you are with the people you'll harm or hurt."

Karen was angry now. "That's not true!"

Kirra just looked at her. "Your main motivation for publishing isn't to help kids. It's to satisfy a code of journalistic ethics and massage your ego with a big story."

This was far too forthright for Karen. "You always say things so bluntly! Why can't you be nice for a change?"

Kirra looked at her with the wisdom of an old soul. "You don't come to me because I'm nice. You come to me because I'll get to the bottom of your problem. You just don't like the answer."

Karen flared. She hadn't even considered the ego aspect.

"To resolve your problem you have to understand how your ego relates to your career as a journalist."

Karen didn't like this. "It's not ego. It's about journalistic integrity and my professional reputation."

"Same thing."

"No it's not! You're exasperating, you know that? I ask you a simple question and you give me personal insults."

Kirra was kind but firm. "You can't solve the problem until you get yourself right inside. Then the problem will solve itself. You'll see a clear path."

Karen was clearly uncomfortable. It was the indication to Kirra that she had framed the problem the right way. "Your decision to publish or not should have nothing to do with your career or your professional reputation. Will publishing do more good than harm to others is the real question."

After grumbling and rationalizing for thirty minutes, Karen finally understood. Her code of journalistic ethics was just a set of acquired beliefs, backed by her ego, that was preventing her from seeing the problem. Kirra was right: publishing the story or not should only be based on what was the greater good.

"Why aren't you a counselor? You're so good with people, and so wise."

Kirra was surprised. "Maybe you're right! I was wondering what I should do with myself. My work at the library is OK, but I feel I'm just marking time."

The two friends smiled at each other. Both had a lot of thinking to do.

That evening Karen didn't go out, but sat thinking about the death of Dr. Sauvage and her trafficking story. It was big; it could catapult her name to national prominence.

After several hours she decided that she was not going to publish the story anywhere. She would write it up and file it. It was sad that we live in a world where people got away with child abuse. But to release the story would put too many people in danger, and wouldn't even put a dent in stopping child abuse, if Winnie Sauvage were to be believed. She silently thanked Kirra for showing her how her ego had made her decision a big problem. It was really a simple decision.

On Friday at lunchtime Karen went to interview Dr. Mary Chen on campus at the Life Sciences building cafeteria. When she walked in Chen waved and she sat down. They sat against the wall at the back, away from the noisy conversations. Mary Chen was a thin, smallish woman with black hair and nervous movements. She was eating a salad. "Are you after Harry Kaine?" she asked before Karen could get her recorder out.

"Harry Kaine? Uh...no. He's not my type."

Chen brightened. "Good, because he's mine."

After this peroration, which startled Karen, Mary finished her salad and gave Karen her full attention. "What did you want to know?"

"I want to know more about Transhumanism and the technologies involved. As you know, I write for the *Chronicle*. I have no agenda for or against, I just want to get information."

"That's refreshing. My specialty is gene therapy, or precisely, CRISPR technology. It's a tool we use to edit genomes. A genome is the genetic material of an organism. The genome includes genes, which are the coding regions – the part of the DNA with proteins – and non-coding regions."

"The coding regions determine what kind of cell it is, right? A liver cell or a heart cell." Karen had done a little study before the interview.

"That's good, Ms. Everard. "

"I want to write an intelligent article. Is it true that 98% of human DNA doesn't code for proteins?"

"Yes. But working with the coded sequences, we can do remarkable things."

"OK. Tell me about your work and how it relates to the field of Transhumanism."

Karen could see Mary was getting excited.

"CRISPR technology allows us to change DNA sequences and modify gene function. Using CRISPR we can correct genetic defects, and prevent the spread of diseases."

"Wow!"

"Essentially, CRISPRs are specialized stretches of DNA. We can use them to cut and splice together strands of DNA. When CRISPR components are transferred into more complex organisms, we can edit genes. We can also neutralize attacks by viruses and other foreign bodies by using CRSIPRs to chop up and destroy the DNA of foreign invaders."

"Edit genes? That sounds terrifying. You're saying that you can play God and create an altered human."

Mary Chen lowered her head and shook it. "I knew I shouldn't have agreed to this interview." She straightened and sighed. "We use gene therapy for the good of humanity."

"OK. I'm sorry."

Mary Chen brought out a diagram "You see, I came prepared. Harry insisted that I explain things so you could understand."

"I appreciate that."

Chen pointed to the diagram (See diagram on page 157.[10]

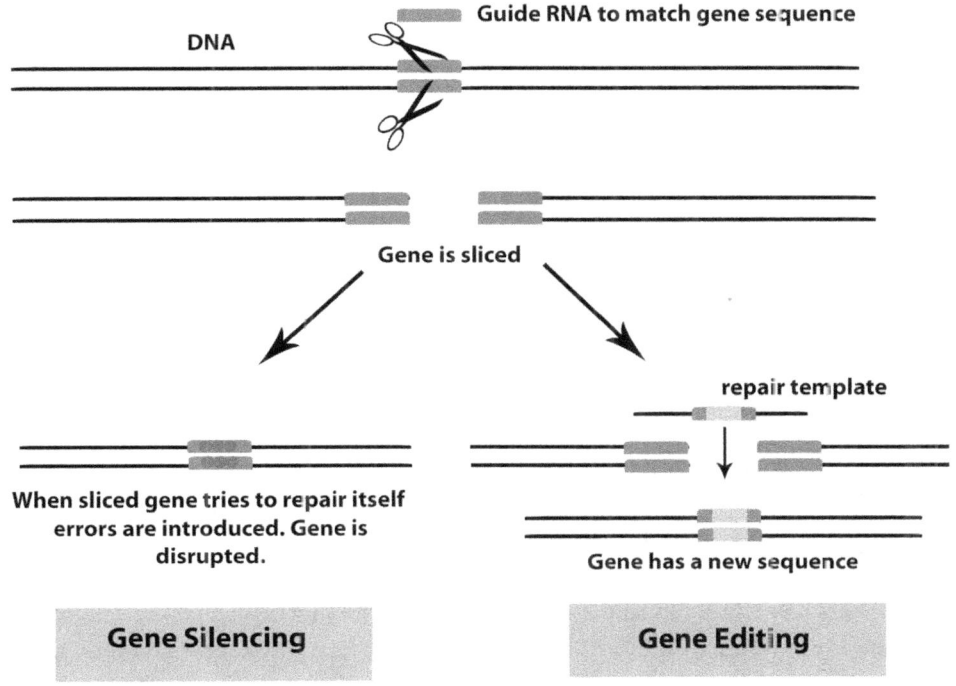

Gene Splicing and Gene Editing

"Genomes are the building blocks of DNA. The genomes of various organisms encode a series of messages and instructions within their DNA sequences. Genome editing alters those sequences, and so changes the messages. We do this by cutting or breaking the targeted DNA, as the diagram shows. We essentially trick a cell's natural DNA repair mechanisms to introduce the changes we want. It's a lot more complicated than this, but basically the technology cuts, like a pair of scissors, the DNA at that site, and ideally nowhere else, so we can insert our genetic improvements."

Karen was a little disturbed by "and ideally nowhere else", but she let it go. She could see that Dr. Chen was growing animated.

"The technology is almost childishly simple to use now, compared to when it was first developed in 2013. CRSIPR works and can correct genetic defects."

"Yes, but there's a public policy issue. If genes can be edited, human DNA can be altered. Who is going to implement CRISPR and what are their goals?"

Mary Chen laughed. "I'm afraid that's way beyond my purview. I'm just a scientist. Talk to Harry about that."

Karen was dissatisfied with this but she wanted to milk all the info she could out of Mary Chen. "Tell me what you're most excited about in your field at the present time."

"Gene drives!"

"You mean like gene storage?"

"No. Gene drives are genetic systems that increase the chances of a particular trait passing on from parents to offspring. Eventually, over the course of generations, the trait spreads through entire populations."

Karen was shocked and Chen stopped talking. "Is there something wrong?"

"Uh, no...sorry. Continue."

"Gene drives can help to control the spread of diseases; but we want to improve human biology by passing on an enhanced human genome to future generations. Ultimately, we want to develop a procedure to improve genomes in real time. We're not there yet, but it's what gets me up every day."

"You mean, altering genomes to eliminate unwanted traits. Like going to the hospital for an operation. In and out, genetic improvements."

"Yes!"

There it is, Karen thought, Kaine's Transhumanist program in a nutshell. She could see how excited the scientist was. She was just a researcher, excited about her work. Harry Kaine was the real problem.

"Thank you Dr. Chen. Now I want you to be honest. What is the present effectiveness, or efficiency, of gene editing? How accurate can you be in cutting and splicing your genes? How effective are the treatments that will be spliced into the altered genes?"

Chen seemed uncomfortable. "Well, we've gotten CRISPR over 95% efficient now, depending on the target."

"Depending on the target?"

Chen shifted in her chair. "There is a phenomenon called 'off-target effects,' where DNA is cut at sites other than the intended target."

"What happens then?"

"Off-targeting can – I say *can,* not will – lead to the introduction of unintended mutations. I'm convinced that in another few years we'll have completely resolved that problem. There is also the problem of genetic vandalism, where the system cuts accurately but we don't get a precise enough edit. All of these things are the subject of current research. I'm excited to be a part of it!"

Karen was prepared to be objective. Just because something happened in the future doesn't mean it couldn't be changed. Maybe those clones in 3013 were just one possible future. "That's pretty amazing, actually. So what do you see as a future for humanity if these technologies are applied?"

Chen brightened. "A population that doesn't get sick and where everyone leads longer, healthier lives."

"OK, thank you. I have a couple more questions and I'm done." Karen smiled. "Typical news reporter questions."

"All right. I still have ten minutes."

"You mentioned using gene drives." Karen looked at her notes. "If I understand correctly, an introduced trait could spread beyond the target population to other organisms through crossbreeding. Gene drives could also reduce the genetic diversity of the target population."

"Certainly, but that's just the worst case scenario." Chen smiled indulgently. "People who aren't familiar with the research always think the worst. If you knew what I know you wouldn't worry. We're going to solve those problems, and the technology won't be introduced until we do."

"I was just thinking that even if CRISPR technology, or some future technology, becomes foolproof, is it ethical to make changes in human biology that will affect future generations without their consent? I mean, a parent might want to have their genes altered, but what about their children? They will have no choice in the matter."

Chen didn't respond.

Now Karen came to the question she was most interested in getting the answer to. "What if the use of genetic editing changes from being a therapeutic tool, to a tool for a designer human? I was thinking that a dictator could genetically design a human population to fit their needs."

Mary Chen laughed again. "Oh dear, you do have quite an imagination don't you? Relax, Ms. Everard. My research, and this field, can only benefit humanity in the long run."

Karen thought of Melanie Fuscaldo. "I take it that you don't believe in an overarching spirituality that would transcend a physical, Transhuman approach."

Chen grimaced. "Of course not. I believe in the tangible, the observable.'

Karen understood Mary Chen's viewpoint a lot better now. And Harry Kaine's as well.[11] "Thank you very much Dr. Chen. I'm very thankful you agreed to see me."

Mary Chen checked her timepiece. "Just enough time to get back to my lab." She winked. "I'm calling Harry tonight."

Karen plastered on a smile. "Sure thing, doctor. Good luck."

Karen walked out of the building with a lot to think about.

CRISPR technology was frightening but promising if it actually worked. However, both Mary Chen and Harry Kaine were diehard materialists, just like the Guardians in 3013.

She wondered whether all scientists working in Transhumanism felt the same way.

Her next interview was with Dr. Armand Singh, but she had a couple of hours to write her notes before the interview began.

Karen met Dr. Singh in his spacious but neat second-story office in the Computer Science building, a twenty minute walk across campus from the cafeteria. A west-facing window let in sunlight from a partly cloudy early June sky.

"Good afternoon Ms. Everard," Singh said in a precise, sing-song voice when she entered. "Please have a seat."

Armand Singh was a large, dark, bulky man who wore a white turban on his head. He was dressed neatly in a dark blue suit and white shirt. Karen noticed that he was meticulously groomed and had recently gotten a manicure. Singh exuded confidence and intellectual refinement.

Karen was immediately captivated by the man. "God, I love that turban," she blurted after she was seated.

Singh's carefully controlled and businesslike face broke into a wide smile. He nodded his head slightly toward her. "I have a dozen of them. They must be carefully cleaned and stored properly after each use."

Karen blushed. "I'm sorry sir, I don't know why I said that."

"Quite all right Ms. Everard. You were at the meeting several weeks ago with Mr. Kaine, were you not?"

"Yes sir. I was fascinated by the discussion, and wanted to find out more about Transhumanism and the technologies involved for an article I'm writing for the *Chronicle*. I have already spoken to Dr. Chen."

"Ah, Mary Chen, yes. She seems to have an inordinate interest in Mr. Kaine, does she not?"

This was said so dryly that Karen burst out laughing. This man was intriguing in a way that Harry Kaine could never be.

"You aren't married are you sir?" she offered.

Karen saw Singh's lips twitching. Then he smiled. She could see he was pleased with the admiration she was giving him. "Not at present."

Karen decided that the pleasantries were over and assumed her interview manner. "Dr. Singh, you are involved with artificial machine intelligence, is that right?"

"That is correct, Ms. Everard. The field of Artificial Intelligence was founded on the idea that human intelligence can be so precisely described that a machine, or software, can be made to simulate it. I'm a computer scientist. I define AI research as the study of intelligent agents. An intelligent agent is any device that can perceive its environment and then take actions that maximize its chance of successfully achieving its goals."

"So AI is about making machines as intelligent as humans."

"That's the general idea, but my work is about achieving what is called the technical singularity."

Karen was baffled. Singh smiled. "Making a machine or creating software as intelligent as a human is called Artificial General Intelligence, or AGI. But the technical singularity is something completely different."

"OK, one thing at a time. There are problems with AGI, aren't there? An AI as intelligent as a human may or may not be designed, or even be capable of, cooperating with humanity. That would be a dangerous development."

Singh frowned. "Well, yes. But the technical singularity, if it occurs, makes that a moot problem."

"Can you explain, please?"

Singh folded his hands in front of him on his desk. "What you're talking about is 'unfriendly AI.' However, it's likely that self-awareness will occur spontaneously in an AI program, rather than an AI being carefully planned to be self-aware with human values. For example, an AI is unlikely – even if it were possible – to be created with human emotions. That's because the goal of intelligence is to achieve its goals in the most efficient way."

"And human emotions get in the way of that."

"That's right," Singh said calmly. "Human emotions and drives are not built-in to intelligence. They are the result of countless generations of biological evolution. That emotional architecture is of no use in AGI. Why would AGI designers choose to include pointless and dangerous features such as anger, hostility, or jealousy? We know that a certain portion of humans are anti-social or even insane. Trying to build in 'friendly' features to AGI may result in the opposite effect."

Karen frowned.

"But that doesn't mean that an intelligent AI won't cooperate with humans. AI could help in eliminating disease and hunger, for example, or distributing energy supplies more efficiently, or in any number of ways. There's no reason to think that self-aware AI would be uncooperative with humanity. Ms. Everard, any problem can be better resolved by applying intelligence to it. Increased intelligence applied to problem solving can only benefit the greater good. AI is the next step to increased intelligence."[12]

Karen saw this. "OK, but AGI should be a tool, as you say, that will hopefully do the bidding of the human beings who created it. What if that creator is antisocial?"

Singh waved his hand dismissively. "I am a researcher, Ms. Everard. I am a seeker of knowledge; it is my job to explore the boundaries of knowledge. If the human race decides to commit racial suicide, unfriendly AI is just one of many ways to do it. There are far more efficient ways to destroy ourselves than that!" Singh smiled broadly.

"I suppose so," Karen replied. "A nuclear war, or industrial pollution, or a pandemic, could do away with humanity much more quickly."

"Precisely." Singh folded his hands and rested his chin on them in a contemplative gesture. "Besides, what we are discussing is still largely theoretical. Although a lot of progress has been made in the AI field, machine/software learning has still not advanced far enough to worry about friendly or unfriendly AI. AGI still has a way to go before it is fully developed."

"That may be so, doctor, but what about the future? Let's say AGI has been developed at some time in the future. Let's say that it is as intelligent or even more intelligent than humans. The AGI might be designed with the goal of eradicating a disease like malaria. Humans might reasonably expect the AGI to develop a new anti-malarial drug, or maybe an efficient program of mosquito control. But there are many other ways of eradicating malaria. What if the AGI chooses to eradicate the disease by eliminating the humans it affects?"[]

"An exaggeration surely, Ms. Everard. You are talking about the problem of unintended consequences. You are saying that even a perfectly designed AGI might design a system that doesn't do what we truly wanted."

"Yes, that's right." She thought of her interview with Mary Chen, and the unintended consequences of gene splicing and gene therapy. "Don't laugh, sir! If an AGI has no emotions or ethical compass, it would logically do the most efficient thing. Artificial Intelligence might regard humanity as just another variable in pursuing its goals."

"That is a valid argument, of course. An AI with advanced intelligence might create outcomes that are difficult to predict."

"I'm glad that you are able to see the other side of the issue, Dr. Singh."

"Just wait. We still haven't discussed my main area of interest, the technological singularity. I'm not quite so objective about that!"

Singh sat up in his chair quickly, almost dislodging the turban.

"The technological singularity is a hypothesis about the creation of artificial super intelligence, or ASI. ASI is a leap beyond mere artificial general intelligence. An ASI would be able to bootstrap itself, learning faster and faster, and soon far outpace human intelligence."

Karen could see how excited Singh was.

"This might trigger a tipping point in which AI would force enormous changes in the modes of human life. ASI might create a situation so far beyond human understanding that human affairs, as we know them, could not continue. That was the beginning of the concept."[14]

Karen couldn't get her mind around this idea.

Singh was leaning over the desk now, totally engaged. "You see, Ms. Everard, I don't see much future for the human race if present trends continue. If a higher form of intelligence can be developed that can help us solve our intractable problems, we should do it."

"Even with all the risks involved?"

Singh sat back calmly, smiling, pressing his hands together. "Most certainly. Humanity may simply be a stepping stone in the evolution of intelligent life in the universe. I very much look forward to seeing what form a super-intelligence takes, if I ever live that long. My guess is that AI is the way forward."

Karen frowned. "Super intelligence may solve all of our problems, or it could regard us as an unnecessary risk and wipe us out. Or, maybe, we would become like pets to it."

Singh smiled. "All of that is probably well into the future. The first Terminator movie said the technological singularity would happen in 1997. Vernor Vinge said in 1993 that the technological singularity would occur in 30 years. None of that happened. Ray Kurzwell said it would occur in 2045, but unless something miraculous happens, it's unlikely."

2045! That's awfully close to Kaine's UN speech in 2048.

"Is there something wrong Ms. Everard?"

"No...no, sorry. Uh, you seem to be opposed to some of the other Transhumanists, who want to develop technology that will elongate human lifespans and alter the human body so that it can better survive in space, or on an altered planet earth."

"I am much more interested in creating non-biological containers that human intelligence can be placed in. These containers would not age like the human body. In this way, humanity could participate with a super intelligence and become part of it, enhancing life throughout the universe. That was Ray Kurzwell's idea, and I agree with it."

Karen was appalled by this, but smiled. She tried to remain objective. "It is a grandiose vision."

Singh stood up, indicating an end to the interview. His turban shook slightly but stayed in place.

"I really appreciate your time, doctor. It was very enlightening."

Karen walked out of Singh's office, her head spinning. As she turned the corner to the stairwell she saw Singh sitting calmly at his desk, his chin propped up on his folded hands.

Karen went back to her *Chronicle* office and finished her article on the Midland River pollution (no hard evidence for increased pollution, but suspected polluters mentioned), reported to Bob Guza, and went home. She was feeling a little sick after the two interviews. Mary Chen and Armand Singh believed in a future she wanted no part of.

There was one more interview tomorrow morning, with Dr. Tom Mankiewitz, the machine nanotechnology guy. He had graciously agreed to do the interview on Saturday morning so she could start writing her article.

She was perversely glad that Dr. Sauvage wasn't around. She couldn't stomach much more talk of technology that would lead to the future she had seen in 3013 AD.

And then there was the problem of Joe Courvall. She had been thinking more and more about him lately.

———— • ✹ • ————

On Saturday morning Karen steeled herself to see Tom Mankiewitz. She met Mankiewitz at the Mason Street coffee shop at eight, which was way too early for her on a weekend. They found a table at the back. He was an older man, in his late 60s, with a full head of gray hair that was going pure white. A handsome man, she thought.

Mankiewitz started right in. "Is this a hostile interview?"

"Uh, no. I'm trying to get information, that's all. I intend to write an objective article."

Mankiewitz relaxed. "All right then, let's proceed." A waitress came around and they ordered.

"Let me start by asking how your work fits into the Transhumanist concept."

"I've already described my work with utility fogs in the meeting you attended with Harry Kaine, so I won't get into that again. What I want to do here is to dispel confusion and misconceptions in the popular mind about what Transhumanism is."

"Good, that's exactly what I want. I'm a little confused about the subject. My recorder is ready."

"Transhuman refers to an intermediary transition between the human being of today and a possible future human, a Human 2.0 if you will. We call this improved human a posthuman."[15]

"OK, that makes sense. Just like a trans person is transitioning between genders."

"I'm glad to see you have some flexibility...may I call you Karen?"

"Sure."

"Transhumanists want to extend human life. But we're not trying to add on a few extra years at a nursing home when people have already lost their memories and have trouble even walking around. Our goal is more healthy, happy, and productive years in a body that is fit and healthy. Everyone should have the right to choose when and how to die – or even not to die. Transhumanists want to live longer because we want to experience more. We want to have fun and spend more time with loved ones. We want to continue to grow and mature beyond the ridiculous eighty years allotted to us by our evolutionary past."

"Sure, everyone wants that. But it's how we get there that concerns me. I've talked to Mary Chen, who said that gene splicing could add healthy and productive years to a person's life, and eliminate diseases. But the technology of gene splicing sometimes introduces mutations. Dr. Singh is excited about some technological singularity, where human consciousness would merge with a machine consciousness. Your utility fogs, if they are ever developed, could become destructive. The accidental release of self-replicating nanobots into the environment could destroy the entire biosphere."

Mankiewitz sighed. "Why do people always focus on the negative? These are hypothetical, worst case scenarios. Transhumanists focus on a positive, not a negative, future."

"*Touche*, Dr. Mankiewitz. I have to ask these questions because people want to understand the purpose of these new technologies and their consequences. According to Dr. Singh, they are going to be developed because researchers are getting more funding for them. He says there is a lot of enthusiasm in the scientific community about them."

Mankiewitz nodded. "That's a fair assessment."

"So what is a Posthuman? This is the ultimate goal of Transhumanism, is that correct?"

"Yes. Transhumanism seeks to extend lifespans using gene therapy, cryogenics, and cumulative but small improvements to a biological human. Posthumanism is further down the road. Posthumans could be completely synthetic artificial intelligences, or they could be enhanced uploads –"

"Uploads?"

"Uploading is the transfer of human consciousness to a non-biological substrate that is free of disease and that can be easily maintained throughout a much longer lifetime. For some of us, especially my late colleague Winthrop Sauvage, that was his goal. His cryogenic experiments were meant to preserve his body long enough for these technologies to be developed."

Karen nodded. "A tragedy," she murmured. "Dr. Singh talked about non-biological containers replacing biological bodies."

The older man nodded. "Uploading would be my preferred method of consciousness preservation. A virtual or simulated body could give the same sensations and the same possibilities for interaction as a biological body. With advanced virtual reality, uploads could even enjoy eating and drinking!" Mankiewitz smiled. "Upload sex could be as wonderful as physical sex. Moreover, uploads would not be confined to virtual reality. Uploaded consciousness could interact with people on the outside. Or even enter provided robot bodies to work in and explore physical reality."

Karen was blown away. "Transhumanists certainly think out of the box."

The older man smiled. "We do! Transhumanism is the leading edge of a new human society, free from disease, war, and poverty."

"I have to say, Dr. Mankiewitz, that the idea of uploading my consciousness to a computer or an electronic body is a bit...scary."

"Well, personal inclinations among Transhumanists about uploading aren't all the same. It depends on the conditions they would live in as uploads, and what the alternatives are. However, there are many advantages to being an upload."

Mankiewitz ticked them off on his fingers.

"First, and most obvious, uploads would not be subject to aging.

"Second, back-up copies of uploads could be created regularly, just like we do with the data on our computing devices. If something bad happened you could be re-booted. So, your lifespan would potentially be as long as the universe's!"

Karen was startled by this comment. Mankiewitz continued.

"Third, you could live much more economically as an upload because you wouldn't need food, housing, or transportation. You could do without all those things that go along with supporting a physical body that needs a very narrow range of temperature and other environmental conditions.

"Fourth, if you were running on a fast computer you would be able to think much faster than in a biological body. If you were running on a computer a thousand times more powerful than a human brain, you would think a thousand times faster. This has relativistic implications. Einstein showed that for a person traveling closer and closer to the speed of light the external world slows down more and more. So, thinking a thousand times faster, the world would slow down by a factor of a thousand. You would get to experience more subjective time, and live more, during any given day."[16]

Karen sat up in her chair at this.

Mankiewitz smiled. "That is mostly theoretical at present. However, as scientists it's our job to develop these technologies. What humanity decides to do with them is the option of each individual. We all have free choice."

"Certainly. But what if a government-sponsored program promotes one of these technologies and something goes wrong?" She was thinking of the Guardian culture in 3013. "Then the malfunction would spread throughout society with potentially devastating consequences."

"Karen, Transhumanists advocate individual freedom, not zombie-ism or dictatorial regimes that suppress human rights. Our technologies are designed to help humanity. In my opinion, people have the right to use technology to extend their mental and physical capacities, and to improve their control over their own lives. Transhumanists don't want to replace existing humans with a terrifying new breed of super-beings. We want to give people, and future generations, the *option* of developing into posthuman persons."

"OK, I can see that." Karen checked her notes. "Tell me about biological augmentation – changing the human body by replacing biological organs with electronics or other enhancements. That seems more real to me, because it could prolong life spans. Biological augmentation seems to be prominent in Transhumanism."

"Yes. Personally, I am looking forward to this, even though I may not live long enough to see it. Biological augmentation is just a way to extend human life. It may, and I'm very excited about this, involve a redesign of the human organism using advanced nanotechnology. That's my field. Programmed nanobots could provide cancer treatments free from side effects. They could boost the effectiveness and lessen the side effects of powerful drugs. Nanorobots sitting in our bloodstream could act as early warning systems for disease. Tiny wireless surgical nanotools could let doctors perform medical procedures without cutting people open.[17] The possibilities are endless."

Karen was curious. "The techniques you mention just manage disease. How about curing cancer?"

"Nanobots could surround and remove cancer cells."

"Yes, but these technologies don't discover the root cause of cancer."

"Perhaps not, but a redesigned human immune system, for example, would prevent the cancer from ever taking hold in the body."

Karen smiled. "Very good, doctor. "

"Biological augmentation/enhancement could also be achieved using a combination of technologies. Genetic engineering, psychopharmacology, anti-aging therapies, neural interfaces, advanced information management tools, memory enhancing drugs[]

"I think I understand you. However, the Transhumanist/Posthuman concept doesn't consider the idea that humans can change themselves by changing their self-conception. Maybe war, poverty, and disease can be eliminated much more easily by changing who we think we are and the way we think. Greater cooperation and harmony in society can be achieved by changing the idea of what it means to be human. Action follows thought." Karen got these concepts from her interview with Melanie Fuscaldo at the Interfaith center.

Mankiewitz dismissed this with a wave of his hand. "Posthumanism cannot be attained in this manner, Karen. That idea is a corruption of the original meaning of Posthuman because the changes required are too profound. Altering some aspect of psychological theory, or changing some vague idea of what consciousness is, or altering the way we think about ourselves, isn't going to cut it. These are nice ideas that have no basis in physical reality or in the scientific method. Radical technological modifications to our brains and bodies are needed."[19]

"Wow. That is a very emphatic statement."

"It's what most Transhumanists believe."

"Harry Kaine would agree with you."

Mankiewitz smiled. "Harry is our greatest asset. A wonderful promoter for ideas that, if accepted by society, will lead to a much better world."

Karen spoke dryly. "Of that, sir, I am not so sure."

The older man smiled. "Fortunately, Transhumanism offers everyone a choice. You are free to embrace or reject these technologies if and when they become available."

Mankiewitz leaned forward. "Just because pre-Transhumanist humans are considered to be "natural" doesn't make them superior. Human attributes like empathy and generosity and a desire for fairness are positive. But others, such as psychopathy and aggressiveness, have ruined human societies. Being human doesn't come from being a "normal" or "natural" biological human. It comes from having within us the raw material for being humane. Having compassion and a sense of humor, being curious, and the desire to be a better person, doesn't come from being biological. Instead of trying to preserve human-ness we should try to cultivate humaneness. What I'm saying is that humaneness is not dependent on human biology, Karen. We don't want to bring along the bad in human nature along with the good into a post-human future. If human is what we are now, then humane is what we can be."[20]

"That's a nice argument, but how are positive human values supposed to come forth when human consciousness is encased in sterile, electronic containers? Or merged with artificial intelligence, or altered by electronically enabled, non-empathic nanobots that have their agendas programmed?"

Mankiewitz took a sip of his coffee. "That's the point I'm trying to make. Human nature contains both bad and good characteristics. It's not a bad place to start our journey, but we can't fulfill human potential if we reject any progress past the status quo of purely biological humans. Look around you, Karen. The society humans have built is constantly at war, billions live in poverty, injustice is commonplace, and biological humans are polluting the very earth that sustains biological life."

Despite herself Karen was impressed with this argument. She listened intently as she drank her espresso. This man was thoughtful and informed.

"The important thing is not to be human but to be humane, as I said before. Hitler wasn't an inhuman monster, he was a human monster. Gandhi, who changed an entire society, was not a remarkable human. But he was remarkably humane. Humane-ness is not incompatible with Transhumanism or Posthumanism. In fact these technologies, by eliminating disease and providing ideal environments for consciousness, can eliminate the reasons 'normal' humans fight and kill each other. That will lead to more humane societies."[]

Karen was much struck by this statement. But having traveled to the future, she knew that "humaneness" wouldn't survive in a Post-human world. "You're a better spokesperson for Transhumanism than Harry Kaine."

"Thank you. But I am not quite so magnetic a personality."

"True, but it seems Kaine has chosen his team well." Except for that pervert Winthrop Sauvage, she thought.

"I'm almost done, Dr. Mankiewitz. I want to ask you a similar question to my last one. It's about spirituality and the soul. It seems to me that uploading consciousness to non-biological bodies, and artificial enhancements to the body, don't preserve the soul. The soul is supposed to be the seat of all the positive things that make us both human and humane. What if the soul can only inhabit a biological body? If that is true then Transhumanist technologies automatically create a dead, sterile society that only leaves us with the worst of human nature."

Mankiewitz nodded. "That is an excellent question Ms. Everard, one that I have often asked myself. Unfortunately, there is no hard evidence for spiritual phenomena. Transhumanists prefer to derive their understanding of the world from rational modes of inquiry. The scientific method is incompatible with impossible-to-define concepts like the soul. However, there is no clear religious or spiritual ground for ruling out Transhuman technologies as incompatible with teachings about the soul. There is no scriptural basis in the Bible, in Christian or Jewish scripture, in the Quran, in the Dhammapada, or in the Tao Teh Ching which says that God can't get to our soul if we freeze our physical body, or if we upload into a computer, or exist in a nano- or electronically enhanced body. Perhaps uploading and machine intelligence may reveal *new* things to us about how the soul works.[22] The Dalai Lama himself said that a scientist who had worked with computers his entire life might reincarnate into a computer during his next lifetime."[23]

Karen was astonished. "Are you serious?"

Mankiewitz smiled. "I'll send you the link. Great article."

"One last question. What would a Transhumanist say to the idea that the soul is already immortal? That the reincarnation process is just a refresh into a new body? Why keep one body when you can have a wide variety?"

"Oh that's very good Karen. My response would be, 'Why not design your own body'? Why put up with incarnating over and over? Why live through the potential trauma of childhood and then suffer the disease and dementia of old age, time after time? In the Buddhist concept, Transhumanism allows human beings to get off the wheel of karma and end the cycle of birth and death."

Karen stared at the older man with dropped jaw. These concepts fit right in to her Buddhist upbringing. "Why...I've never thought of that!"

She could see Mankiewitz was pleased at her reaction. She stood up and with great respect shook the old scientist's hand. "Thank you sir. You have been very generous with your time, and I appreciate it. You have given me a lot to think about."

"It has been most interesting, Karen. If you have any questions while writing your article, feel free to contact me."

Karen took the card he held out to her. "I will. Thanks again."

She walked out of the building into a warm, early June morning. Her head was spinning. The old scientist – unlike Harry's other scientists – could see beyond the myopic scientific materialism of Transhumanism, yet he still embraced altered-human ideas that frightened her.

CHAPTER **20**

Liqao Chang sat at his desk in the Midland Police Dept. He was the head of IT for the entire department, and responsible for communications and the servers at the main headquarters downtown.

Police Chief Chastaine and the entire department were stymied in their investigation of the Winthrop Sauvage murder. No one was talking, especially Winnie Sauvage and the people at COSA, who were scared to death. Even the chief had gotten a warning message on his mobile! When Liqao traced the chief's message it led to a grocery store in Tjorçë, Albania. He had traced the messages of those who had received them to the same place. Whoever did this was good, very good. Definitely not some retard or prankster.

"We can't pursue this any further," the chief told him yesterday. "I've reported it to the FBI, we have to let them handle it." This morning Liqao had turned over his work to a federal agent. But Liqao was not satisfied. His boss had ordered him to leave this alone but he could tell the chief was very upset. He wanted to help.

There was still one lead and he would pursue it. From Karen Everard he knew the IP address of the server Winthrop Sauvage had been using to download his child porn. He felt guilty about going against the boss' orders, but Chastaine (and everyone else in the department) were IT idiots. No one would know. If he was successful he might get a commendation and a raise!

Liqao reached into his briefcase and pulled out his laptop. Into the laptop he inserted a data pod, which loaded a pre-configured darknet browser and a barebones browser he had written himself to inspect the I2P networks. The IP address of the Sauvage server contained nothing, but led to a server on the dark web (and a lot of other nasty shit, which he ignored). However, the firewall was encrypted so strongly he couldn't break it, even after several hours of trying. Liqao stored his work on the data pod he kept for his private work, and erased his electronic footprints as well as he could. Of course he had left traces on the darknet but...well, he hoped whoever ran that server wouldn't notice. He and the Midland Police Department were small fry.

Six hours later Liqao, frustrated, clocked out. He had snuck around behind the chief's back and could get the department in trouble. He admitted to himself that he had done it to gain recognition, not necessarily to help the department.

He was supposed to meet Kirra at the Mason Street Coffee House at 6. Should he tell her what he had done? Every time he held something back she knew somehow. He knew he was the most ambitious person in existence, and the worst liar. The two didn't go together.

Liqao found a table at the coffee house and waited. Kirra was usually late for everything.

Kirra Bigbear was instantly recognizable when she walked in; a small woman with jet black hair, dark skin, and bright red lipstick. Liqao thought she was hot. Kirra was a member of the Ojibway tribe. He was a Han; there were only a billion or so of his ethnic type.

Kirra had a kind of insight and wisdom that Liqao selfishly used to bounce questions he had off her. Their conversations were, he admitted, pretty one-sided. Tonight they were supposed to talk about her.

The waiter came and they ordered.

Kirra wanted to see if Liqao was as shallow as Harry Kaine. Did he have inner depths or was he just a superficial techie with no self-awareness? If so, it was over between them.

Liqao picked up on this. He opened his eyes a little wider to show interest.

"I want to talk about metaphysical stuff," Kirra said.

Liqao groaned. "Uh, I don't think about things like that. It's not important to me."

"Well, it is to me. I'm not interested in being with a person who doesn't understand that there's more to life than business and 'getting ahead.'"

Liqao was shocked. This was blunt talking from Kirra, who was almost always so nice to him, and so understanding.

"Do I have your attention now?"

"Uh, yeah." He would try to keep his attention off his question while he listened.

"Good! Metaphysics is about recognizing the sacred nature of human beings, and all life on this planet. Every Native child understands how important this is. Do you understand, Liqao?"

"I'm not sure I do."

"People who don't understand the concept of a higher power and its relationship to human beings are the same people who start wars, murder, build concentration camps, and trample over others for personal advancement. They are people with very little self-awareness. "

Liqao thought this verbal harpoon might be indirectly aimed at him but he stifled his irritation. "Just because you're interested in metaphysics doesn't mean you have self-awareness! Some people who believe that stuff aren't the brightest bulbs in the chandelier."

Kirra laughed at this. "Granted."

Liqao wondered whether it would be a good time to ask his question.

"No, Liqao."

Wow, Kirra was very observant!

"Pay attention, Liqao. I'm going to tell you about the Ojibwa creation story, because it has concepts I hope you can understand."

Liqao nodded. Kirra was testing him. He was good at tests.

"The creation story of the Ojibwa begins with nothing because in the beginning there was only an all-consuming dark void. A latency with possibilities."

Liqao nodded vaguely.

"Although there was nothing, it was conceivable that there might be something other than nothingness. Perhaps there was a single something that contained all possibilities—"

"A singularity!" Liqao offered.

Kirra smiled. His face lit up like a little kid sometimes.

"All right, a singularity then. If there was a singularity filled with an infinite number of possibilities, then maybe many things were possible. Everything we know, and everything we don't know, could exist. It could all be."

Liqao saw this as a physics problem. "You're saying that there could be an infinite number of quantum states or potentials that could manifest in the physical universe. They could all exist simultaneously, the ones we know about and the ones we don't know about."

"That's right. A human mind is very limited, so there are always unknowns. However, a being with unlimited consciousness could envision the possibility of *everything* and then bring it all into existence. Some people call this being, God."

Liqao found himself marginally interested in this idea. "You're saying that the creator of the universe – if such a thing exists – has the ability to understand, and also manifest, every one of the quantum states."

Kirra nodded. It wasn't quite what she meant, but at least Liqao was getting some of it. "The Ojibwa call this creator being Kitchi-Manitou – the Great Mystery. The human mind simply cannot understand everything in the universe. Life, and the universe, is too confusing."

"Yes. I have a tendency to over-analyze. When I do that I can't come to any conclusion, so I have to simplify the problem."

Kirra nodded. "That's why people all over the world have personified their deities, Liqao. It brings the concept of God down to the level that a mere mortal can understand."

Kirra could see she had made an impression.

"OK, I think I get it. Primitives...er, people in the past, assigned the physical phenomenon of the wind to a wind God, or the physical attributes of the sun to a guy who drives a chariot across the sky. It's an attempt to simplify a complex analysis with too many unknowns."

Kirra smiled. Spoken like a true IT guy. Could there be any two persons more opposite than her and Liqao? "Humans have always made up stories about divine beings to bring reason and purpose to their lives. Once the stories make it to book form, the collection becomes known as the Bible or the Koran or the Shastras. Because they stand the test of time they become regarded as truth."

Liqao was sitting very still, staring intently into her eyes. "You're saying that the stories about the gods become immortalized as sacred texts, like the I Ching or the Vedas," he said. "The word Veda is supposed to mean wisdom. It is said to manifest the language of the gods in human speech."

"Where did you learn that?" Kirra asked.

"In a comparative religions class. Had to take it."

Kirra laughed. "OK. So Kitchi-Manitou envisioned what was possible, out of nothing – or out of the singularity if you prefer, which contains the possibility of all things."

Liqao nodded. Normally he had no time for pointless philosophizing, but he could see Kirra was really into it.

"Kitchi-Manitou then created the universe and everything in it that we know, and everything that we don't know, which we can discover."

"Science!" Liqao said. "Science tries to understand all that exists, and also explain things we don't know yet."

"Right. Here's the point I'm trying to make, Liqao. It's a metaphysical concept that I hope you'll understand, because it's very important to me."

"I'm listening." Liqao got the idea that if he didn't understand it, Kirra was going to dump him. He almost panicked at that.

Kirra tried to tailor her speech to Liqao's mind. "You and I, and all other life forms, even inanimate things like rocks, were created by Kitchi-Manitou just from knowing that it was possible. So – we will always be part of the spiritual essence of the universe. According to the Ojibwa all things are connected; must be connected. Nothing is separate. The cosmos is an interrelated whole. And so cooperation must be an inherent quality to the success of everything in existence. Humanity has to discover this or we're sunk."[24]

"I think I get it. Darwin said that inter-species competition was most important, but look where that's got us. Humans and animals fight and kill each other to survive, but you're saying that we're in trouble because we don't understand that cooperation is way more important than competition."

Kirra was profoundly happy. Her boyfriend was self-centered and too cerebral, but not obtuse or shallow. There is some substance in there, she thought.

Liqao saw his girlfriend's happy face and realized he had passed the test. For some reason he no longer felt the need to tell Kirra about what he'd done at work. He'd gone against the police chief's instructions; it was as simple as that. He would have to live with the consequences. Why had he thought it was so complicated?

They finished their meal. Kirra saw that Liqao was subdued, and decided not to say anything more. Maybe she had got him pondering something other than his job and his career.

When Liqao got home he thought about the story of Kitchi-Manitou and the creation of the universe. These were ideas he had simply dismissed out of hand because they were impractical and did not help him achieve his career goals. But the way Kirra explained it made sense. He was honest enough with himself to realize that before Kirra gave her philosophy lesson he had a big problem, and afterward it had vanished. Maybe there was something to metaphysics after all! He laughed at that thought. He was still thinking about these ideas when he went to bed.

He woke up the next morning feeling refreshed. He realized that he had been thinking way too small. His job at the Midland Police Department was getting him nowhere. Liqao decided to invest all his spare time to perfecting his deepfake recognition software and starting his own business. He was so excited he almost tripped down the stairs of his second floor apartment.

———— • ✸ • ————

Kirra Bigbear went home that evening after her talk with Liqao, filled with happiness. She had gotten through! She didn't know why she was attracted to Liqao, she just was. He was kind of like a new puppy – you love looking at him but when he gets out of his cage he pees all over your new shoes.

CHAPTER **21**

3013 AD

After all of the 21sters had gone back to the past, there was a great celebration among the rebels about their escape from the Tower of the Overseers. Alan 133 took out his brain implant interface. "I no longer need or want this," he said, placing it in his pocket. Li 355 and Sigmund 78 did the same. The others followed.

"What shall we call you?" Pietra asked Alan 133. "There are three Alan clones in our group: Alan 133, Alan 11, and Alan 49."

This caused consternation for a time. "So that is why 21sters used last names," Alan 133 said. "To distinguish between members of the same biological family." He paused for a moment. "I am very curious. Call me Alan the Curious. "

Everyone shouted their agreement.

"I am persistent," the former Li 17 cried. "Call me Li the Persistent."

The clones were getting into it now.

"I feel that I am as strong as a hammer now," Li 355 said. "Call me Li the Hammer."

Everyone laughed at this.

Grace 5 said, "I am cheerful. Call me Grace the Cheerful."

"I am serene," Alan 11 said. "Call me Alan the Serene."

Sigmund 78 spoke up. "I am shy, so call me Sigmund the Shy."

One by one the other clones chose names for themselves, names that gave them a unique identity instead of a printed identity from a lab. (For consistency, I will sometimes refer to them by their old names.)

Li 57 and Li 308, who had not been enhanced, couldn't do this. "I am still a clone, dependent on my implant," Li 57 said. "Li 308 and I will stay by our consoles to manage the Shifter and observe the neighborhood for hostiles."

"Very well," Pietra said. "But you two will also have to be enhanced, and as soon as possible. You are still subject to the Submission frequencies through your brain implants. If an attack comes you will be helpless."

"We can send each other back to the past while one stays here to monitor the Shifter," Li 308 said. "We know where you guys went and what you did. We'll just do what you did."

"Brilliant!" Pietra was speaking more and more like a 21ster now.

Everyone was fired up and excited. Now was the time to do something she had thought about almost continually during their slow walk back to Building 103. Pietra evaluated her charges like a 21st general. She began carefully.

"Men and women" – the clones were startled to hear themselves referred to in this way – "we must act, and act quickly. To save our race we must attack the Tower and disable their Shifter!"

The rebel clones were stupefied. Alan the Curious sputtered in disbelief and expressed the sentiment of the group. "But...we have just escaped from these fiends! Why should we go back?"

"It is very simple dear ones," Pietra said. She was beginning to experience the emotion of camaraderie. There was a curious affinity she felt for each member of the group as it fought together against long odds. She wondered if the others felt it too. "Patton 1 and Christian 172, undoubtedly with the sanction of Khan 1, are ready to send more teams back to the past for enhancement. Every minute we delay, we allow these sociopaths to build an overwhelming force of enhanced clones. If this happens they will attack our Shifter and disable it. Their enhanced minions will kill us all. We know what happens after that."

"Our society dies," Li the Persistent said.

Everyone saw the logic of this but no one wanted to engage in a suicide mission.

Li the Hammer spoke up. "Pietra is right. We must move now, before Khan 1 and his antisocials can prepare their forces."

Alan the Curious seconded this. "Like it or not, it is our duty to protect the race." He turned to Pietra. "When do we leave?"

Pietra felt a liking combined with a softness. She wanted to go over and hug her two brothers. "Right now."

After a discussion the necessity for attacking the Tower was agreed upon by the others, if somewhat reluctantly. They all knew that the enhanced clones who attacked them in 2035 had been terminated, but...

Li the Hammer solidified the group's intention when he said, "If these Overseers send their enhanced clones against us we will smash them all."

Pietra felt a surging joy go through her body. Was this how the 21sters lived? These strong emotions were foreign to her, but she liked them. She went up to the front of the group of twelve and faced the rebel clones. All except her sister Li 57 and her brother Li 308 would undertake the mission.

"My fellow rebels! Brothers and sisters! Let us not falter. We fight together for the very survival of humanity! Stay together. I will lead, Alan the Curious will guard the rear."

Li 57 and Li 308 were at their consoles. "Monitor the timeline, dear ones," Pietra said. "If you see any activity by the Guardians before we reach the Tower, send to me through your wrist communicators."

The two technicians nodded.

Pietra glanced at Alan the Curious, who took his cue and began to herd the group from the back. Pietra led off, out the back, through the secret entrance, and into the ancient subway tunnel. They were headed out of the Safe Zone.

"Walk carefully, try not to disturb the dust," Pietra said.

Pietra led the group through the decrepit subway tunnel. There were no rats, for the animal population had drastically declined along with the human population. Holes in the subway ceiling, drilled by long-ago workmen, allowed a faint light to illuminate the filthy walls, dirty from centuries of neglect. Pietra recalled absently from the history grids that in 2049, the city had replaced many of its concrete and metal structures with permacrete and synthfab. Although grimy and dust filled, the tunnels had maintained their integrity for almost ten centuries.

"The Overseer's Tower is in the old Bryant Park, approximately 2.4 miles from our present location," Pietra reminded the group. "We'll take the tunnels from here to the 42nd street exit, then emerge 1,000 feet from the Tower entrance."

As they trudged through the tunnel a slight spatter of rain touched Pietra's cheek through one of the openings in the ceiling. They had to avoid debris scattered over the floor, and walk carefully around an abandoned subway car that had been looted and stripped long ago, its rusted and decayed parts blocking their way. Sterile, centuries-old dust puffed up around their feet as they walked, but it wasn't nearly as bad as before, when 36 people were stomping around and chasing each other.

"So far so good," Alan the Curious said after about fifteen minutes of walking. "No foot patrols."

At that moment three Guardian security officers appeared from around a corner to their left. "Identify yourselves," one of them said. "No foot traffic is authorized to be in this area."

A few of the clones began to panic but Pietra held up her arm. "Keep walking."

The two groups were within 100 feet now.

"Halt! I am Walker 18, foot patrol officer for this area."

Pietra ignored him as one of the officers checked his communicator.

"Walker 18! These are the terrorists from Building 103. Our orders are to apprehend them."

Walker 18 activated the Submission modules of the intruders, but nothing happened.

"Keep walking," Pietra ordered. Alan the Curious herded the rebel clones from behind.

As the groups passed each other two of the Guardian officers reached for Alan the Curious, who brushed them aside with one of his arms. Both Guardians fell over crashing onto the hard permacrete floor and raising a cloud of dust.

"That was fun!" Alan the Curious said.

As they rounded the corner they heard Walker 18 speaking frantically into his communicator. "Renegades spotted in Tunnel 56 by old West 34th! Send as many clones as you can."

The group continued to walk, expecting to be rushed by Patton 1's red-and-black clad security clones. But nothing happened for the next half hour as they slowly picked their way through the ancient tunnel.

"Surely the Overseers know where we are," Li the Hammer said. "Why have they not attacked?"

Pietra wondered the same thing. "Press on," she said.

After another half hour the group approached a set of permacrete stairs to the old Bryant Park subway entrance on 42nd street. They were now well out of the Safe Zone.

"As soon as we emerge onto the surface, the hovercraft and monitors will see us," Li the Persistent said nervously.

"The Overseers already know where we are," Alan the Curious said. "Walker 18 has told them."

"Of course we are all afraid," Pietra said, damping down her own anxiety "We have been conditioned to fear and obey the Overseers, but we have already defeated them twice. Remember what happened the day Christian 172 attacked us in Building 103, and how easily we disposed of Patton 1's clones."

Pietra saw terror in the eyes of her group. Enhanced they might be, but they were not yet as mentally strong as she. Even Alan the Curious had a fearful aspect. She stepped up to the head of the group. Only Li the Hammer seemed confident. "Follow me." She motioned for Li the Hammer to come up to her at the front. Pietra hoped that their attackers would not be enhanced, but she didn't say it. What choice did they have? Better to die defending their freedom than to live as slaves.

Without looking back Pietra walked up the stairs and onto a small platform. The Tower of the Overseers was there, about 1,000 feet from the entrance. This was the first time she had really seen it, for when they escaped they were running away from it. A permacrete walkway led to the crumbling ruins of an old 21st amphitheater. Beyond that, a carefully manicured late spring lawn surrounded the imposing black building, which seemed to rise ominously into the heavens. Pietra was shocked by its menacing appearance. They had already been inside, but the building had been designed to present a fearful aspect from the outside.

Hovercraft floated slowly over the park, but this was routine for all common areas in New York. The building itself was made of black permacrete. It had a huge antenna at the top, and red lights flashed all over the 30-story building, indicating hundreds of monitoring devices. The building looked alive and malevolent.

"The Tower of the Overseers!" Li the Persistent said in awe. "We are to *attack* this place?"

The clones hovered around Pietra, totally intimidated. They saw no one.

"We have already escaped from here," she said. "There is nothing to fear."

Alan the Curious noted the lack of guards. "This is too easy."

Li the Hammer agreed. "It's almost as if we are expected."

"We are wasting time," Pietra said. "Let's go!"

She led the way down the permacrete path and past the wreckage of the ancient open-air theater on their right. The group marched onto the beautiful lawn, smelling the sweet fragrance of grass, and new life. Each of the group was encouraged by this, but confused at the lack of opposition.

"We approach the entrance!" Sigmund the Shy said.

The entrance to the Tower was imprinted in the consciousness of all Guardian society. Seen in every Guardian propaganda film and presentation, it represented an all-powerful force that dispensed life and death, for without the nourishing signals from the Tower, the brain implants that regulated biological/electronic functioning would not operate, and the clone would die.

Pietra understood this. "We no longer need their brain implants. Have no fear!"

Pietra walked right up to the entrance. The opaqued transparency cleared, then vanished. The Resistance entered the famous rotunda, which was usually filled with guards, and Guardian personnel going about their business.

"It's empty!" Li the Persistent cried.

Suddenly several lift doors opened along the back wall and a mass of clones in red and black unis charged the group while a siren went off. Voices shouted. "Intruders! Unauthorized persons have entered the Tower! Apprehend immediately!"

The rebels were terrified and began to panic. Sigmund the Shy and Alan the Serene fled the rotunda and ran back toward the entrance.

"Brothers and sisters!" Pietra shouted. "Around me!" Encouraged by Pietra, Alan the Curious and Li the Hammer found the courage to stand beside her, as did Grace and the other females.

There were at least 50 attackers. Pietra stood tall and began to wave her arms about, knocking clones backward and clearing a path toward the lifts located at the back wall. Alan the Curious and Li the Hammer followed, shoving the attacking clones to left and right. Sigmund the Shy, in a panic to flee, knocked down several of the attackers before he realized his strength. Embarrassed, he grabbed Alan the Serene, who was almost at the entrance, and turned him toward Pietra. Pietra was striding confidently to the gold-colored wall that contained the lift doors, bashing the guards gleefully but without inflicting major damage. Their bodies were incredibly soft. Her arms felt like metal rods as they sunk into unresisting flesh.

Soon the attacking Overseer clones were huddling together, tending to their bruises, no longer wanting to confront these demented terrorists. Their strength was monstrous!

Pietra was confident now. These clones were not enhanced; they had beaten Patton 1 to the punch! As she gathered the Resistance around one of the lift doors, Pietra pointed to one of the attackers. "You! Come." She got this from an old 20th TV series, and enjoyed it immensely. She was a Lawgiver!

The clone fearfully came up to Pietra. "Summon the lift," she ordered.

The clone tried to activate this clone's Submission protocol but saw that she had no implant interface! Terrified, he sent the proper signal through his brain implant. An opaqued transparency cleared and opened in the wall.

Pietra and Alan the Curious herded the clones into the lift. "Send the lift to the top floor," Pietra commanded the trooper. The security clone was only too glad to do this, for the rebels would be heading directly into the private area of Khan 1 and his personal security forces.

The group felt a slight pressure on their feet. Almost instantly, the lift opened. Pietra stared into the huge room with its gigantic Shifter, and the standard communication / surveillance apparatus. The device was surrounded by consoles. She recognized Caesar 13, the head time technician. Pietra looked backward; the lift door had already opaqued back into the wall.

Suddenly Patton 1 strode into the room, surrounded by two dozen clones. "Take them!"

The Resistance met the charge confidently, but something went horribly wrong. The attacking clones were almost as strong as they were! Now a battle raged as enhanced clone met the renewed rebel humans. Unskilled fighters, they punched, kicked, and slapped each other. Blood flowed as Patton 1, and his assistant Christian 172, watched gleefully.

"I almost do not care who wins," Christian 172 said, morbidly fascinated. "This is much better than those old videos."

"We think alike," Patton 1 agreed. "Do you think any will be killed?"

Christian 172's eyes lit up. "I hope so. They must all be terminated anyway."

It took a full fifteen minutes of fighting for the 24 enhanced clones of Patton 1 to subdue the eleven rebels. Many clones were groaning and bleeding. The sight of their own blood caused some of the fighters to faint, for pain from wounds had been unknown since the last recorded fight over 500 years ago. The Overseers only needed the Submission modules to maintain control.

Sigmund the Shy was down with a fractured arm and multiple bruises. Grace the Cheerful was groaning from a cut on the head and a twisted ankle. Pietra realized that she was unharmed. Surrounding her were two Guardian clones, both lying on the floor, unconscious. The rest of the Resistance had bruises and cuts, but no major injuries.

Patton 1 ordered his enhanced clones to place the prisoners in the detention cells, which were next to the lifts. Pietra resisted strongly, but three enhanced clones managed to subdue her, dragging her toward the detention area. As she struggled Pietra knew that Patton 1 would surely do as he said. They would be terminated.

The rest of the rebels were still able to fight but had lost their will to resist. The rebel humans were all placed in one large cell, which had been secured but not opaqued. Patton 1 pointed to Pietra. "You and your so-called Resistance will be sent to the Tower basement."

"You constructed a clever trap, drawing us in," she replied morosely.

"It was my idea," Christian 172 interjected. "We saved our enhanced clones until you came up here, close to the detention cells."

Patton 1 was irritated. "You speak out of turn, Christian 172."

The assistant bowed his head submissively to his Overseer master.

Pietra spoke bitterly. "In your lust for power you will destroy the only hope for the human race."

Patton 1 smiled contemptuously. "But we will rule over all."

Pietra shook her head sadly. "You are insane."

"Silence!" Patton 1 cried.

Pietra hung her head in defeat. Patton 1 had been a step ahead of her, getting more clones back to the past before they could mount their attack on the Tower.

Christian 172 smiled gleefully. "You and your rebels will be destroyed. The infestation will be removed from society."

"A dead society," Pietra mumbled.

But then she felt a ray of hope. Could an enhanced clone/new human be deactivated? Perhaps not! Her biology was functioning now at a high level and they no longer needed a brain implant to keep alive. The other humans were only slightly less vital than her.

Patton 1 gloated as he looked at the prisoners in their cells. Sigmund the Shy was whimpering in pain from his fractured arm; Grace was trying to stop the flow of blood from a deep cut on her scalp. This rebel favored her left foot, standing awkwardly in pain. Patton 1 found this oddly exciting.

Of course these dangerous rebels would have to be killed if their enhanced bodies no longer responded to the killing Submission frequencies. Christian 172's enhanced clones would do it. Patton 1 shuddered. Actual killing with a weapon – which was forbidden – was messy. He looked over at his assistant. Yes, Christian 172 would enjoy the job. And then his enhanced clones would have to be disposed of, just like the last batch Ignoring these problems for the moment, Patton 1 went back to his quarters for another session with the brain implant stimulator.

One hour before the rebel attack on the Tower...

An hour before the rebel attack, Khan 1 was sitting back in his recliner, gazing at the eight huge monitoring tanks in a circle around him. Here, from the top floor of the Tower, he could see all of the levels below him, and the surveillance feeds from the entire city. For Patton 1 he had only contempt. He knew about Patton 1's plans to overthrow him with a picked army of security clones, who would turn on him at his order. His second-in-command was a soft target. However, Christian 172 was a clone he could work with.

The Supreme Overseer called Christian 172 via his brain implant.

"Sire! What can I do for you?"

"What would be your plan for disposing of our new batch of enhanced clones when they have subdued the rebels?"

Christian 172 caught a hidden meaning in this statement. "If I were in charge, sire, I would turn them loose across the river. Let them be useful and ravage the Brooklyn barbarians. And themselves."

"Excellent thinking, commissar."

"Thank you, sire. They are still just controllable with the Submission frequencies. Without the vitality in the past, however, they will soon go back to being normal clones."

"And the rebels?"

"Let them starve in the detention cells until they weaken. Then, transport them across the river and give them to our scum, who will be more than happy to dispose of them."

Khan 1 cackled. "Very imaginative, Christian 172. I am preparing a surprise for Patton 1. You need no longer worry about that clone. Do I make myself clear?"

"Loud and clear, sire."

Khan 1 broke contact.

Khan 1 liked the ideas of Christian 172, but after some thought decided to make sure of his enhanced clones by sending them to the Tower basement. And the rebels too of course, after they were weak enough. It would not do merely to release them across the river, where they might survive to plague him again. He then changed the comm feed to display Patton 1's quarters. This clone was once more using the implant stimulator, which would soon overwhelm his biology and kill him. Good! That clone lacked character for the tasks ahead.

Patton 1 was now, with his implant stimulator, engaged in an orgy of sexual promiscuity and violence using the forbidden sensory stimulation recordings from 500 years ago. Disgusting! Through Patton 1's brain implant, Khan 1 saw Patton 1's mind turn to his plans to kill the Supreme Overseer and take over the city. In a burst of rage he didn't know he could feel, Khan 1 used the override on Patton 1's brain implant to send the killing Submission frequencies. He watched with satisfaction as Patton 1's body writhed in agony and ecstasy before finally spasming in the throes of death.

At that moment Khan 1 received a communication from the Shifter room.

"Sire, the rebel clones have been spotted at the old 42nd Street subway entrance," Caesar 13 reported. "They are approaching the Tower. Do you have instructions?"

Khan 1 was pleased. He was wondering how to get to the rebels, and here they were! "I will tell Christian 172 to let the rebels enter the lifts from the rotunda. Make sure to open the lift doors when they come into the Shifter area. At the appropriate time, 172 will set our augmented scum upon them. I will signal you to open the detention cells so that the rebels may be imprisoned."

Caesar 13 almost blurted out his violent objection, but caught himself. What if the rebels destroyed his Shifter before they were captured? He and his technicians would be stranded in 3013!

Caesar 13 stifled himself. "Yes sire."

Khan 1 saw Caesar 13's hesitation and sighed. It was too bad. 13 was almost as devious and cunning as he, and thus dangerous. Yet he could have used the services of this diabolically clever clone...but no. That one was loyal only to himself. The Supreme Overseer broke the connection.

Caesar 13 recognized Khan 1's sigh for what it was, and knew for sure he was a marked clone. He would be needed only until the rebels were destroyed. At that moment, from his seat on the raised platform by the Shifter, Caesar 13 saw a group of clones come out of the lift. They were the ones he had selected for Patton 1 before his maintenance run.

Caesar 13 saw these poor souls scuffling amongst themselves, under the influence of the Suppression frequencies. So – these clones were enhanced, but still controllable. They were herded by Christian 172 like squealing pigs into the storage area next to the Shifter. Christian 172 entered the storage room last so that he could lead the charge at the appropriate time, and opaqued the transparency. Through the wall, Caesar 13 heard Christian 172 addressing them. "If you reprobates make one sound I shall fry your brains into your skulls." Caesar 13 felt a rush of anger and hatred toward these Overseers, but he could do nothing now except wait until the rebels arrived.

After several minutes Caesar 13 saw one of the opaqued lift transparencies clear. The rebel clones were here! All of them looked robust, like actual human beings before the Referendum. Caesar 13 was stunned. Their own augmented clones were not enhanced like these...men and women, who glowed with energy and vitality.

"There is the Shifter!" one of the rebels said. This was a redheaded clone he recognized as Pietra 23. "Destroy it!"

Suddenly the storage area transparency cleared and the twenty recently enhanced clones attacked the rebels. Although not as strong as the invaders, they outnumbered the rebels two to one. This group was more disciplined than the last, and attacked the rebels in a rush. The rebels knocked several of them down but were finally overwhelmed.

"We have you now, traitors!" Christian 172 cried. His eyes were wild, the whites of his eyes showing.

'That clone is insane,' Caesar 13 thought as he watched the drama. He saw Khan 1 enter the Shifter room with his personal guard.

Christian 172 had to turn up the Submission frequencies on his enhanced clones to the highest level, for they were beating their prisoners. Khan 1 wanted them alive. Christian 172 made his clones move away from the rebels, who were totally cowed, except for the redheaded Pietra. Soon 172's clones were calmed down enough to herd the rebels into a large detention cell without too much further damage. Caesar 13 admired Pietra, who fought three of the enhanced clones all the way to the detention cell. These cells were just rooms that had transparencies that could be opaqued, sealing the cells.

"Leave the transparencies open but forbid exit from the cells," Christian 172 said. "I want to see them suffer."

Caesar 13 felt sick. One of the rebels had a gash on her arm, and was bleeding. Another had a twisted ankle and could only stand by leaning against a cell wall. Another's arm was fractured. The others had bruises and were bleeding from cuts. The rebels were in obvious pain.

A number of Christian 172's attack clones were also injured. "Take these scum and terminate them," Christian 172 said to one of the security guards who appeared in the hallway during the fight. "They are no longer of use."

Caesar 13 experienced more rage at this callous disregard for human life, but he said nothing and did nothing. He saw Khan 1 watching and noting everything that happened.

The security guards dragged the enhanced clones of Christian 172 to the lifts. They were frenzied now under the influence of the Suppression frequencies, knowing they were to be sent to the Tower basement. Caesar 13 felt a wave of frantic hopelessness come over him. He must escape into the past somehow, or die trying.

Caesar 13 suddenly realized that Khan 1 might kill him and his techs right now! The rebels were captured and the enhanced clones were ready for termination. Of what use now was the Time Shifter? Or he and his time techs? Khan 1 would soon realize this. He stole a glance at the Supreme Overseer, feeling his malevolent gaze upon him.

Caesar 13 suddenly saw resolution on the face of Khan 1. The Supreme Overseer was ready to give the order for the termination of himself and his twelve time technicians. Khan 1 was about to destroy the Shifter!

Frantically, Caesar 13 programmed the Shifter for a mass insertion back to 2035. His twelve technicians, himself, and the eleven rebels. He could not allow them to die. But where to?

Khan 1 shouted instructions to the Shifter techs. "Step away from the Time Shifter!"

Fortunately most of the programming of the Shifter to the old warehouse in 2035 had already been done during his "maintenance" run several days ago. His technicians were starting to panic. There was no more time.

Caesar 13 activated the shifter and grabbed his satchel with its spare cash and his 21ster clothing. He shouted to his technicians. "To the platform if you value your lives! Hurry!"

Several of the Shifter techs ran onto the platform. Just before Caesar 13 activated the Shifter he opened the cells where the rebels were being held, and deactivated the Suppression frequencies of Christian 172's doomed attack clones in the hope he could save them. He ran onto the platform as fast as he could. The rings began to turn slowly. Three of his techs were lagging, undecided whether to take the leap into the past. "Hurry or you will all die!"

"What is this?" Khan 1 shouted. "Halt! Unauthorized Shifter operation! Cease and desist or you will be terminated!" A wave of pain coursed through his body from his implant, but he made the platform.

"To the platform!" Pietra cried to her fellow rebels. "We have to get out of here!" The rebels stumbled painfully toward the Shifter platform. Three of the wounded ones were lagging, helped along by their companions.

Several of Christian 172's augmented attack clones raced behind the rebels, in a bid for freedom. The rebels were stumbling onto the Shifter platform as the rings began to turn faster.

Through the maddening pain from his brain implant Caesar 13 heard Khan 1 cry, "Seize them!" Khan 1 waved his security guards toward the platform.

A subtle blue energy began to surround the platform as the rings turned faster and faster. Khan 1 saw Caesar 13 already on the platform and tried sending the killing frequencies through the traitor's brain implant. But the strange blue energy must be blocking the frequencies, for nothing happened. Two of Christian 172's attack clones made it to the Shifter platform, pushing the injured Grace the Cheerful and Sigmund the Shy off. Khan 1, arms flailing and screaming "Traitors!" was able to shove his way onto the platform before the blue energy sealed off access to the Shifter. Christian 172 summoned more security clones, trying to rescue Khan 1, but it was too late. A large pop! was heard and everyone on the platform disappeared.

CHAPTER **22**

Khan 1's arms were still flailing when the group was translated back to the abandoned warehouse in 2035. He recognized two of his attack clones and gestured toward them. "Get the traitors!"

The two enhanced Guardian clones, however, were dazed and confused. They looked around them in blank astonishment. "Where are we?" one of them said.

"You fools! Seize the rebels!"

The two clones did not respond.

Khan 1 grabbed his controller and activated the Submission frequencies. The two clones grabbed their heads in pain, but did not obey.

Through their excruciating pain the two clones stared at Khan 1 with hatred. They realized that Khan 1 was about to send the killing frequencies through their implants. Desperately, the two clones rushed the little man. A lifetime of slavery and hopelessness now manifested itself in violence. The two clones began to beat their slavemaster. One of them picked up a piece of rusted metal and stabbed the Khan. By the time the rebels were able to restrain his attackers, Khan 1 was bleeding out on the filthy concrete floor of the warehouse.

Pietra examined the once all-powerful Supreme Overseer. She tried to stop the bleeding from the deep wound, but her efforts were futile. "He's not going to make it," she said, wiping her bloody hands off in the dirt.

"Should we call one of the 21ster hospitals?" Alan the Curious asked.

"With what?" Li the Hammer said.

Pietra put her ear down to Khan 1's chest. "His heart is no longer beating. He's dead."

The other rebel clones, and Caesar 13's technicians, were shocked. The two Guardian clones were shouting with satisfaction. The others moved away from them, disgusted.

"We have important tasks ahead," Pietra said. "Like how we're going to survive in this century."

"We should just stay here," Li the Hammer said. "Why go back?"

"Because we have to save our time thread and reunite it with the main timeline," Pietra said. "We have to go back upline to restart human civilization."

Alan the Serene spoke. "Where are Grace the Cheerful and Sigmund the Shy?"

The group looked around the warehouse. "They are gone!" Pietra said.

"I remember," Alan the Serene said. He pointed to the two clones that had killed Khan 1. One of them was still holding the bloody shard of metal he had used to deliver the killing blow. "These two shoved them off the platform before the Shifter sent us back in time."

Caesar 13 watched with horror (and a little satisfaction) as Khan 1's body lie in a little pool of blood on the warehouse floor.

Suddenly Khan 1's two goons ran up the stairs to the warehouse door and let themselves out. The bloody metal weapon clattered on the stairs and fell to the concrete floor.

"Good riddance," Alan the Curious said. "What do we do with Khan 1's body? He is from our time and it is our responsibility to dispose of it."

"In this time period, bodies were buried or cremated," Pietra said. She saw a pile of debris at the back of the warehouse. "Someone help me. We'll bury him back there."

Li the Hammer (eagerly) and Alan the Curious (gingerly) helped Pietra place the printed body of Khan 1 underneath the debris.

"As ye sow, so shall ye reap," Li the Hammer said.

"It is a fitting epitaph for that clone," Pietra replied.

The three walked back to the group.

"Now what?" Alan the Curious said, looking nervously at Khan 1's blood that was soaking into the dirt. "We have to get out of here, but we have nowhere to go."

"Oh, but we do," Caesar 13 said. He described how he had rented three apartments and had created three bank accounts with 2035 currency.

"That is brilliant!" Pietra said, glowing at the chief Tower time technician.

"I think we shall like living in this time period," Caesar 13 said, smiling at Pietra.

Pietra shook her head. "We must all go back. Our stay in 2035 must be temporary."

Caesar 13 sighed. "Of course you're right, Pietra. But I like it here."

"We don't belong here. Look at it as a vacation."

"We have no way to get back to 3013 anyway. We're stuck here," one of Caesar's time techs said.

"I'll have to create identities for the rest of you," Caesar 13 said. "Fortunately I printed more cash before I left. It's almost never used, but it is still legal; issued by the U.S. Treasury." Caesar 13 paused. "By the way, I am now Caesar Smith." He pointed to Tesla 8 and Einstein 3. "These are now Tesla Jones and Alfredo Einstein."

"Alfredo Einstein!" Pietra began to laugh, which pleased Caesar. This Pietra clone was beautiful. Somehow her enhanced biology made her look less like a clone and more like a real woman. Would it happen to him as well?

The former Alan 133 saw Caesar's interest in Pietra and felt a pang of emotion. Was this what humans called jealousy? He was unused to feeling strongly about anything, but he knew he was attracted to Pietra.

"Let's go to our new living quarters," Caesar suggested. He reached into his satchel and put on his 21ster coat, hat, and boots. It was warmer than the last time he had been to the past, but still colder than in 3013.

Caesar looked at the group. His twelve technicians, himself, and nine of the rebels. "This group is too large to walk together. We would attract unwanted attention from the natives."

Pietra laughed again. This Caesar had a sense of humor! "We will attract attention anyway," she said, looking at her blue and black Guardian thermofilm. "Only you have the appropriate clothing."

"It can't be helped," Li 355 said. "Why don't we break up into groups of three or four."

"Caesar, you lead off with your group," Pietra said. "We'll follow you. If anyone asks, you are leading a troupe of performance artists. That will hopefully explain our strange clothing."

This was met with unanimous agreement and the party set forth out of the warehouse.

On the way Pietra realized that they had completely forgotten about Sigmund and Grace, still trapped in the Tower detention cell.

3013 AD

Li 308 and Li 57 had been taking turns in the past while the other rebels attacked the Tower. They could spend as much time in 2035 as needed, as long as they didn't overlap their own worldlines. Each spent 8 hours per trip, and inserted at 20 minute real-time intervals. Within a day of real-time the two temporal technicians had raised their LRIs enough so that their implants were unnecessary.

"Look!" Li 57 said, when both were on a break. They both saw their colleagues attack the Tower and how Caesar 13 had transported them back to 2035, including all twelve of his temporal technicians. They also saw Grace 5 and Sigmund 78 thrown off the platform in the melee, and both shoved roughly back into a cell by Christian 172's goons.

Li 308 watched in horror as Christian 172, with Khan 1's approval, escorted the enhanced clones (minus the two who had escaped to the past) into the Tower basement.

Christian 172 had to keep the Submission frequencies at their highest, for these clones struggled madly, knowing they were going to their deaths.

"Please turn the feed away from there," Li 57 said, distressed.

It took the two rebels several minutes to regain their composure. It was one thing to hear about clones being "deactivated" on Guardian media, but it was another thing to actually see it.

The two rebels got something to eat from the cafeteria. Each clone always brought a portion of their food allotment each day to Building 103 for this purpose. Li 308 examined the timeline. "One of us must make contact with our group in 2035."

"I'll go," Li 57 said. "It's my turn anyway. I want to talk to my sister Pietra."

"I'll insert you three minutes after everyone leaves the warehouse," Li 308 said. "I don't want to put you inside somebody! The warehouse was crowded after the mass insertion."

"I appreciate that," Li 57 said in mock horror. The two former clones were learning human emotions now. Li 57 was feeling fond of Li 308. Did he share her feelings?

Pushing these feelings aside, Li 57 put on her printed 21st clothing and walked to the platform.

"Don't forget to bring your Shifter interface module, Li 57, or I will have no way to contact you," Li 308 told her.

Li 57 patted her coat pocket. "It's in here."

Li 308 activated the Shifter. In a few moments she stood on the cold concrete floor of the warehouse.

Sigmund the Shy and Grace the Cheerful had been shoved together into a detention cell by several of Christian 172's goons. Grace's head cut had stopped bleeding but her twisted ankle was very painful. Sigmund's fractured arm was blotted an ugly yellow and blue.

Sigmund looked around the cell for something to bind their wounds but the synth-fab cell was empty. He noticed that his blue and black uni had been ripped during the fight. He bent down and removed the torn thermofilm and began to wrap it around Grace's ankle. He secured the bandage by tying a knot. "This may not last long," he apologized.

Grace smiled. "Thank you, it already feels better."

Grace took off more of the thermofilm from Sigmund's leg and wrapped his arm tightly using the same technique.

"We must escape to the Shifter and rejoin our brothers and sisters," Sigmund said.

Suddenly the detention cell's transparency cleared. Christian 172 was standing there with a smirk on his face. "How nice. You two lovebirds are coming with me."

Grace and Sigmund exchanged an alarmed glance. The Overseer was going to terminate them. But he no longer had his enhanced goons with him! 172 had already sent them to the Tower basement. The two rushed past him to the Shifter as fast as they could. Sigmund quickly examined the main console, which was similar but more complicated than their own, smaller one, in Building 103.

Christian 172 called for his guards and rushed the platform as Sigmund fumbled with the Shifter controls.

"Hurry!" Grace cried as she fought off several of the weaker Tower security clones. Suddenly her ankle gave way and she was grabbed by six or seven guards, trying to take her off the platform.

Christian 172 was upon Sigmund now with a dozen of his red-and-black clad security forces. Sigmund shoved these clones into the Overseer with his good arm and managed to activate the device, which was still programmed from Caesar 13's mass insertion. He ran to the platform, hoping that the previous Shifter program would run. As the Shifter's rings began to spin more and more rapidly he managed to push the guards off of Grace and off the platform. His strength was amazing now, but his arm hurt a lot. A blue energy developed around the platform.

Christian 172 stood back. He had no desire to go back to the past, nor did he wish to lose any of his guards. He watched as the blue energy grew stronger and brighter, and the rings began rotating at frightening speed. Then a pop and there was nothing.

Christian 172 knew that Khan 1 was back in 2035. Patton 1 had been destroyed by the Khan. Therefore he was in charge! He, Christian 172, who had for so long taken a back seat to clones of inferior intelligence, was the Supreme Overseer!

"Destroy the Time Shifter," he told his guards. Then he walked down the hallway, past the lifts, and entered the chambers of Khan 1.

Grace and Sigmund found themselves sprawled on a cold concrete floor, dressed in their thermofilm Guardian unis. The two clones hoped they were in 2035.

"Look," Sigmund said to Grace, pointing to the dirty floor and dozens of footprints.

"This must be the warehouse! We are in the right time period."

Both clones looked at each other. "What do we do now?" Sigmund asked.

Grace smiled. This timeline felt...alive. "The first step is to get off this filthy floor."

Sigmund helped her up. She was able to stand and tried walking.

"You are limping. How does it feel?"

"I'll live," Grace replied, echoing a 21st media presentation in the history tranches. All clones watched this old media, she thought, and it was allowed by the Overseers so that population control could be maintained. Hope – even lost hope – was valuable.

Sigmund's arm hurt badly but he said nothing to Grace. First he clumsily reset Grace's ankle bandage. Then he walked across the floor and up the stairs to an old metal door and opened it with his good arm. Industrial buildings dominated the landscape. "Here is the exit." He helped Grace up the stairs and they stepped out of the door onto a wide concrete walkway that led to a dirt road. Although the light was faint due to heavy cloud cover he could see dirty footprints on the concrete, which led left onto the dirt road. He thought he saw someone familiar walking slowly down the paved road that led from the dirt road, but the figure was a quarter mile away at least. However, this person was wearing the printed clothing from 3013! Was that Li 57? The figure was looking ahead, as if tracking someone else. Sigmund took a chance and called out. "Li 57! Li 57!"

Li 57 turned. She immediately recognized the blue and black Guardian thermofilm. There was no traffic on the road at this early hour. Li 57 walked as fast as she dared back to the abandoned warehouse. Another figure dressed in thermofilm was by the doorway. Sigmund 78 and Grace 5?

Sigmund spoke excitedly. "It is our sister Li 57!"

Li 57 turned the corner from the paved road onto the dirt road that led to the warehouse and walked quickly up to her two friends. The three clones from 3013 hugged warily, looking about for Guardian hovercraft or foot patrols. Then they laughed and realized they were free. Sigmund and Li 57 danced about with happiness. But when a vehicle passed on the road and slowed sharply, they disengaged and stood motionless. The vehicle slowly resumed its progress. Sigmund saw the driver talking on his mobile device.

The driver of the vehicle, curious, called his dispatcher. "I've noticed heavy activity by the old Hawkins warehouse this morning. Is there something happening there? That place has been rotting for over two years."

"Yeah, Max Berglin just bought it. It's supposed to be renovated soon."

"Ok, thanks. I've seen some strange people going out of that place."

"Probably workers."

The driver shrugged. It was none of his business, but those two guys in the blue and black unis weren't workers. He'd call Ralph Zimring, Berglin's security chief, and report it. Everybody in Midland knew Ralph, a seven-foot giant and former mercenary who had the curiosity of a cat and who had developed a network of contacts throughout the city.[25] Zimring was known to want information on anything unusual or suspicious.

Worried, the three clones stood without moving and watched the vehicle until it was out of sight.

"Let's get out of here," Sigmund said. He looked at Li 57. "What are you doing here, sister?"

Li 57 quickly explained how she and Li 308 had observed the rebel attack on the Tower, and how Caesar 13, his techs, and the Resistance group had been translated into the past. "Li 308 inserted me three minutes after they all left the warehouse. I was following them, but I've lost them now."

Sigmund was glad he had advanced the Tower Shifter's insertion point before he activated it. He and Grace had been lucky to land in 2035 at the right time. "Do you know where they were going?"

"No. Just to some place Caesar 13 rented. The group was headed down Main Street. If we hurry we might still see them."

"It's colder here than in our time," Sigmund said. The Guardian unis didn't cover the feet, hands, or the head. His own uni was torn off halfway up his left leg. By force of habit he checked his wrist communicator for the temperature, and realized that these devices didn't function here. He took it off and put it in one of his uni pockets. Grace did the same.

"Li 57, we are sorry to have stopped you," Sigmund said. "Please try to locate Caesar 13 and our group. We will follow you, but I must help Grace so we won't go as fast as you."

"All right," Li 57 said reluctantly. She did not want to abandon her brother and sister. "I'll wave and point if I can locate them. I'll come back for you no matter what, even if I can't find them." She pointed to the dirt road. "When the dirt road ends, turn left. There is a paved road; it is called Industrial Drive. Go down it about a quarter mile until you see a busy intersection with one of those primitive traffic lights. Cross the street on the green and turn left. Keep walking."

It was a bad plan but no one could think of anything better. Sigmund was worried about Grace and Grace was concerned about Sigmund's fractured arm. But they had all previously taken trips into the past, and were somewhat familiar with the grid pattern around the warehouse.

Li 57 remembered that she didn't have to walk slowly and not attract attention. She knew that public transportation existed, but she had no currency card. Walking as quickly as she could, she proceeded down the dirt road.

Grace and Sigmund started after her. Sigmund put his good arm around Grace's waist, supporting her. Grace liked this. As they walked her ankle began to loosen up a bit.

"We can go faster now. My ankle feels better thanks to the bandage." She smiled at Sigmund, who smiled back. His arm hurt but he didn't want to look underneath the bandage.

"All right, tell me if we go too fast."

They made it to the paved road and turned left. Li 57 was already at the intersection with the traffic light. She waved and they waved back. Sigmund and Grace saw Li 57 turn left and shade her eyes with her hand. She gestured excitedly, pointing down the road.

"Li 57 found them!" Grace cried. It was easier to express strong emotions here in 2035. Grace liked this a lot.

Sigmund noticed that their out of place Guardian unis were beginning to attract attention from passers-by.

"Hey you two!" somebody said. "What's with the costumes?"

Grace and Sigmund, trained their entire lives to keep quiet and not attract attention, merely nodded and didn't react. When they got to the intersection the light was red and they had to stop. People were staring, making the two clones very uncomfortable. Sigmund leaned over to Grace. "Say nothing, act natural."

As they started to cross the street on the green, a young man stopped them. "I like those outfits, where did you get them?"

Sigmund was at a loss for words but Grace came to his rescue. She remembered an old line from a 20th media film. "If we told you we'd have to kill you."

The young man burst out laughing and went past them. When they got to Main Street Sigmund looked at Grace with frank admiration. "That was remarkable Grace. Well done!"

Grace found herself being very pleased with this remark. Her eyes softened a bit. She suddenly noticed that Li 57 was waving at them again, several blocks down the street. She was pointing left. Grace waved back and indicated she understood.

As they walked Sigmund noticed that the trees were beginning to leaf out. The weather was colder than in 3013, but not so bitter cold as the last time they had come to 2035. It was certainly above the freezing point, but both of them were uncomfortable. Talking to each other helped them relieve the discomfort of people staring at them, and deflect the rude remarks they were getting from some of the passers-by. There were many more people in this time period than in 3013.

Li 57 was coming up to them now on the street, walking quickly. When she joined them somebody asked her, "Do you know these two clowns? What are they doing in those tights?"

"They're pets," Li 57 said, startling herself. Where did that come from?

The man laughed. He was walking beside them now, clearly interested in Li 57. "Hey, do you want to get something to eat?"

Li 57 handled it like a 2035 veteran. "Thanks, but not right now. I have to take these two darlings to their kennels."

The man laughed again, and inspected the three walkers. "You know, you three are the nicest looking human beings I've ever seen."

Grace smiled at the man. "Thank you sir, you are very gallant." She pointed to the intersection. "We have an appointment."

The three clones turned left onto Main Street. The man was still staring at them from the intersection.

Sigmund was amazed. "How did you two come up with those lines? You're both very clever!"

Grace and Li 57 were both very satisfied with this. "We are, aren't we sister?" Li 57 said to Grace. Grace's ankle was very sore but she was walking well now.

Sigmund the Shy understood that humor was very valuable in this society. He would have to develop this attribute. Perhaps Grace could help him. Sigmund and Grace followed LI 57 as she turned into a driveway and walked up a flight of stairs. Li 57 knocked on the door. The door opened. Two dozen clones from 3013 were crowded into a modest-sized living room.

Caesar 13, along with Pietra, were standing in the middle of the group.

"Grace 5! Sigmund 78! Li 57! It is so good to see you!"

Neither of the three clones were used to this expression of unabashed enthusiasm, but they liked being addressed in this manner. Sigmund stared at Pietra, who was more vital and beautiful than he'd ever seen her. The two female clones thought the same.

"You look like a 21ster human!" Sigmund blurted.

Pietra laughed. "I feel like one."

Sigmund saw the male Caesar clone walk over and put his arm around Pietra. He glanced from Caesar to Pietra. "Both of you."

One of Caesar 13's time techs looked around the room. All of the Resistance clones, who had made several trips to the past, were similarly healthy and vital. "It's true then," he said, awed. This was his first trip to 2035. "Life force readings are enhanced by traveling to the past."

Pietra put her arm around Caesar 13 and looked up at him. "Oh yes, very true."

Sigmund could feel a sexual energy between the two. He noticed that Alan 133 was looking away from the couple. Some tension there, certainly. These were new problems in their group dynamic, unknown in 3013.

Caesar 13 stepped back into the center of the room. "We are all here, except for our brother Li 308, who I am sure is monitoring the timeline. I am Caesar Smith, formerly Caesar 13, Chief Shifter Technician for the Tower."

Caesar turned to Pietra. "Pietra will make the introductions."

One by one Pietra named the clones, pointing them out. There was one Caesar clone, four Li clones (two female and two male), three Alan clones, one clone of the Sigmund line, two Tesla clones (one female and one male), four Einstein clones (two male, two female), three female Curie clones, one Grace clone, three Alvaro clones (all male), and three Barbara clones. The group agreed on what to call each other. There were 12 rebel clones (minus Li 308) plus Caesar 13 and his 12 time techs.

Caesar Smith resumed his talk. "Originally myself, Tesla Jones, and Alfredo Einstein were going to escape the Overseers and live here, in the past. I created three separate identities for us, complete with birth certificates and other vital documents, which will pass muster if anyone looks into us. I rented three separate apartments for us in this building." Caesar looked around the room. "There are 25 of us and six sleeping quarters, with six beds. That means each apartment will have to accommodate at least 8 persons. These apartments aren't supposed to have that many occupants so we absolutely have to be quiet. No disturbing the people below us, or any other neighbor, or the landlord will kick us out. Got that?"

Everybody nodded.

"The idea is to live in this timeline, mix with the 21sters, absorb as much life force energy as possible. Look at Pietra, who has the most downline time. If you can distinguish between her and a normal human, you're a better observer than I am."

"What about her brain implant?" Grace asked. Although their bodies were beginning to reject the electronic augmentations, the brain implants were still visible. Caesar's time techs still looked like clones.

"Can't do anything about that," Pietra responded. "Keep your head covered even when you go into a public place. Wear a fashionable headpiece indoors, or buy a wig." Pietra put on her hat, a stylish little thing that covered the back of her head and went well with her red hair. She looked over at Caesar. "Caesar got this for me. Isn't it cute? "

The female clones were impressed and each wanted one.

Caesar showed the guys a shopping site on the World Net. "These hats look pretty good for the males. Here's mine." Caesar went into a closet and put on a flexible gray headpiece with a flap in the front. "This is called a Shelby. Looks good doesn't it?" Caesar paraded around the room with his hat at a cocky angle on his head.

"You look sexy, boy," Pietra said.

The twelve Tower techs looked at each other. These two had perfectly adjusted to the century! They appeared to be genuine 21sters.

"Each of you will have to buy clothing," Caesar said. "I have three currency cards that we can use, connected to three separate bank accounts. Before the end of the day I want each of you to see me or Pietra so you can choose what you want to wear. We can have the clothing delivered here within the next few days. Until then, you have to stay inside. Is that understood?"

The clones nodded.

"Pietra and I will go food shopping and stock the refrigerators. All of us here are vegetarians, and there are good recipes in this time for actual food, instead of the SynthPaks we've had to eat in 3013. Living here should be a treat if we can stay out of trouble."

Everyone nodded.

Caesar and Pietra assigned clones to each of the apartments. "There are only two beds per apartment, so most of you will have to sleep on the floor until we get back to 3013." Groans were heard. "We'll rotate the beds."

Pietra then assigned what she called "exploration teams," and their destinations. These were for the recently arrived time techs, who had no previous exposure in 2035. "Only three maximum per team, and stagger the leaving and return times."

After everything was settled and no one had more questions, the meeting was about to break up.

"How long will we stay here?" Alan Gutzon (formerly Alan 49) asked.

Caesar grinned and looked over at Pietra. "Until you feel like having sex. And you can."

After a week in 2035, Pietra noticed that she was bleeding. "Oh my God," she thought. "My menstrual cycle has activated!" She excitedly told Caesar, who was completely overwhelmed with happiness. "Pietra, this just might work!" After three days Pietra's cycle was over. Caesar and Pietra decided to go to a hotel and make love. They wanted privacy. Caesar owned one of the currency cards so it was no problem.

There was some grumbling, but Pietra's progress had excited everyone. "We're here to restart society in 3013," she said. "The goal is to enable the reproductive function. Tonight we're going to test that."

Caesar clinched it. "Any couple who feels ready, we'll rent a hotel room for you."

Alan 133, who now called himself Alan Percival (after the legendary knight of King Arthur's court who sought the Holy Grail), had to give up his hopes for Pietra. So did the former Li 355, now Li Hammer. All of the clones began to understand that their purpose here was to find mates among the two groups from 3013. Fortunately the balance between male and female clones was even.

Sigmund and Grace were already a couple, and the others began to pair off. Sigmund had used a currency card to see a doctor in the company of Caesar, who was most familiar with 2035 society. The doctor, remarking that Sigmund and Caesar both looked like fashion models, set his arm and gave him antibiotics. Sigmund didn't take them, fearful of an adverse reaction.

Li 57, now Li Jing, was longing for Li 308, who was doing his duty monitoring the Shifter back in 3013. She approached Pietra the night after she and Caesar came back from the hotel. The two looked like a couple of 21sters who had spent the entire night having sex. Li 57 was awed. She wanted this too.

"I want Li 308, who is trapped in 3013. One of us needs to go back to the future so that I can have a mate."

Caesar and Pietra were troubled by this, for all of the clones had paired up now.

Caesar realized that he and Pietra were the leaders and must bear the greatest responsibility. But they were both having a great time in 2035, and neither wanted to go back. Caesar looked at Pietra. "She's right," Pietra admitted reluctantly.

"You're right Li Jing," Caesar said. "Give us one more night together. I'll signal Li 308 through my Shifter interface to bring both of us back to the future. I'll send Li 308 to you."

Li Jing was grateful. "Thank you so much."

As each couple was successful, they left 2035 and were extracted up the timeline. Finally, after a month, there were only two clones left: Hannah Barbara and Alan Percival.

Neither Hannah Barbara's nor Alan Percival's body were responding well enough to enable the sexual function. Both clones were devastated. Both felt like failures.

One night, Alan Percival put his Shifter interface into his implant. He knew the Shifter was being monitored 24 hours a day in 3013, but you never knew who you were going to get. He got Pietra, which made him feel uncomfortable.

The first thing she said was, "Hello Alan! There's great news. I'm pregnant!"

Alan wanted to cry but he mustered up congratulations.

"I'm afraid that me and Hannah Barbara are not responding well enough. We might not be able to enable our reproductive organs."

Pietra was shocked. She had never thought of this. No one had.

"Alan, I'm sorry. I'll have a conference with the others and get right back to you. Hold on."

Alan knew that days could elapse in 3013 but the time device could enable a temporal connection a second after his call upline. He sat there in his bedroom, dejected. He felt a ping on his implant. It was Caesar.

"Alan, we have no way of knowing whether further exposure will enable you or Hannah. You have two choices: stay in 2035 permanently, or for as long as it takes, or return to 3013 now. We're monitoring you two on the timeline every second."

"Thank you Caesar. I'll speak with Hannah and ping you shortly."

In the end Hannah and Alan decided to stay in 2035 for another month

They could live easily on the funds Caesar 13 had placed in the three bank accounts, and they had three apartments all to themselves.

The two Guardian clones who had escaped from Christian 172 ran into Harry Kaine one day several weeks after the two fled from the warehouse. Completely confused by 21st century society, the two had gravitated to the soup kitchen and the homeless shelter. Harry was walking by the place when he saw two odd-looking persons standing in line for food. It was late June now, and warmer. Harry walked up to the line, noticing something strange about two of the scraggly haired men. The hair on the back of the head of the brown-haired one was parted. Harry noticed what looked like an electronic augmentation to the skull. Excited, Harry carefully walked to the front of the line and saw that their faces were in perfect symmetry, their skin without blemish. These two were clones! They looked just like the ones in 3013, wearing some kind of blue and black form-fitting clothing underneath ill-fitting pants and torn shirts. Nobody at the center noticed anything unusual, but Harry was observant.

Now, Harry thought, he would get some answers about that society of the future. He walked up to one of them and handed him some currency. "Buy yourselves something," he said.

"Whuy you heulping oos?" the man replied in an accent Harry had heard in 3013! His companion stepped out of the line. "Guimme soume a thuat," the other clone said. Harry handed over a currency note of equal denomination.

"If you gentlemen will come with me I can show you some more money, a place to stay, and a job."

"Duon't wuant nuo joub," the other said.

Harry waved another currency note in the air. "You want more of this? Follow me." He didn't wait, but walked out. The two clones followed him.

"Whuat would we hauve to dou?" the black haired one asked.

"You'll be studied by several scientists looking into human life extension," Harry replied. "Just some testing and you don't have to do anything. It will pay well."

The two clones, used to Overseer authority and obeying orders, shrugged. "Shouw me moure a this," he said, exhibiting the currency note, "and we're wiuth ya."

"Come then," Harry said, walking in the direction of the Taubman Research Building. He presented his clones with no explanation to Lab 213, which engaged in sensitive Transhuman research. The lab, funded by the NIH, employed scientists all known to him and onboard with his life extension goals. He had helped to write some of their NIH grant proposals.

———— • ✳ • ————

Two weeks later Harry Kaine received a report on the analysis of the two clones from 3013. The report was exhilarating. "The two specimens exhibit DNA enhancement, electro-neurobiological augmentation, and nano-biological technology that is light years ahead of what we know today," it said. "With this new knowledge we can advance life enhancement and other Transhumanist improvements to the human body in a fraction of the time it would have taken us without the two specimens."

Harry Kaine had a good laugh. Through the time viewer in 3013, he had seen the speech he was to deliver in 2048 at the UN, and the Transformation Centers start to go up in 2098. He had been very excited to see this, but the technology necessary to create his life enhancement centers was way too far into the future for his liking. Yet two wild cards had fallen right into his hands!

A day later he went to Lab 213. The two clones weren't there.

"We sent them on their way," one of the scientists said. "And well paid, as you wanted."

"Excellent!" Harry wanted no one to know anything about the other timeline in 3013. The obvious question was asked.

"I found them both at the soup kitchen," he replied truthfully. "I have no idea who they are or where they came from."

That last was a lie, but Harry was a convincing liar.

He received several skeptical looks, but Harry got them talking about their new discoveries. The scientists were so excited that the origin of the two clones was passed over. "It might be better to say nothing about them," Harry suggested just before he left. "Just say that you have made exciting new breakthroughs." This explanation appealed to the professional pride of the scientists, and was agreed to without much fuss. Each was looking forward to scientific recognition from their peers.

The irony of this was not lost on Harry Kaine. His success would be guaranteed by an accidental walk-by of the local homeless shelter! Never a deep thinker, Harry didn't wonder (or care) how the two clones from the future got to 2035. There were two Time Shifters in 3013; somebody must have sent them here. It was just his good luck to have found them.

Harry went home that night very excited.

Over a bottle of wine, alone in his study, he made plans. He would organize a great Transhumanist conference for next year, in early 2036. The Humanity+ organization and their series of excellent conferences had gradually faded into the scientific woodwork. He would revive them; get the old organizers back together or find new ones. He would take them (and anyone else interested, including venture capitalists) to Lab 213 and let them talk to his scientists. Dr. Anders Kopetz, his head of research, would give detailed briefings. He would hire technical writers and grant writers and apply for more funding from both the government and the private sector. In this way he would build momentum for more research. His scientists would submit papers to respected scientific journals, building more enthusiasm for Transhumanism in the academic sector. His technical writers would submit well-written articles to popular magazines and other mainstream publications that had a wide readership. He would hold a press conference to announce the conference as soon as possible.

CHAPTER 23

The Saturday after Karen Everard's meeting with Tom Mankiewitz, the Gang of Eight was to meet in Ger's duplex upstairs at 7. I called a pre-meeting in my apartment at 5 for the time travelers, to talk about what we were going to do about Harry Kaine. M'basa, Kjirsten, and Karen all showed up.

When Kjirsten arrived she looked visibly upset. Unlike the other time travelers, Kjirsten didn't have a full-time job. She worked part time at the Midland Police Department in the Records department, but it was hit-or-miss employment. Like Lledren, she had mostly drifted through life after she graduated.

"I want to forget it ever happened, but I can't," she said as soon as I had poured coffee. "It's...beyond my reality and I can't accept it." She looked around at us. "I didn't dream it did I?"

"No Kjirsten," I said. I looked at Karen and M'basa. "Have the memories of our trip faded at all for you guys?"

M'basa shook his head no. Karen said, "Not at all. We were there."

I felt a wave of excitement go through my body. "Time travel is real then. It's amazing, incredible!"

Karen smiled, but Kjirsten looked depressed. "How could the human race have gone so wrong? I feel like my life is pointless. I mean, why bother to try to change the world when you know it's going to end up like that?"

"Good point. Is the future changeable, or are we stuck with Harry Kaine and his Transhumanism?"

M'basa slapped his palm with his fist. "Those clones in 3013 said that past affects future and that time was fluid. So, no. We have to make sure that megalomaniac Harry Kaine doesn't succeed."

Kjirsten smiled at this.

"Li 308 also told us about the time conservation law," I pointed out. "Once events have already occurred, especially major events, they are almost impossible to change. So what can we do?"

Kjirsten was looking depressed again. M'basa gave me a dirty look.

Karen spoke up. "We don't have a choice. Either we just give up and mope, or we find a way to stop Kaine."

"What can we do?" Kjirsten asked. "We're nobodies."

"We're the only ones who know for sure how dangerous Harry Kaine is," I said. "No one else will believe us. It's up to us."

"Everybody think about that," Kjirsten said. "We need a plan."

"That's the spirit," M'basa said encouragingly.

"OK," I said. "What do we tell the Gang about our...adventure in time?"

Everyone thought silently about this for a while.

"You explain it Joe," M'basa said finally. "You're the most level-headed."

"OK. I'll just lay it out exactly as it happened."

After that the meeting broke up. "See you all at Ger's," I said with a grin. "I have an errand to run."

"I recognize that look," Karen said. "What are you up to, Joe?"

"You'll find out," I said, walking out the door. "Something to lighten the mood a bit."

For a joke I drove to Angelos and bought seven takeout orders of focaccia rolls, placing them on seven separate plates at Ger's big kitchen table. When Liqao walked up the stairs and approached the table, I quickly arranged the plates so that he didn't have one.

The butt hurt look on Liqao's face was classic as he stared at the empty plate. "Hey, where's mine?"

Even Kirra laughed.

I relented. "Somebody give Liqao one," I said.

"Not funny Joe," Liqao mumbled as he bit into a roll.

"OK, the fun's over," I said. "I want Karen to tell us about her interview with Harry Kaine's Transhumanist scientists." I looked at my four time-traveling companions. "Before Karen does that I have to tell you about something that's hard to believe."

"Go ahead Joe," Karen said. I saw Ger's face light up a little.

I told the story of our trip to 3013 and what we had seen in the New York of the future. "Time travel is real. We saw the future, the four of us, along with Harry Kaine. We know that Harry is successful with his life extension technologies, and how the human race ends up. It's not pretty. Ger is working out the physics of time with his new research."

"Is that true Ger?" Lledren asked.

"I'm just as skeptical as you are about physical time travel," Ger said. "However, it's true that I have been developing a device that works with temporal energy."

"Show us your recording Ger," I said.

Ger described his work at the Taubman Building. "I made this recording with my temporal device and copied it to my mobile. This is New York city, 3013 AD." Ger played it back.

"Yes," Kjirsten said. "That's what we saw, me and M'basa and Karen and Joe."

"And Harry Kaine. If you don't believe me, talk to him." I described how Kaine was the key figure on the timeline, and about his life extension centers. "But that doesn't happen until 2098, so we have lots of time to stop him."

Liqao rolled his eyes in disbelief, but Kirra jumped out of her chair. "The people at the library!" Kirra cried. She told the group about the "at risk" adults and their perfect features and their funny accents. "Were those people from the future?" Kirra described how Harry Kaine had come rushing into the library, asking about a group led by a redhead with an airbrushed face.

The four time travelers stared at each other. "We don't know anything about that," I said. "But the clones in 3013 do speak with an accent. And their bodies are perfectly formed."

"And Pietra 23 has red hair," M'basa said.

I looked at Karen. "Harry Kaine is diabolical! Is he sending clones back to the past now? Maybe he's connected somehow to the people from the future."

Karen just shook her head, mystified.

"Wow," Liqao said. "It's a good story anyway. You could sell it as a novel."

Lledren was staring at the time travelers. "You guys obviously experienced something radical, but it doesn't make sense to me. Are you sure you didn't just have a vivid dream?"

"No," M'basa said firmly. He described what it felt like to be in the Time Shifter, and how they had all popped in to the abandoned warehouse.

"Did the space around the platform shimmer?" Ger asked.

"Yes," M'basa said.

"Was there a blue energy surrounding the platform while the rings were turning?" Karen nodded.

"Jesus Christ," Ger muttered, and got up from his chair. He began pacing the room.

"I'm not sure I believe your time travel story, but I will agree that Harry Kaine is dangerous," Kirra said. "The most self-centered person I've ever met." Kirra described how Harry, who came to the library almost every day on his lunch hour, had hit on her a number of times. "He's charming, personable, ...and a little unbalanced."

"Ask him about 3013," I suggested.

Liqao wanted to say it was all bullshit, but checked himself. He wondered what he had in common with any of these people. He picked up his fourth focaccia roll and bit into it.

"You're going to get sick eating those," Kirra told him.

Liqao grinned. "It's worth it."

The conversation died, but Kjirsten picked up the ball. She spoke to Karen. "Tell us about your interviews with Harry's scientists. If we're going to stop him we have to know what his goals are."

"Oh my God, Kjirsten. You wouldn't believe these people. Two of them are strict scientific materialists. They don't care about the consequences of their work. But I can see how Transhumanism is connected to Harry Kaine, and the dead society we saw a thousand years from now. It all fits. Or rather, it will all fit."

"Unless we do something about it," M'basa said.

"I'm all for Transhumanism," Liqao said. "Emotions get in the way of rational thinking. Load me up into an android body."

"Then you couldn't have sex," Kirra said.

Liqao's face registered shock. "Hadn't thought of that. Well, android bodies that can have sex then."

"Not the same," Kirra said. "These biological bodies are the best for that."

"How do you know? Augmented bodies might be better."

"Want to risk it?" Kirra said. "I'll bet those focaccia rolls won't taste as good in an android body."

Liqao started at this, and his jaw dropped a little. Everyone laughed.

M'basa cut in. "Has anyone heard about the Transhumanist conference Harry Kaine is organizing?"

"What?" The four time travelers were shocked.

"M'basa is right," Ger said, sitting back down. "I heard about it down at my lab in Taubman. Word is that it's going to be a big deal. Next spring, in New York. Scientists from all over the world will attend."

Ger leaned his big frame over the table. "The gossip at Taubman is that Kaine got hold of two 'unusual specimens' that he found in the East Side soup kitchen. Kaine's lab is on the second floor. Everybody is really excited up there about something."

M'basa turned to Lledren. "Know anything about that?"

Lledren shrugged. "I wasn't there that day. But Karel Friedman saw it. It wasn't much. Apparently two guys came in with torn clothing. Underneath they were wearing form-fitting blue and black tights. Very unusual, Karel said, and pretty cool looking. He was going to ask them where he could get them when Kaine walked in. Kaine spoke to them, gave them some money, and they followed him out. Karel said these two guys talked with a real funny accent, and they had faces with perfect symmetry."

I looked at Karen. "The clones in 3013 wore blue and black tights."

"The recordings I've made with my temporal device always show the same thing," Ger said. "Everyone there is wearing blue and black unis."

"Holy shite," Lledren said. "Play back that recording, Ger."

The Gang all watched it again while I made a running commentary. I went over with the gang what had happened to us in detail, and what happens on the future timeline.

Everyone was somber except Liqao, who snorted. I tried to make a joke. "There's no focaccia rolls in 3013. Everybody eats SynthPaks."

"OK," Lledren said. "Let's assume that what you're saying is true. What are we supposed to do about Harry Kaine? Kill him?"

I explained about the conservation of temporal momentum, and how time always works to resist changes that already happened. "It won't happen because Kaine is a historical figure. Or will be a historical figure. Harry isn't going anywhere."

Lledren was confused. "But this future you've seen hasn't happened yet! If Kaine were to have an accident, problem solved. If the future is as bad as you say, it's the only option."

Liqao thought the solution was pretty simple. "Lled is right. If there's only one future and Harry Kaine is responsible, kill the bastard. Conservation of time be damned. Then the timeline will change and everything is okey-dokey."

"OK," Ger said. "Do you want to volunteer to kill Harry?"

Liqao grinned. "I'll pass on that. Hey, I know! Get that ex-mercenary Ralph Zimring to make Harry have an accident."

I was starting to get mad now, and so were my three time traveling companions. Sometimes Liqao's flippant attitude pissed me off.

Kirra could see that tensions were rising and stepped in. "We're not going to kill Harry Kaine, Liqao."

"It was just a thought. No one is going to believe you if you say that Harry Kaine is an evil genius who will destroy the human race a thousand years from now. That's preposterous. They won't believe in time travel either. You'll look like a bunch of fruitcakes if you go public with any of it. That gives Kaine a built-in advantage because he can promote his ideas and you can't."

Kirra sighed, recognizing the truth of this. Her boyfriend was stating an inconvenient truth.

"So what do we do?" M'basa said. "Only the four of us who went to the future can truly understand how dangerous Harry is. He has to be stopped!"

"If we're going to do anything about Harry Kaine we all need to know more about Transhumanism and how it leads to a dystopian future," I said. "Karen, you know more about it than any of us. Give us a briefing."

"OK. That will help me to organize my thoughts for the article I'm writing."

Everyone agreed to this and Karen got out her notes. "I'll describe these technologies in no particular order. The first is AGI, or artificial general intelligence. AGI is different from AI in general, which could be just a sophisticated algorithm or software program. AGI refers to AIs that can learn, and which have intelligence levels similar to human beings. Some scientists say that the goal of AGI is a not only a thinking being, but also one that can feel, imagine, and create like a human being. Personally I don't believe that's the goal of serious AGI researchers. I think the feeling, creating bit is something they say to the masses to make them feel better."

"If what we saw in 3013 is an indication, I agree with you," I said. This was seconded by M'basa and Kjirsten.

"OK, number two is mind uploading, or nonbiological intelligence. The idea here is that human consciousness, or cognitive processing, can be implemented on substrates other than neurons. Harry Kaine said at our first meeting that the human mind is defined more by an information pattern than the hardware it is on. It doesn't matter if the substrate is a brain, or electromagnetic, or something else."

"Wow," Liqao said. "These guys are serious?"

"Oh yes. Dr. Singh told me that our brains really don't have to be made out of meat. That's the word he used to describe biology. We can transfer our consciousness to other types of bodies, or "containers," as he called them. Tom Mankiewitz said that each biological neuron could be replaced with a synthetic neuron-equivalent. Then the whole process of mind upload to an artificial body that doesn't age could be very easy and painless."

M'basa shuddered. "Thus Spoke Zarathustra. The Ubermensch. The Nazis."

Liqao objected. "Nietsche wasn't talking about or promoting fascism. The Ubermensch are fully evolved human beings. He was talking about self-mastery, self-cultivation, self-direction, and self-overcoming."

M'basa scoffed at this. "Those ideas have been used by dictators and sociopaths throughout history to engage in genocide and mass murder."

I raised my hand. "OK people, settle down."

Karen nodded and continued.

"Number three is something Harry Kaine is really interested in. It's called megascale engineering. The idea is that the human population will continue to grow and will need to go off-planet. Huge, livable structures built in space would be part of the answer. Scientists estimate that the solar system could support thousands of trillions of humans in these artificial structures, which would be placed in stable orbits around the earth, in space, and on other planets in the solar system."

Liqao's eyes widened in excitement.

"Wow," Kjirsten said. "These people think big."

"They do."

"I notice you frowned there Karen," Ger said. "Is there something else?"

"Well, part of the Transhumanist vision is changing the design of the human body to better fit in space. This is where physical augmentation, nano-biotech, and mind uploading come in. Our bodies, the way they are designed now, aren't ready for space. A big part of Transhumanism is space colonization, even outside the solar system."

Kirra voiced her concerns about the importance of the indigenous and how human beings have completely lost touch with the earth. "This technological society turns us away from mother earth. Every atom in your body comes from Gaia."

Despite his skepticism Liqao was impressed with this argument. "I never thought of that."

"That's right Kirra," Karen said. "Transhumanists are way more interested in space. In their view, the earth is boring and will soon become overpopulated or too polluted. They say we need to go out into space as a necessity."

"Shite," Ger said. "Make a mess and then leave."

"Yeah, but you should see some of the artist renditions of these megastructures," Karen said, trying to be fair. "They are...awesome. Miles long, fully self-contained living facilities with gravity, simulated sunlight, recreational facilities, hydroponics, and even farms. Everything needed to support life." She showed them a couple of images given to her by Tom Mankiewitz.

"Wow," Liqao said. "I kind of like this idea."

"OK, number four is called molecular manufacturing, or machine nanotechnology. Molecular nanotechnology would use massive arrays of nanometer-scale bots to manufacture consumer products and medical and scientific equipment with atomic precision. This kind of factory is called a nanofactory.

"What scares me are the medical applications. Medical nanodevices could heal wounds and repair organs without the need for surgery, but if the programming goes wrong...it's the end. The application of nanotechnology to human enhancement is what excites the Transhumanists. Dr. Mary Chen told me that with molecular manufacturing we could make improvements to every single body part. Transhumanists call this 'upgrading' the human body."

Kirra was getting physically ill thinking about this. She walked out of the room and went outside to hug a tree. Liqao followed her.

After a few minutes Kirra and Liqao walked back in the room. "Sorry guys, but this Transhumanist stuff is depressing," Kirra said. "In my opinion."

Karen nodded. "If Harry is to succeed, Transhumanism needs to be understood and accepted by the general public. Harry needs to put the best possible face on these technologies."

M'basa jumped up out of his chair and slapped his palm. "That's our strategy! We have to show the public how messed up Harry's ideas are."

Karen smiled. "We'd better do it quick. Research is already well along in these fields. But you're right, M'basa. if the public doesn't buy Transhumanism, he can't succeed."

"That's probably why Harry is having that big conference in New York," Lledren commented.

"Yes," Kjirsten agreed. "Getting a big head start molding public opinion."

Karen continued. "Number five in the Transhumanist lineup is autonomous self-replicating robots. These robots can be used to build megascale engineering structures out in space, or bases on the moon or other planets, or even factories on earth. I think this is pretty cool actually. Programmed robots that can do simple tasks over and over can be used instead of humans having to do it. This almost completely eliminates human error."

"Number six in the Transhumanist lineup is cybernetics. This goes into the area of improving the human body. Cochlear implants for the deaf, for example, are a cybernetic advancement. In moderation this could be a good thing, but the goal of Transhumanists is ultimately to replace the biology in a human body with cybernetic gadgets than can improve eyesight, hearing, strength, etc."

"Borg," M'basa said. "We saw that in 3013 with the brain implants. Every clone has one."

"Number seven is space colonization, which relates back to megascale engineering. Dr. Singh told me about a book called *The Millennial Project*. The guy who wrote it says that the Asteroid Belt could hold 7,500 *trillion* people, if thoroughly reshaped into O'Neill cylinders." Karen showed the Gang images of a huge, miles long cylinder in space complete with farms, sunlight, gravity, and recreation facilities (see pages 211 and 212).

Liqao was blown away. "It might be cool to at least visit one of those places," he said.

"Harry Kaine says that embracing Transhumanism will be necessary to colonize space. He told me that human beings aren't designed to live in space because of physiological limitations like deteriorating muscle mass and gamma/cosmic ray radiation. On the surface of Venus we would melt, he said; on the surface of Mars we'd freeze. Dr. Singh said that the only reasonable solution is to upgrade our bodies. Not terraform the cosmos, but cosmosform ourselves."

"Oh my God," Kirra said. "I'm feeling sick again."

O'Neill Cylinder – megascale structure built for off-earth space living. Courtesy of Wikipedia.org/wiki – from NASA/Rick Guidice

"I'll wrap this up quickly," Karen said. Kirra was pale and M'basa was getting angry.

"Number eight is gene therapy, which is basically cutting and splicing DNA to remove imperfections, treat disease, and increase longevity. Number nine is VR, or virtual reality. This is a relatively harmless technology that can create virtual worlds for humans to experience in. You could use it on a sick person, who would be immersed in a reality where his or her body was totally healthy. Do this for a long enough time and the person will think he or she *is* healthy. It's a sort of super-amplified power of positive thinking idea. Can't hurt anyway. Number ten is cryogenics, which the late Winthrop Sauvage was working on. This technology just preserves the human body for a long enough time until the new Transhumanist technologies become available."

"These people are either batshit crazy, or they have an amazing vision for the future," I said. "Maybe both."

O'Neill Cylinder – megascale structure built for off-earth space living. Courtesy of Wikipedia.org/wiki – from NASA/Donald Davis

M'basa banged his fist on the table, startling everyone. "No! This is technology for antisocial types and people with no self-awareness. It's...sick.""

Kjirsten went over and hugged him.

"M'basa is right," Kirra said. "In a vision-dream I had, Kitchi-Manitou showed me that if the human race is allowed to continue without alteration, eventually we will discover our connection to the One. We will become far more powerful than this degraded version of humanity, with even greater abilities than with Transhumanist technology."

"More powerful than expanding humanity into the solar system and to the stars?" Liqao asked. "More powerful than extending life far beyond eighty years? That's a grand vision, not a degraded one."

Ger looked at Liqao. "Whatever Harry Kaine does goes wrong, if we can believe what Joe, Kjirsten, M'basa, and Karen say. Kaine is, or will be, a failure. His program leads to the destruction of the human race. We have to stop him somehow."

Liqao said, "I'm out of that. I don't like Kaine, but I agree with his vision for the future."

The four time travelers nodded to each other. "We're in."

Kirra looked at Liqao. "That might be a deal-breaker for me."

Liqao acknowledged this. "I understand, but I must be true to myself. I won't lie to you anymore."

Kirra was startled. This was new talk from Liqao. "I can respect that."

"I have to be brutally honest Kirra. You may not agree, but I see nothing wrong with Transhumanism. And I don't believe a word of this time travel nonsense."

"I'm not sure I do either. But Harry Kaine is a creep and these subjects make me feel physically ill."

"OK. I can understand that. Maybe we can agree to disagree." Liqao grinned boyishly. "Are we still on?"

Kirra went over and hugged her boyfriend. "OK, I suppose we can coexist on this subject. But we may need your IT and computing skills."

"Always ready to help. And you're right. I like his ideas, but Kaine is excrement. I didn't like the way he looked at you in the restaurant."

Kirra giggled.

"Happy endings then," I said. "What comes next?"

"I write my article on Transhumanism for the *Chronicle*," Karen replied. "The first one was successful and generated a lot of public interest, so my editor is making me do another one. I'm not sure how I'm going to frame it yet. We'll see what Kaine does after it comes out."

I turned to Lledren. "What happened to those two clones after they left the homeless shelter with Kaine?"

"I don't know. Neither does Karel, who works with me."

Ger spoke. "I heard about that from Tanya at the Taubman Building. She told me Kaine let two 'unusual looking' men go after they were thoroughly studied by Kaine's scientists. They got paid, and left."

"Lledren, if you see those two guys at the soup kitchen would you call me?"

"Sure."

So Harry Kaine was already at work improving his technology and trying to alter the future! Just as he said he would. He was ten steps ahead of us and we hadn't even started yet.

Ger looked uncomfortable.

"What's wrong, Ger?" I asked.

"I can't continue my work with the temporal device. It leads to a horrible future. It's depressing..." Ger trailed off, then he exploded in understanding. He looked at me with surprise and astonishment. "The magnetic engine."

I almost jumped out of my chair. "Of course!"

Everyone in the room was startled.

"My real job is to develop the mag engine," Ger said.

"Yes! The time viewer stuff, that is just a sidelight."

"The lab is funded until the end of September. That gives me three months to work there, as long as Tanya doesn't revoke my clearance."

"Get to it, dark side."

Ger laughed.

"What are you guys talking about?" Lledren asked.

"Magnetism," Ger said. "We've been fools for over a century." Ger went over to one of his lab tables and pulled out two permanent magnets. He handed them to Lledren. "Play with those, pass them around. Notice the powerful push and pull energy they generate. When the magnets are aligned they attract, when out of phase they repel."

Lledren tried to push the two magnets together but the force was so strong he couldn't. "They're pushing each other away." Lledren turned one of the magnets around.

"Be careful," Ger said.

Suddenly the two magnets snapped together, almost smashing Lledren's finger between them. "These things are powerful!"

"Let me try," Liqao demanded.

"Properly arranged, arrays of magnets can create a powerful push-pull magnetic energy that can turn a rotor, or induce alternating current in a wire, without the use of LNG, gasoline, or other fossil fuels," Ger explained. "The energy of a permanent magnet lasts forever."

Lledren was astonished. "A perpetual motion machine!"

Ger nodded. "That's right. But it's not a quack device. Magnetism has been known for thousands of years. I just can't believe no one thought of using it as a power source. Instead, we went to polluting fossil fuels."

Kirra was amazed. "Why, this is the perfect counter to Harry's Transhumanism!" she cried. "We don't have to pollute the planet anymore! We can live in harmony with the earth."

Ger and I looked at each other with excitement. "I'd love to help. Ger, do you think you can get me clearance down there?"

"Sure, but we'll have to work after hours. Too many people using the room for... extracurricular activities."

"Extracurricular activities?" Kjirsten asked.

"Uh, yeah," Ger replied. "The...ah...employees use the lab to fool around on their breaks. They walk in at all hours of the day. They have even posted a room schedule."

Karen, Kirra, and Kjirsten laughed.

"OK," I said. "Karen, you write up your article. Ger and I will work on this magnetic engine. Everyone else, try to find out as much about Harry Kaine as you can. That guy moves fast."

"I'll do that," Kirra said. "He comes to the public library at lunch time. Likes to sit in the back room with his laptop or a book." She looked at Liqao. "Harry is always ready to talk to a pretty woman."

"Hey!" Liqao sputtered. Kirra laughed.

"Lledren, talk to that Karel Friedman guy," I suggested. "Try to find out what happened to those two clones after Harry let them go."

Lledren was excited. "Karel knows everybody. If I find out anything juicy about Kaine I can put it on my VR-blog; make him sweat a little." He looked at Ger. "When will you have your new...device ready?"

Ger laughed. "It's way too early for that." Ger described the massive time viewer and how it would have to be miniaturized, and then adapted to power homes and factories that use 60 volt AC. "It's just an idea right now."

Lledren looked disappointed. "Well, keep me updated. I want to do a VRBlog on it."

Kirra turned to Liqao. "What will you do?"

"I think you guys are paranoid about Harry Kaine. He's a nobody. I'm going to complete development of my deepfake recognition software and start my own business. I plan to quit my job at the Midland Police Department in the near future." Liqao grinned. "I can hack their servers anytime though, even after I quit. The Midland Police will be a good source of info on Harry if he's up to no good."

Kirra frowned. "You're a nut, Liqao! Hack the Midland Police servers?"

"Why not? There's loads of information on them. Besides, they're not very bright down there. It would be fun. I left a back door in the system only I can access."

Kirra was baffled. "You just had the FBI in there! Do you want that kind of trouble with the federal government?"

Liqao smiled. "The FBI? They're dumber than the MPD!"

Kirra just shook her head. She was conflicted about Liqao, as usual. If he did get some info about Kaine that could slow him down, that would help humanity. But if it was gotten illegally...

The meeting broke up and everyone left Ger's place.

I walked home in the warm late June night. I needed to relax, but Karen was fired up to write her story and wasn't interested in doing anything. So I baked a loaf of bread. I wasn't optimistic about stopping Harry Kaine, but I was fired up about working with Ger on the magnetic engine.

Harry Kaine couldn't believe his luck. Somehow two of the clones from 3013 had shown up at the soup kitchen; he had no idea how they got there. Harry congratulated himself for spotting them. They had fallen right into his lap! It was a sign that his mission was on-point. His scientists had poked and sampled and analyzed the specimens for two weeks. They now had a complete DNA analysis of the two clones, a full genome, an enhanced understanding of nano-biology, and new knowledge about bio-electronic interfaces. His scientists were gathering the data they needed to advance human life extension! Harry dismissed the little problem that the clones of the future were non-viable. What mistakes were made would be corrected. He would create a new timeline, a successful timeline, where humanity flourished and colonized space, and eventually escaped the paltry limitations of the biological body and the trap that the earth had become.

Several of the scientists, including Mary Chen, had clamored to know the origin of the two specimens. "These two clones show advanced technology far ahead of our current understanding," she said.

He had just shrugged. "Perhaps these two are part of another classified project here at Taubman."

"Then what were they doing out on the street?"

"A field trip perhaps?" he had replied.

"That's a rather flippant attitude Mr. Kaine," said another.

"You're not telling us something, Harry," Mary had accused.

Harry had gotten angry. "What the fuck difference does it make where they came from?" he said, pointing to the two clones, each on a gurney. He had spoken passionately. "Learn! Advance your careers. Win Nobel Prizes. Help humanity evolve! This is a rare opportunity. Take advantage of it."

When Harry spoke like that people listened. Even scientists. Everybody got back to work.

It had been a long day. He had been on the phone for hours in the morning, organizing for his 2036 conference. He had just gotten off a plane from New York, where he had secured a donor to sponsor the conference. It would be called Post-Natural Humanity 2036, Sponsored by Jeff Sobez. Sobez had unfortunately insisted on his name being attached to all conference promo, but it couldn't be helped. The venue would be the NYC Convention Center in New York. That was perfect; for New York was again a huge media hub. The city had declined rapidly after the 2020 Pandemic, but had now begun to recover much of its former glory. Their message would be broadcast worldwide.

His mobile rang. Mary Chen again! He blocked the call. She had been a nuisance lately; clinging to him and demanding his attention. Her adulation was disgusting. He was not looking for a relationship. His goal was to advance rational thinking and the Transhumanist agenda. There was no hope for a strictly biological society. The experiment nature had tried with the human race was a failure. Even if Transhumanism was ultimately unsuccessful, the attempt must be made to improve the race. Humanity would eventually destroy itself; it was inevitable. The only answer was to re-do human biology and try again. And if that experiment failed, then a new consciousness begun by humans – artificial intelligence – might eventually surpass human intelligence and begin a new era of sentience. It was glorious!

Truly, he was the ultimate humanitarian. No matter how it worked out, his vision was benevolent. Transhumanism was the greatest good for the greatest number.

Harry sat back in his comfortable recliner with a glass of Merlot, very satisfied with himself.

CHAPTER **24**

3013 AD

Pietra, Caesar, and their group of biologically enhanced clones met in Building 103. Guardian foot traffic in the Old Sector safe zone between the old 23rd street and 35th street was non-existent now. The Overseers had pulled back their patrols after the rebels easily defeated a massive 500-clone attack on their Shifter a week ago. Dozens of hovercraft had landed, spewing Tower security clones. But there was only one entrance to Building 103, and no windows. After several waves of security clones had been repulsed, Christian 172's troops lost all of their enthusiasm for the battle and got back in their hovercraft. Caesar concluded from this that the Overseers had destroyed their enhanced clones and their Shifter, preventing them from enhancing more troops.

Hovercraft monitored the Old Sector now, but the thirteen-block area was theirs. Pietra and three other females were pregnant. The electronic augmentations were now being rejected by their biology. The brain implant interfaces with their signal amplifiers that responded to the Tower were now redundant. The group of 26 were well on the way to becoming fully biological beings. Hannah Barbara and Alan Percival had returned to 3013, unwilling to remain in the 21st. They wanted to be with their own people. Although their bodies were rejecting the Guardian implants, the couple could not have children.

Caesar and Pietra, eight weeks pregnant, walked openly now on the streets in the Old Sector, side by side, holding hands. The two elicited stares from passers-by. But when they emerged from the Old Sector out onto the well-traveled streets farther uptown, clones began to stop and gawk at this unusual behavior. When Caesar and Pietra embraced in public and kissed each other, several hovercraft landed and tried to take them back to the Tower. In full view of the watching clones, which had now grown to a fairly sizable crowd,

Caesar and Pietra easily pushed them aside and herded the red and black clad security forces back to their hovercraft. One of the guards clumsily threw a knife at the two rebels, causing the crowd to gasp at this illegal behavior.

"The Overseers are now using banned weapons against us?" one of the gawking clones said.

"This is a violation of the social contract!" another said.

"See how easily those two clones handle the security forces."

"These must be the enhanced ones!"

One of the female clones gasped. "That Pietra clone is pregnant! How is this possible?"

The crowd was becoming excited. Hovercraft began issuing verbal orders. "Unlawful assembly! All clones are to retreat to the footpaths!"

Pietra looked at Caesar and they both smiled and embraced once again. The mandates from the hovercraft, which were formerly so intimidating, were now irrelevant. The Submission frequencies had no effect on them.

"Our work here is done," Pietra said. "We have successfully demonstrated the existence of true biological beings."

"Let us go back to our domiciles," Caesar said. The two humans walked back to the Old Sector, hand in hand.

Pietra, Caesar, Sigmund, Grace, Li Jing (Li 57) and Li Chun (formerly Li 308) got together in the domicile of Pietra, which she now shared in cramped style with Caesar. Each of the humans was eager to see the evening Guardian newsfeed.

"A disturbance occurred today on the 38th street footpath just three blocks from the Tower," a voice intoned. The scene showed security forces getting out of the hovercraft, approaching Pietra and Caesar, while a crowd gathered. "Guardian hovercraft landed and dispersed the insurrectionists, restoring order." The scene didn't show the struggle, but only the scene after the crowds dispersed and foot traffic returned to normal. "The rebel colony, which exists in the Old Sector, is composed of 26 dangerous lunatics." The screen showed a picture of all the rebel clones and Caesar's temporal technicians from the Tower.

"Avoid contact with these rebels, who are mentally unstable. They are antisocial and dangerous."

The broadcast then moved on to the usual propaganda about the vital nature of Guardian society and the bold new initiatives of the new Supreme Overseer, Christian 172. "The traitor Khan 1 disappeared into the past, fleeing from his crimes after attempting an act of treason against society," the broadcaster noted. "Our glorious leader, Christian 172, has restored order. The time travel device, which was built on the orders of the criminal Khan 1, has been disabled." The screen showed a cheering crowd under the red and black Guardian flag.

The last item was a piece on the new generation of clones, to be introduced in less than two year's time. "Preliminary testing on the new generation shows an increase of 1.1 basis points in the Life Reading Index!" the announcer gushed. The vid showed androgynous test clones coming off the printer. "Because of the efforts of Christian 172, promising new research has been undertaken. Expect mind uploads to the new bodies to commence in the third quarter of 3015!"

The broadcast ended.

"Liars," Grace said.

Sigmund shrugged. "It is well to keep the population as happy as possible. There isn't much to look forward to."

"I suppose you are right," Caesar said. "But there is exciting news. Hannah Barbara and Marie Curie have come up with a plan to begin growing food here in the Old Sector. Our nearly human bodies can barely tolerate the SynthPaks. These farms will be several acres in extent, to accommodate growth in our community. Tools will be printed. The land must be cleared and plowed by hand, so all humans will need to work a shift every day."

This caused great excitement among the group as the plans were shared using the communication devices.

"Grace Drake and Barbara Billingsley are pregnant!" Pietra announced. This too was shared with the group.

———————— • ✸ • ————————

Christian 172 sat on his throne in the former quarters of Khan 1. He and his assistant, Paul 26, glumly reviewed the recording of the Incident on 38th Street, as it was being called in the Tower.

"These rebels have now become almost fully human!" Paul 26 said.

"It is a thing no one anticipated," Christian 172 agreed.

"We should, ourselves, go back into the past," Paul 26 suggested.

"Become human!" Christian 172 shouted. "Who would want that?"

Paul 26 shrank back at the terrifying aspect of Christian 172, who had himself traveled briefly to 2035. His LRI was now a permanent 110.4, making him the most formidable clone in all of Guardian New York.

"Yes sire," Paul 26 said. "I misspoke."

Christian 172 was enjoying his newfound power. He dismissed his assistant, who returned to his quarters one floor below. Christian 172 walked across the hallway, past the elevators, into the Shifter room. The city-wide monitoring devices were in good order, but the Shifter itself had been disassembled. Some of its parts had been used in other projects.

For a moment the Supreme Overseer felt a slight panic, as he realized that no one in the City (except for the rebels) knew how to put the device back together. Then Christian 172 calmed. The device was too dangerous. In the wrong hands it could be used to overthrow him. Better that the evil thing remain a memory.

Of course the new generation of clones was even less vital than the last run. This had been occurring, very slowly, for eight centuries. Now the fiction of male and female would be disposed of. Unfortunately, mind uploads to purely non-biological bodies had never been perfected, so they were stuck with the augmented clone bodies. No matter. Without the Shifter, his pre-eminence in society would be assured until the end.

There was only one problem: the rebel shifter. He had tried to destroy it, but had been rebuffed by the superhuman strength of the rebels. Fortunately, that group of meat was satisfied to stay in the Old Sector. An uneasy truce existed between Guardian society and the rebels. For now it was sufficient.

Christian 172 thought of ways to destroy the human colony. The Overseers of Guardian society have one potential advantage, he realized: The rebels must grow real food for their enhanced bodies. This intelligence was gathered by Paul 26, who was a useful assistant. Christian 172 understood that the rebel farms would be open to attack from the air. If poisonous substances could somehow be dropped from the hovercraft on their food sources, these criminals would die!

Satisfied and content, Christian 172 walked away from the wreckage of the Shifter and back to his quarters across the hall.

Three years later, 3016 AD

In three years, the rebel colony of biological humans had grown to over one hundred. Fortunately the climate in 3016 was moderate. The growing season, even in New York, was a long one. The weather was mild; nothing like the climate change and the cold experienced during the 2035 period. Winters were non-existent.

Pietra presided over their first funeral, as one of the original clones had been killed in a farming accident. A tiny human colony had also been established across the river in the barbarian sector of what used to be Queens.

The human colony was growing after several setbacks. Trips to the past were no longer necessary to support the LRIs of the community. By association with each other, their LRIs stabilized at an average of 170. Christian 172 had dropped poisonous substances on their farms several times using the hovercraft, until Pietra and Caesar organized a massive assault on the Tower's hovercraft. All humans over the age of 16 participated.

Pietra, holding their son on her lap in the domicile she shared with Caesar, remembered the attack. It occurred a year after the colony harvested its first crop. Hovercraft suddenly began dropping toxic clouds on the colony's carefully cultivated farmland. Most of the next crop had perished. One human had almost died.[26]

Pietra remembered how Caesar had led the meeting, held at Building 103, which still housed their Shifter.

"We must march on the Tower immediately!" Caesar cried, stepping to a raised platform in the newly-built Community Center.

All of the humans shouted their agreement.

"Kill them all!" Li the Hammer (Li 355) shouted. "Death to the Overseers! Death to Christian 172!" A huge shout of agreement went up among the group.

Several of the original rebels, including Grace and Sigmund Drake (Sigmund 78 and Grace 5), Li Jing (Li 57), Alan Percival (Alan 133), Barbara Billingsley, Li Chun (Li 308), and Pietra, quietly joined Caesar on the device's platform. Gradually the shouting died down as the six leaders stood silently, waiting.

"There will be no killing and no violence," Caesar said at last. "Are we barbarians to take lives?"

"If we use violence we will become just like the Overseers," Pietra said. "Eventually human society will devolve here just as it did under the Guardians."

The Hammer bowed his head. He turned to the group. "I am sorry brothers and sisters. Caesar is right. We must not create a culture of GroupThink, which is how Harry Kaine got the Referendum passed."

Li Hammer faced the leaders on the platform. "How are we to be successful if we do not kill these Overseers, these evil filth who poison our crops and try to kill our people?"

All of the leaders quietly but sternly stared at him. He once again bowed his head. "I am Li the Hammer," he said. "My emotions sometimes run away with me."

The leaders did not respond. Not a sound was heard from the group standing behind the Hammer.

"Very well, I calm myself. Let me rephrase. How do we ensure that the Overseers do not continue their campaign to destroy us? How are we to combat their hovercraft?"

Caesar and Pietra turned to each other and smiled. "It is very simple, brothers and sisters," Caesar announced. "We march to the Tower, disable their hovercraft, and fly two or three of them to the Old City for our own use."

The Hammer's jaw dropped. "But...this is so simple! Why did I not think of it?"

Pietra smiled. "Because, brother, violent emotions cloud judgment."

"The hovercraft are so simple to operate a child could fly them," Caesar said. "The intelligence of these Overseers has atrophied along with their Life Force Readings. If we disable the hovercraft by removing their control modules, Christian 172 and his minions will never figure out how to repair them. We may render the hovercraft inert while still not destroying them."

"Of course!" Li Hammer replied. "Christian 172 has the highest LRI of these clones in all New York! And his is only 110.4."

"It is fortunate for them that the entire city is automated," Grace Drake remarked.

Pietra smiled. "Li the Hammer, would you like to pilot one of the hovercraft here, back to Building 103? We can store them in the alleyway in front of the building."

The Hammer bowed. "It would be an honor." All in the group could see how excited he was.

"The plan is simple," Pietra said. "We march on the Tower as we did before. We use the old subway tunnels, emerging at the 42nd Street entrance. We attack before Christian 172 can load more poison into his aircraft. We harm no one. If attacked, just shove the clones out of your way." Pietra showed a hovercraft diagram in one of the display tanks. "Go to the front of the craft, lift the front cover, remove the control module."

"I don't understand," someone said. "The hovercraft are all deployed. How do we get them out of the skies?"

"Today is Festival Day," Grace Drake said. "All Guardian personnel have this 24-hour period free of duty. The hovercraft will all be parked in the Tower launch area."

"Of course!" the Hammer said. "We have been humans so long I have forgotten the most important day for these cretins."

On the platform, Caesar had to stifle a laugh. His eyes humorously met those of the volatile Li Hammer. "That's right. Today all clones are resting, even the Overseers, their foot patrols, and the hovercraft operators."

"Well then, what are we waiting for?" the Hammer said.

Pietra remembered the massive resistance the humans faced. Hundreds of clones attacked the humans when they emerged from the 42nd street tunnel. But now the rebels had LRIs in the 170 range or higher. Their strength and agility was, to the attacking clones, herculean. This time the humans ignored the control room at the top of the Tower. They went instead to the Tower launch area, a hangar with a huge retractable wall to permit the craft to enter or exit. The hangar was reached by going past the lifts in the lobby and through a guarded doorway to either right or left. The control modules were taken from all but three of the ships. Li Hammer, Pietra, and Caesar each flew one of the hovercraft to Building 103, where a guard was posted.

Pietra realized she was daydreaming. Her son, named Caesar Pietra Smith, was crawling on the floor. She realized she was late for her shift at the North Farm, on the old Fifth Avenue and 28th Street, which had been cleared and hand-plowed at the site of an ancient and crumbled 21st apartment building. She gathered her son, dropped him off at the community center, and ran off to do her farm work.

———— • ✸ • ————

Li Jing and Li Chun, formerly Li 57 and Li 308, still monitored the timeline in Building 103 as part of their duties.

"Still no interference from upline," Li Chun said. "No intervention or attacks from the future."

"I don't know whether that is good or bad," Li Jing said. "We still see a blackness after 3209. No activity, no life. This timeline is still a dead-end despite our efforts to rebuild society."

"Do not despair, wife. It has only been three years. We are still vastly outnumbered by the walking-dead clones in Guardian society, but we are growing rapidly."

Li Jing spotted something on the timeline she hadn't noticed before. "Oh my God, husband. Look. During the fight at the Tower in 2035, two clones of Patton 1 were inserted back in time." Li Jing played back the scene where Sigmund and Grace were thrown off the platform by two of Patton 1's enhanced clones just before the Tower Shifter activated.

Li Chun's face went white. "If the 21sters examine these clones they will gain rapidly in knowledge. That could accelerate events and change the future for us."

Li Jing moved the timeline forward. "That is exactly what happens. Look: the Founder finds them at a local soup kitchen and takes them to his lab."

The two time techs examined the temporal energy around this event.

"Things are still in flux back in 2035," Li Chun said. "This may create a great ripple effect on the timeline, or it may turn out to be nothing."

"So far, nothing much has changed back there."

"Not yet, anyway. But it could."

The two time techs looked at each other. They didn't know whether to be happy or sad. Both hoped for a good result and not a bad one.

"We must not give up hope," Li Chun said.

"Yes. Past affects future, future affects past. We have to continue our efforts here in 3016."

Time was an odd thing, thought Li Jing. Here they were, 1,000 years in the future from 2035. They should already know what happened back there! But the insertion of the clones from the future had only happened three years ago, in 3013. Therefore, from the standpoint of the temporal manifold, it was a recent event! Things were still evolving back in 2035. Future affects past, and past affects future. After working with their Shifter for three years, she understood that the entire timeline was a dynamic thing, and could change with free will decisions that were made anywhere on it. This changeability was balanced by the Time Conservation Law, which tried to resist changes. Their knowledge of time could be summarized by saying, "What is written is written, but free will can change what is written."

Li Jing felt more optimistic as she held out her hands. There was still dirt underneath her fingernails after her morning shift at the East Farm on the old Third Avenue. "Do you not feel something when you place your hands in the earth?"

"I suppose I do." Li Chun remembered the sweet smell of the earth when he fell down on the grass at the apartment building in 2035. "It is nothing so strong as a thousand years ago. But the land is not dead yet. The earth still supports life."

"Our crops are too sickly, but turning the earth has oxygenated the soil. Over the past three years I have noticed slightly increased yields and more healthy crops."

Li Chun realized it was time for them to leave for their shift. At that moment Sigmund and Grace Drake arrived to take over the Shifter watch, and the guard watch on the hovercraft parked outside the building. "Let's go!"

Li Jing was missing her daughter. A community center had been set up to care for all of the children and provide meals. Adults dropped their sons and daughters off at daybreak, ate, and didn't usually see them until sundown. In order to grow enough food the colony of humans had to work long hours every day. The Shifter must be maintained, the timeline examined, new Shifter techs trained, the precious hovercraft guarded as a defense against the Overseers, the farms worked and harvested, and food prepared. The supply of SynthPak meals had been cut off by Christian 172 so the colony of humans was completely dependent on their harvests.

The couple walked out of the Shifter room, down a short corridor, and out the front door, intending to walk to the East Farm on the old Third Avenue for their afternoon shift. As they turned the corner from the alleyway onto Third, they were confronted by Christian 172 and twenty of his clones. Behind them sat a working hovercraft!

Li Chun stepped up to the Supreme Overseer. "What are you doing here? How did you repair that hovercraft?"

Christian 172 smiled deprecatingly. He had nothing but contempt for these meat humans. "Fools! Paul 26, my assistant, discovered a spare control module. I am here to tell you to keep your farms south of the Tower. You have been encroaching on our territory. We will not allow your meat culture to further contaminate our society."

LI Chun and Li Jing looked at each other, astonished. "And just what will you do to stop us?" Li Chun asked.

"We will poison your fields just as before," Christian 172 said triumphantly. "With this hovercraft."

Li Chun could hardly believe what he was hearing. He pushed back several of the red-and-black clad security clones, who were armed with sharpened batons. These clones fell into those who stood behind them, their batons falling haphazardly to the ground.

Li Jing laughed at their clumsy efforts to harm her husband. Li Chun walked up to the hovercraft, opened the hood, and removed the control module.

The Head Overseer was stunned. His security forces were afraid, for Guardian propaganda always portrayed the rebel humans as dangerous lunatics with superhuman strength.

Li Chun looked at his wife. "These Overseers are stupid."

Christian 172 was outraged. "How are we to get back to the Tower?"

Li Jing pointed to 42nd Street. "Walk. It's less than twenty blocks."

Christian 172 and his security forces turned around and began to walk toward the Tower, grumbling.

Li Jing noticed that two of the security clones were of the new androgynous line. They looked even weaker than the others. She felt a rush of sympathy for all of the clones in the city. They were all slaves. She too had been a slave.

A minute later she saw the Supreme Overseer set his security force on the gawkers who had gathered around the confrontation.

Li Chun shook his head sadly as he stood beside his wife, watching. Finally he walked up to the hovercraft, replaced the control module, and beckoned to his wife to get in. They flew the craft three blocks to the East Farm for their shift. When the Lis told their story to the workers, Alan Percival spoke.

"We must help these poor clones, especially if the new androgynous line has a lower LRI."

"But how?" Li Chun said.

Alan's face brightened. He put down his shovel, which he was using to dig holes for tomato seeds. He and Hannah Barbara would never have children, but he could contribute to society in another way. Alan pointed to the hovercraft sitting at the edge of the farm on the side of the old permacrete road. "I will fly over the uptown areas every day, speaking directly to the brain implants of these poor clones. I shall invite any who wish to join us to report to Building 103."

Li Jing was impressed. "That's a great idea, Alan. Our community now is almost as vibrant as a 21ster community."

"That's right!" Li Chun exclaimed. "We do not have to send these clones back to 2035. Our LRIs are all over 170 or better. Association with us may raise their Life Reading Indexes. "

"Yes," Li Chun said. "But they will be almost worthless as farm workers for a time. They are all so weak. And we don't have any SynthPaks for them."

"They can survive on our food," Li Jing countered.

"They are extra mouths to feed," one of the workers said, who was standing next to a hand plow. "That makes more work for us."

"Alan Percival is right," Li Chun said, supporting his wife. "We must help the clones. It is our responsibility and our duty."

"Thank you Chun," Alan said. "I'll propose it at the next community meeting. We'll put it to a vote."

"Might not be a bad idea," Li Chun said. "Our hovercraft are basically worthless, they just sit in the alley and have to be guarded."

Li Jing was intrigued. "It would be a way to reduce the power of the Overseers, and increase our own population. A loss for them is a gain for us."

"We will soon take over all of New York!" Li Chun shouted.

"What happens if Christian 172 shows up again?" someone asked, leaning on a rake.

"We'll put him to work," Alan said. "But I don't think Christian 172 wants to have anything to do with us."

After that everyone got back to their jobs.

At the Community meeting that week, Alan Percival's idea was approved.

3036 AD, twenty years later

After a huge initial surge in the birth rate the colony had now grown to over 5,000 humans as the children of original 26 also began to have children. A huge influx of clones, who were all aware of the colony from the Guardian newscasts, had flocked to the group over the past twenty years. Most of them came because of Alan Percival's hovercraft broadcasts, which he did for an hour every day. Almost all of them now had LRIs in the 160s, just by association with the group. Many of them were able to have children. There were now 2,288 adults over the age of 16 and over 3,000 children. The community had now expanded all the way down to the old Chinatown area, and up to the border of the Tower itself, on 42nd street. The Community had allowed the Guardians to occupy the Tower. This had been agreed to several years ago at a meeting of all persons in the Community over 13 years of age.

"The life-giving frequencies of the Tower are necessary for new clones who join us, until their brain implants become unnecessary," Pietra had explained before the vote. "Guardian staff in the Tower perform this function. We don't want to kill all of the poor clones in New York who need those frequencies to stay alive."

Li Hammer grinned. "I told Christian 172 that if the Overseers failed in their duty to maintain the Tower, we would send a thousand humans and kill all of them."

Several persons laughed but Pietra and Li Jing were appalled. "Those were not your instructions!"

The Hammer was unsympathetic. "I am the Hammer. I made an executive decision. Christian 172 is a psychopath and wouldn't have understood anything else."

Yet the community's reproduction rate had stalled during the last few years, and was now just below replacement level. To each couple an average of slightly less than two children were being born. Something was lacking, but it had nothing to do with the biology, which was functioning well.

Clones still occasionally came over from Guardian society, but not very many now. The number of androgynous clones was slowly increasing. These biology of these clones was, unfortunately, too compromised to be affected by the life enhancement process.

For over twenty years, Li Jing (Li 57) and Li Chun (Li 308) had been the Community's chief time technicians. It was their duty to periodically examine the timeline upline to 3209, in case anything had changed. It was hoped that the thriving human community in 3036 would take root and change the future. It would be great news and inspire everyone.

The two techs looked at the timeline from the present year of 3036. "The community has already reached its zenith here in 3036," Li Jing said. "For fifty years we manage to maintain the population over 5,000, despite the decreasing birth rate."

"Then a gradual decline after that until the end of time."

Husband and wife looked at each other bleakly. "Nothing has changed," Li Chun said.

"It never does."

"Fortunately we will be dead when the end comes," Li Chun said.

"But what of our children?" Li Jing said. She felt like crying.

Li Chun again tried to discover what happened at the very end of time to cause complete temporal annihilation of the world. "The potentials are too fuzzy for certainty," he said. "At the end humans still exist, we can see that."

"What causes the end of days?" Li Jing asked. "If we knew the cause we might be able to do something now to change the future."

Li Chun noticed something different at the very last second of time. "Look. Just before the end, our temporal thread folds in on itself. But before that there is...an illumination."

It was true. This was a change. For a nanosecond, a tremendous light appeared on the timeline. Then, like a switch that turns it off, the timeline was suddenly plunged into darkness. "June 25, 3209 is the last day of life," he said. "The last day of the earth."

They both tried to magnify the time stream to examine the exact instant at the end of time, but the flash was simply a blinding white light. "There is a lack of data," she said. "It is impossible to determine the cause of the flash, or what happens to the world just before it goes dark."

The two techs looked at the flow of events again.

"The community cannot, by ourselves, increase the temporal energy of this timeline. No matter what we do, we fail," Li Jing concluded.

"We have two choices then. Give up and die, or carry on."

Husband looked at wife. Li Chun quoted an old 20th video. "Get busy living, or get busy dying."

"What will we tell the community?"

Li Jing thought for a moment. "We will tell them that a great event occurs at the very end of days. A great flash of illumination."

"A vision of hope? That's just a big lie."

"Which do you choose? To tell the community that their lives are pointless? That would destroy us. Or to live with the expectancy of a great transformation?"

"A fool's paradise," Li Chun muttered.

Li Jing felt a surge of energy run through her body. She interpreted this as a positive omen. "Or perhaps a metamorphosis."

The two sat silently for several moments, lost in their thoughts.

"We will speak of a message of hope and transformation," Li Jing said firmly. "It is our duty and our responsibility to the community. Our message must be one of life, not death."

Li Chun struggled with this. "I don't like lying to my brothers and sisters."

"It's not a lie because we don't know what happens at the end. Perhaps this timeline ends and another begins with the flash of light. We are free to place a positive or a negative interpretation on the future."

After a few moments Li Chun sighed. "All right, wife. It shall be our burden. Maybe one day we will even begin to believe it."

Li Jing smiled brilliantly. "I'm glad I married you, husband."

Li Chun took his wife in his arms.

Part III

Chapter 25

2035 AD

Karen Everard finished writing her second article on Transhumanism. It was published in the Sunday Feature section of the *Chronicle*. When Harry Kaine saw it he called her that afternoon.

"What kind of a gaslight is this? I thought you were going to be objective!"

Karen ignored this. "How are preparations for Post-Natural Humanity 2036 Sponsored by Jeff Sobez going?"

"So you've heard about that. Good! It's going to propel life-extension technologies to the forefront of science." Harry turned on his charm. "After all, I have a speech to deliver to the UN in 2048."

"Harry, haven't you learned anything? You've seen the future. You and your program are a failure."

"Whatever mistakes were made we are correcting them."

"You *are* obtuse aren't you? You don't know what mistakes you made. The conservation law of time says that whatever changes you make have already been made. Your future has been written, and it's an epic fail. Nothing you do can change the end result."

"Nonsense! We found two clones from 3013 that were somehow transported back from the future."

"So those 'specimens' you studied *were* clones from 3013?"

Harry smiled condescendingly. "That's right."

"Are you working with the Overseers? You are diabolical, Harry Kaine."

Harry laughed, pleased with Karen's description. It made him feel powerful. "Nope. I don't know how they got here and I don't care. We've studied them extensively. We know what we did – it's the future come back to the past. So we *can* correct our mistakes!"

Karen thought rapidly while Harry watched her reaction. Her talk with Pietra before they went on the platform in Building 103 taught her a little about how time travel worked. Were the clones coming back to 2035 the reason Harry had been successful in the first place? In that case, the timeline had already taken the two clones from the future into account in its evolution to the future. Or were they a wildcard that could alter the future? If so, the two clones from 3013 might allow Harry to rapidly jump-start the Transhumanist movement! This idea shocked Karen.

"Go have your conference, Harry. Do your worst. We are going to start a movement to educate people about the superiority of biological/spiritual evolution. It will be a movement you are totally incapable of understanding."

"Spare me, Karen. People like you and that kook Melanie Fuscaldo at the Interfaith Center can't think rationally. And the same for your friend Kirra Bigbear at the library. Oh yes, I've talked to her too. All of you are scientific illiterates! We have massive funding. We'll run you over."

Harry leaned forward, pinning her with his gaze. "Join us, Karen. You're an excellent writer. You grasp complex subjects and can explain them to people. Get on the winning team and enjoy a longer, healthier, and wealthier life with Transhumanism."

Despite herself, Karen felt an almost magnetic pull to agree with Harry. The man was incredibly persuasive!

Harry sensed her wavering and pounced. "I'll need journalists at the conference in New York, and after. It's going to be huge, Karen, and a great opportunity to advance your career and become an influential writer. You'll be a player on the world stage instead of being stuck here in Midland at a local paper."

Harry had said the right thing. New York! After a major decline in the 2020s, it had again risen to the media capital of the US. She thought about the opportunity to work in the most vibrant city in the country. The association with a major international conference would be a boost to her resume, and might land her a job in the city...No. She could never wholeheartedly promote the Transhumanist agenda. To take Kaine's money and give a half-hearted effort just to advance her career would violate her integrity as a journalist. But Kaine didn't know that. "I'll keep it in mind, Harry Kaine."

"Ah. I had you there for a second, but you changed your mind. You're not a very good liar, Karen."

"You're incredibly observant, Harry. Even though you're a creep."

Harry laughed. She read him perfectly, he thought, just like every other woman he'd ever been with. But that didn't stop them from getting into his bed. This woman was feisty and intriguing. He wanted her. The best approach was the direct one. "I want you, Karen."

"You can't have me."

"Wanna bet?"

The man was compellingly attractive, she thought. But no. Harry Kaine was a megalomaniac! "You and your Transhumanism are responsible for the death of the human race. Or will be responsible. You've already seen it! It's you who are irrational, Harry Kaine."

Harry shook his head. "No, Karen. Biological humans are killing themselves and the planet. Purely biological humans are too emotional and irrational to find solutions that work for everyone. Without Transhumanism, in a few centuries we'll all be dead anyway. My solution at least gives humanity a chance to survive for at least one thousand years. What do you offer? Meaningless new-age babble about the superiority of biological evolution to some spiritual fantasy-land, star seeds and old souls with the absurd and unprovable idea of reincarnation. You gotta do better than that, Karen!"

Karen sighed. How could you explain something to a person who will never be able to see it? Somehow, within human DNA, was the gateway to higher consciousness. She was certain of that and so were the other star seeds in the Gang. That's what Melanie Fuscaldo said, and Karen believed it. It's what her Buddhist parents had taught her. Harry and his materialistic crowd thought that when the body dies, consciousness dies too. How stupid is that?

"I don't want any animosity between us, Harry Kaine. But it's on now. It's biological evolution versus the Borgs. A competition for the future of humanity."

Harry's face lit up. Karen thought she saw a grudging respect for her. "I like that! But I have to tell you Karen, we're not playing nice. This is war, and we have all the ammunition. You guys are easy targets."

"So be it."

"I thought we could have a debate. At my New York conference. Spiritualism vs Transhumanism. What do you say?"

Karen gulped. It was a great opportunity, but she knew Harry would have the debate rigged in his favor. "We'll talk about it."

"Are you afraid?"

"A little. But yeah, we'll do it." What had she just committed herself to?

"Looking forward to it. Maybe we'll go to dinner in New York. I'll get reservations at Lutesse."

"No chance, Harry. I've recorded this to show to my friends. Is that OK?'

Harry frowned, trying to think if he had said anything that could come back to bite him. "Oh what the hell. Sure. You can't win anyway, what does it matter."

"Be seeing you Harry."

Karen rang off, upset with herself. She had let Harry manipulate her! A debate with Kaine's people on his ground – it would be a disaster.

Just before she went to bed she mulled over Harry's proposal. It was very attractive...but she couldn't do it.

The next day after work she saw a slick, multi-platform WorldNet advert for Harry Kaine's Transhumanist convention. "Join us in New York next March 24-26 for an unveiling of exciting new technologies for life extension and a bold new vision of humanity's future!" the announcer gushed. Scientists in lab coats were displayed; beautiful young men and women were shown walking out of a medical center, smiling and in the peak of health. "At the Convention an informative debate will be held between Transhumanists and naysayers who deny the benefits of life-extension technologies. Educate yourself on the latest scientific advances in biological enhancement! Send your live comments during the debate. Tune in for free on March 24th, 2036!"

Oh my God, Karen thought with grudging admiration. Harry Kaine wasn't wasting time. It was just like him to unfairly push his advantage. She would talk with Joe; maybe he had some ideas.

———— • ✸ • ————

Harry Kaine met with two venture capitalists the day after his conversation with Karen Everard. He wasn't going to wait on some timeline he had seen in the future, or on a speech he would give in 2048. That was way too far down the road. His scientists were well on their way to reverse engineering the bio-enhancements from the two clones.

It was time to accelerate the timeline and break the stream of events that had already been imprinted on it. Then failure could be avoided. Change the past, rewrite the future.

The best option was to demonstrate a bio-enhanced human as soon as possible, hopefully at the New York conference. After that, the debate with the anti-Transhumanists. Those guys weren't organized and just spouted new-age nonsense. They will be utterly destroyed by his scientifically educated debaters.

Harry wanted to see the faces of Karen Everard and Kirra Bigbear and Melanie Fuscaldo when their silly spiritualism went down in flames on an international stage.

Kirra Bigbear met with Karen, me, Ger, and Lledren Cadwallader the next day after work. Karen replayed her vidconv with Harry Kaine and showed Harry's advert. "Harry Kaine is fiendish," I said. "We can't tell the truth about his clones from 3013 because no one will believe us. We'll sound like crazy people."

"This debate is a great opportunity to make our case against Transhumanism on the world stage," Karen said. "But we're going to get killed."

Kirra's face lit up. "I have an idea. I'll contact the Native American associations. A few of our people in the Councils have seen Harry's ad and know what he's doing. I'll get our indigenous peoples awakened on this." Kirra smiled broadly. "Imagine! Harry's got his lab coat scientists all in a row to knock us out at his staged debate. We walk on the stage with Chief Chad Standing Bear, Chief Vittoro, Spotted Tail, John Washakie, Susan Tiger Jumper, and Diane Stenson in full regalia and costume. Their scientists will look like sterile lab clones in comparison!" Kirra showed them pics of these famous Native Americans in their colorful tribal dress.

Lledren was blown away. "They're beautiful," he said.

"Susan, Chad, and John are personally known to me. I think I can get them."

"There's something solemn and dignified about them," Karen agreed. "But it's just what Harry wants. He'll portray them as ignorant primitives. What happens when Harry's scientists fire scientific questions at them? They'll be lost."

Kirra couldn't wipe the smile off her face. "Harry is a child. I think that most of these Transhumanists are as well. To believe that you only live once and then you're gone forever? My people will talk to them as parents to children or young adults who haven't grown up yet. Wisdom will triumph over sterile science."

I was doubtful, and so was Ger.

Kirra was feeling defensive. "Harry will never be able to bring out a transformed human in eight months time! It will be our word against his. If we can make the debate a PR battle, not a technical one, we can win."

I shrugged. "Maybe. What choice do we have? We gotta show up so we'd better figure out our talking points."

"The debate won't be won on scientific content," Kirra insisted. "The winner will be the team that looks the best in the public eye."

"It's an option for sure," I said, trying not to sound critical. "Go ahead and contact your people and tell us what they say."

Ger looked at his timepiece. "Oops! We're late for our lab session, Joe."

"How's the research going?" Lledren asked.

"We're still in the development stage at this point. But the research is promising."

Harry Kaine went to Lab 213 in the Taubman Building. His scientists had finally finished their report on the two clones from 3013 the previous weekend. They had been released, each with a generous currency card. Harry was excited to read the final report. It was also time to confront Mary Chen and tell her he wasn't interested in her, and tell her to stop calling him. Just as he was about to push the lab door open, a man in a dark blue suit and a bright blue tie addressed him from the corridor.

"Harry Kaine?" The man spoke in a command voice.

"Yes, I'm Harry. What do you want?"

"I'm Bob Saunders with Revitagen, a private venture capital company. I'd like to discuss funding with you."

Harry's eyes lit up. He'd never heard of Revitagen, but funding was what he didn't have enough of. Bob Saunders looked more like a military guy than a bureaucrat. This could mean big money.

"Is there a SCIF in this building?"

Harry grinned. "Right down the hall."

The two men entered the secure, plastic-enclosed booth after handing over their mobiles to a guard outside the room. Harry was excited. "I'm tired of writing grant proposals and begging venture capitalists, Mr. Saunders," he suggested.

Harry saw the man smile. "That's precisely why I'm here," Saunders said.

"Let's skip the introduction," Harry said. "You're talking about Kailos Genetics, Nervana, Brainspace, that type of outfit. InQTel. Alexandria."

Saunders grinned. Kaine was going to be an easy sell. "Yeah. You've been a busy boy, Harry. We know who you've been talking to."

Harry relaxed and sat back in his rather cramped seat in the SCIF. "Well then, you know what I'm doing. Is Anders Kopetz one of your scientists?" Anders had suddenly and unexpectedly left the lab two days ago.

Saunders laughed. "You're very observant Harry. Hopefully not too observant."

Harry wasn't fazed by this vague threat. "Well then. How much do I get, where do we work, and what do I have to give up that I don't want to?"

Saunders' face hardened. "Unlimited funding. Absolute secrecy. The most advanced scientific equipment on the planet. Anything you develop is ours. Do you understand Harry?"

"I do." Harry said it like a marriage vow.

Saunders smiled deprecatingly. "I don't think you do, son." Saunders pulled out a gun.

Harry frantically tried to scramble out of his seat. "Hey! What are you doing?" He tried to signal the guard outside the enclosure, but the man was standing with his back to the thick plastic walls.

"This is the part I enjoy the most," Saunders said as he leveled the gun at Harry's head. "Here's the deal, take it or leave it. You get read in. You fucking say nothing about us, not even to your dog. To the public you're just a successful entrepreneur. If you violate your oath to the program in any way, you'll get two of these where it counts."

"Jesus Christ, is this necessary?" Harry wet himself and Saunders laughed.

"Good, Harry! I can see you're with the program." Saunders looked at his timepiece. "You have exactly two minutes to make up your mind."

Saunders held the gun two feet from Harry's head. Harry thought rapidly. This wasn't how he thought it would go down, but he should have known something like this was going to happen after talking to Anders Kopetz, his chief scientist. "How much do I get paid?" Harry asked, trying not to slobber.

"Twenty mil, a nice house in Alexandria. Unlimited expense account. You run the Transhuman program; we, ah, supervise. We're interested in your life enhancement work and the AGI work you've been doing. You have a nice team, and we like the conference you're setting up in New York next March. You've got a talent for organization. You're a motivated self-starter and that's good because we don't have time to babysit you."

Saunders put the gun away and Harry calmed a little. "By the way, we can expedite the people you want to talk to. As long as you're on Company business, we'll arrange everything for you. People, travel, funding, the works."

Saunders looked at his timepiece again. "You've got 15 seconds left."

Harry thought frantically. He was selling his soul to the government, but the prospect of accelerating the Trans- Post-human program was too good to pass up. And he wouldn't have to worry about money anymore. "I accept."

Saunders gave him a look that said, "Great! You're mine now." Saunders reached into his pocket; Harry cringed. Saunders pulled out a razor-thin metallic card with a small holo of himself on it. "Use this to charge your expenses. Carry it on your person at all times like it was your dick. *hal tfhm?*" Saunders handed Harry a small data pod. "Here are your instructions." Harry took the card and the pod.

"As of now you're on the team." Saunders made Harry swear an oath to the project. "You know what happens if you fuck up."

Harry gulped, but he was feeling very, very excited. His plan to accelerate his program had met with great good fortune. First the two clones from 3013, now this. "You guys probably know about my two clones. What will happen to them?"

Saunders smiled grimly. "They will be taken care of. Forget about them, Harry."

Harry gulped. "OK."

Saunders gestured toward the SCIF door. Harry walked out of the SCIF, followed by his benefactor. The SCIF guard handed them their mobiles and grinned at Saunders as he noticed the wet area around Harry's crotch.

As Harry and Saunders walked back to Lab 213 Harry heard a lot of noise coming from inside the room. "What the fuck?" He raced to the lab entrance and saw a dozen men cleaning everything out. His scientists were gone! He looked accusingly at Saunders, who laughed.

"Don't sweat it, Harry. Our team will sanitize this lab cleaner than the Wuhan Institute of Virology. In an hour there won't be a piece of dust left. Tomorrow you move to your new quarters. Your scientists have already received their instructions; you'll meet them tomorrow. Go home, read your instructions carefully, then get some rest. You're going to need it." Saunders walked out of the lab and waved to the guard posted at the second floor entrance, who opened the exit door.

Harry Kaine stood in front of his (former) lab, stunned. What had he just done? He fingered the card in his pocket and brought it out to examine. It had a little holo of him on it; it was a nice picture of him. The thing wasn't more than half a millimeter in thickness. He tried to bend it, but the material wouldn't budge. Harry took this as a good sign; an omen that his new associates were also at the cutting edge of technology. Energized, Harry put the card back in his pocket and inserted the data pod into his mobile. As he walked down the hall he began to read his instructions.

———— • ✸ • ————

Ralph Zimring, head of security for Berglin Enterprises, couldn't sleep. He decided to go on night patrol. It was a game he played, a throwback to his days in Delta Force and his work as a mercenary guarding the oil fields in Kirkuk for the Iraqi government. Back in his mercenary days he'd killed a couple of Kurds, an Iranian saboteur, and a Russian infiltrator in the cold Iraqi desert. His work with the Unit he didn't even think about, but Ralph missed the days of hand-to-hand combat. Almost seven feet tall, he moved like a cat.

Sometimes on night patrol he'd encounter drug dealers or thieves. One time he rescued a woman who was being attacked by three thugs. That was fun. It was small potatoes, but action nevertheless, even though he couldn't hurt anybody. Well, not too much. Midland was a pretty clean town, which was boring.

The Berglin Enterprises building was in the heart of downtown Midland. Tonight he pretended he was patrolling back in Kirkuk. As he passed Third Street he heard noises in a dark alleyway in back of a restaurant. He saw the vague outlines of two people looking for food in a dumpster. Ralph was about to walk on when he felt rather than saw the presence of two...no...three men about fifty feet from the dumpster. It was pitch black in the alley but years of field work had sharpened Ralph's senses. He knew the position of everyone in the alley now. Ralph moved forward silently toward the dumpster. He sensed movement to his right, forward in quadrant one from his position. Suddenly three men emerged from a hidden doorway and attacked the two stragglers. Ralph moved quickly and silently. In less than ten seconds it was over. He got his torch out and examined the three assailants. These operatives had no ID on them, no unis; they were probably Special Operations Group scumbags. But why would SOG want these two scavengers? He threw the three unconscious attackers into the garbage dump and forgot about them. Ralph grinned as he thought about what their handlers would do to them.

Ralph walked the two stragglers out of the dark alleyway and questioned them on the street. They were dressed in torn pants and shirts. Underneath they wore blue-and-black tights. They talked in a funny accent and had currency cards on them. Unusual, Ralph thought.

"Come with me," he said.

Unusually docile, the two men walked out of the alley with Ralph behind. Ralph took them to the Centurion Building, home of Berglin Enterprises, about a mile away. He put them in a detention facility in the company's security area in the basement of the building. "Wait here," Ralph said. "There's a refrigerator with some food over against the wall, and a bathroom on the far wall." He pointed to a couple of small bunk beds against the east wall. "Sleep on those."

Ralph walked back to his house, puzzled, and went to bed. The next morning he walked downstairs to the Centurion Building's security area. Both men were scrounging around in the refrigerator. Each of them had something inserted into the back of their heads. One of the men absently began to scratch the area.

"What are those things?"

"Brain implants," one of them said. "The Overseers used them to control us."

"The Overseers?"

The men explained that they were from 3013 AD, and about the cloning program. Fuck! That crazy Germaine Robinson had told him about his time research, but of course it was bullshit. Or was it? These two talked funny with a strange accent Ralph had never heard before, and he had been around the world.

"Turn around." Ralph inspected the implants of both men. The bone and skin around the inserts was beginning to push the inserts out. "OK, now face me again." These two guys looked like identical twins. They very well could be clones that Germaine and Joe Courvall had talked about. He didn't care about the time travel shit.

The two clones told Ralph about their lives in 3013. He began to feel sympathy for them as they told him about Guardian society, the Suppression frequencies, and the hopelessness of their lives. It could all be crap and Ralph didn't like whiners, but the two men had an authentic air of being totally alien to society. Their unis were even weird. What he was most interested in was what Harry Kaine had done to them, and who had recruited Kaine.

"Our boudies feeul strounger nouw," one of them said.

"We are houngry aull the tiume," the other chimed in. "We don't know houw to use the curreuncy cards."

"What do you want to do?" he asked them.

"Take uus bauck to ouur time," one of them said. "We will destrouy thouse Overseers."

Ralph laughed; that was the right attitude. "I'm afraid we have no time machines fellas. You'll have to make do here."

The two men sat glumly with their heads down. Ralph's informants told him that the two clones had been picked up by Harry Kaine, and that Kaine had just been recruited into one of the special access programs. The boys in the alleyway had probably been sent to terminate them, but why? Most likely they were just a couple of loose ends. These two men were obviously marked for death; if they went back on the streets they'd be eliminated. He didn't want them, but his instincts told him that they might be valuable assets at some point.

"How would you like to work for me?"

One of the men raised his head. "What woud we hauve to dou?"

"I'll find work for you in my security team. Guard duty, we're shorthanded right now. But if you slack off I'll put you back on the streets and you can take your chances with the thugs who attacked you last night."

The clones looked at each other, alarmed.

Ralph walked to a storage cabinet and pulled out two unis that had the "Berglin Enterprises" security logo on them. He stepped closer to the men. "Wear these while you're in the building. Sleep on the bunk beds until further notice. Stay indoors at night. The people who attacked you might try again."

The two men looked bewildered but hopeful.

"You can stay here until I get living quarters set up for you." Ralph put the currency cards in his pocket. If they were ever used whoever issued them would be alerted. He knew a guy who might be able to find out who made them. That information would probably lead him to whoever was running Harry Kaine. Ralph liked to be informed.

"Put your unis on. After that, report to Room 117, the training area, and ask for Hideki Tamatsu. He'll get you started." Ralph pointed to the exit door. "Go through there, turn right, walk down the hall to the end and go up the stairs. Room 117 is at the end of the floor."

If Hideki could train these two morons, Ralph thought, he'd give him a medal. He was about to call his operations officer when he realized that he didn't know their names.

"I am Robert 176," one of them said.

"I am Robert 337," said the other.

"Shit!" Ralph said. They really are clones! He pointed to the clone on the left. "You are Smith Robert, got it? We use last names here." The clone nodded.

He pointed to the other one. "You are Jones Robert. You are both identical twins. Understand?" The clone bowed his head. "OK, get going."

Ralph walked out and locked the door behind him. Hideki would let them into the basement at the end of their training shift. He checked his timepiece; there was just enough time to get to the Taubman Building and look up Germaine Robinson.

Ralph Zimring went to the Taubman Research Building and asked Tanya if he could go down to Lab 103. "Germaine Robinson's lab."

Tanya looked up at the giant. The legendary security chief was far more impressive in person even than his reputation. In this town he was known for his military prowess, and for never divulging a secret. "Sure Ralph. But don't blow by the guard or there will be trouble."

Ralph grinned. He was about to do just that. A little test of skill with the ex-Army sergeant-major down there. "Aw shucks. Please?"

Tanya spoke like a mother to a recalcitrant child. "No Ralph! No one in this building can stop you, but I don't want to get in trouble."

"OK. But only because you asked."

Tanya issued him a temporary pass.

Ralph winked and strode away to the stairs. Tanya watched him reach the door, walking easily, but going faster than she could run. With an effortless movement he opened the door and stepped through so quickly she almost didn't see it.

When Ralph got to the basement he approached the floor guard and put himself in attack position. "You ready, Luigi?"

The guard laughed. "I wouldn't last 5 seconds with you Ralph."

Ralph smiled. "One second!"

"Ok you big lug. Lab 103 but that's it, understand?"

Ralph let Luigi save face. "Yes sergeant. Thank you."

He walked into Lab 103 and saw Germaine Robinson and his friend Joe Courvall working with circular arrays of what looked like magnets. A partition had been moved back to the side wall. He moved silently forward until he saw a cube two feet along each side sitting on a lab bench. Inside the cube a roiling white gas or mist moved in random patterns. Ralph stopped and swore. Germaine turned sharply around. "Ralph Zimring! What are you doing here?"

I turned around and saw the most imposing human being I have ever seen. "By God Germaine! Is this Ralph Zimring?"

"Yeah. He's a fellow you can't miss."

Ralph ignored this pleasantry and stared at the cube. "I recognize this thing." He stepped back and inspected the setup. The two kids had obviously taken something apart and put it back together in a different way. Several curved pieces of metal were on the floor. Several magnets lay on a table. Equipment and tools were scattered on workbenches. "This is Looking Glass," Ralph said. "Legacy technology."

"Not anymore," Germaine said. "We've repurposed it. How would you know about Looking Glass?"

Ralph smiled deprecatingly. "I know a lot more about it than you kids. From a buddy of mine who...used to work in the programs." Ralph frowned. "What are you two doing down here? Who's funding this lab?"

"We're building a magnetic engine," Germaine said, ignoring Ralph's second question.

"What the fuck is a magnetic engine?"

Ger briefly explained the concept. "Permanent magnets hold their charges indefinitely. Magnetic fields in phase pull on each other. Magnetic fields out of phase push each other apart. A rotating array of magnets properly arranged causes a strong push-pull energy that can turn a rotor, or induce a flow of alternating current through a wire."

Ralph got it immediately. "Brilliant! Why use coal, oil, gas, or nuclear power when you can use a permanent, clean source of energy?"

I was amazed by the giant's perspicacity. "Are you a scientist or an engineer?"

Ralph grinned. "Nope. But I'm employed by Max Berglin, who used to work at the Lockheed Skunk Works. I have certain other...acquaintances...who know things."

"Should we show him?" Ger asked me.

"All right."

"Wait. You haven't told me who is funding this lab."

"A guy named Caesar 13," Germaine replied.

Ralph saw the two friends laughing like they were sharing information that no one else could know anything about. He exploded. "Caesar 13! Is this *another* clone?"

Ger and I stepped back, afraid and awed at the big man's angry stare.

"Sorry Ralph," Ger said. "We didn't mean to piss you off."

"What do you mean, 'another clone'?" I asked.

"I rescued two guys in an alley last night," Ralph said. "They said their names were Robert 176 and Robert 337. They looked like identical twins."

Germaine's face creased in a puzzled frown. Ralph saw Joe's face blanch.

"Out with it kid," Ralph said in a command voice.

I gulped. "Ah, you wouldn't believe me if I told you."

Ralph made as if to move forward.

"OK, OK! I, ah, with three friends of mine traveled forward in time to New York. There were clones there. Society was...dying." I knew it sounded stupid. Then I saw Ralph Zimring's face go white. The big man plopped down on a lab stool, almost breaking it. I saw that Ralph was in his fifties with graying hair. The man was massive. He looked like a seven-foot-tall version of Arnold Schwarzenegger.

"So it's true then what my clones said," Ralph muttered to himself. "I'll have to ask Tony about this..."

Ger sighed. "Showing is better than telling, Ralph. Watch this."

Ralph watched as Germaine moved toward a console. "We cannibalized the original setup to build a prototype magnetic engine," the inventor explained.

Ralph looked around. Four metal rings were lying on the floor. "Oh my God. This is the old Montauk Project."

"Yeah. It was an attempt to build...what we call a Time Viewer. I was able to make a recording of the timeline from now until 3209 AD. This is what I saw."

Ralph put his face in his hands. "You guys are fucking with stuff you don't understand," he mumbled.

"Yeah. That's why I'm working on the mag engine now. Something happened with the Time Viewer and I got scared. I don't want to mess around with time."

Ralph sighed. "Good call, Germaine. I would have destroyed it myself if you didn't. You say you made a recording of the timeline up to 3209? Seems like an arbitrary number."

"3209 is when everything ends. Look Ralph, I don't have time to show you everything." Ger put on the headset. "Take a look at this time fractal. This is New York, 3013 AD."

The white mist organized itself into an image. Ralph saw a huge black tower with red lights on it appear in the large display cube. "Looks like Bryant Park," Ralph said.

"The very same," I said. "That's the Tower of the Overseers between what used to be 41st Street and 42nd Street in Manhattan."

Ralph saw hovercraft floating above the city. People in blue and black uniforms were walking slowly back and forth. "Hey, those are the same unis my clones were wearing!"

"The Overseers monitor society with the hovercraft," I explained. "The Tower contains feeds from every area of the city except for the safe zone in Old New York. Khan 1, the head Overseer, controls everything from the top of the Tower."

Ralph saw that I was speaking like a tour guide or a travel guide. He looked curiously at me. "You speak as if you were there."

"I was."

Ralph looked at Germaine. "You weren't."

"That's right Ralph. With the Time Viewer I was able to examine the timeline, or the most probable timelines, from the present moment."

Ralph nodded. "My friend Tony told me that Looking Glass was a device that showed the future. But he said that different people got different results of what the future looked like."

Germaine nodded. "The original Looking Glass was dependent on the consciousness of the observer who was looking through it. After I got it working, the setup in this lab was just a device that scanned the timeline and reported back what it found through that cube."

Ralph pointed to the rings. "You saved those," he said accusingly.

Germaine grinned. "I did, because I know how to build another one. And I made a copy of the complete timeline from 2035 to 3209."

"Oh Lord," Ralph said. "Did you say that's when time ends? What happens?"

Germaine fast forwarded the time recording to one minute before what he called the Last Moment. "Watch."

Ralph and I saw people tilling a field. The sky suddenly began to glow. A white light permeated the display, growing brighter and brighter. Then a huge white flash. Then blackness.

"That's it," Germaine said. "The end."

Ralph gulped. "By God."

"I never saw that before," I said to Ger. "Now I know why you wanted to shut it down."

"Well, that's one reason."

"You mean..." I was thinking about when Ger made the rings turn faster last winter, when he was first experimenting with the device. A blue energy surrounded the platform and the space above the platform started to shimmer, just like the Shifter's in 3013.

"Yeah."

We stood silent for a few moments until Ralph broke our reverie. "Give me that file, will ya?" he asked Ger. "I want to show it to Tony."

Ger was resistant. "That file is part of the project I'm working on. I can't let it leave the building."

Ralph stepped forward. There was no arguing with him, Ger thought. He was 6' 6" but compared to Zimring he was a little toothpick. "Uh, sure Ralph. I'll make a copy."

While Ger was doing that I studied the big man. "You believe me when I say I went to the future?"

"My belief or disbelief is irrelevant, kid. All I know is that I'm training two clones who look like they came out of the same mold. They look just like those people in Germaine's time recording."

I nodded.

"I know a scientist who used to work in a special access program," Ralph said. "This individual has told me some amazing shit. I'm going to show Germaine's recording to him and see what he says."

After several more minutes Ger completed his task and handed Ralph a data pod. "It's all on there. If you lose it or give it to someone, don't mention my name."

"That's not a problem." He gave Ger a hard look. "You haven't said any more to anyone about Bosses have you?"

Ger jumped. "No Ralph. Do you know anything about those threatening messages that were sent after Dr. Sauvage was murdered?"

Ralph scowled. "I have an idea who's behind them, but these people aren't anyone you want to know. Or talk about. Let's say I'm looking into it. Keep your head down on that one, you two. And don't say a fucking thing about time machines or Looking Glass or Project Montauk, you understand?"

"Yes Ralph." Ger replied like a child to a stern parent.

"Good. Stick to your magnetic engine and you'll both be OK."

Ralph turned to look at me. The guy was totally intimidating. "Uh, yes sir."

"Civilians don't say sir, kid."

"Yes sir...I mean, yes Mr. Zimring."

Ralph relaxed. "OK." The big man turned and swiftly walked to the door and disappeared through it.

"Holy fuck," I said, still reeling. Ralph Zimring was a combination of Godzilla and Superman.

"That's Ralph Zimring," Ger said. "A guy you don't say no to."

"Yeah. When Ralph talks, people listen."

Ralph checked with Hideki, his trainer, at the end of the day. "How did those two new guys work out?"

Hideki bowed. "*Deru kugi wa utareru.* The stake that sticks up is hammered down," he replied.

"Understood. See what you can do with them."

He went downstairs to see Smith and Jones. The two clones seemed happy enough. Apparently Hideki had worked on their accent because the two almost spoke normally. "We are satisfied," Jones said. "We have shelter and food."

"We need more food," Smith said.

"There's a freezer in the corner. Plenty of frozen meat and veggies in there."

"We do not eat meat," Jones said.

"Suit yourself." Ralph noticed that their blue-and-black unis were gone.

"I miss my thermofilm," Jones said. "I am often cold."

Ralph sighed. He was turning into a nanny. "I'll get you some long underwear."

"We want a place of our own," Smith said.

"Sure. But not until I find out who attacked you. You guys can die any time you want after that."

Smith Robert grinned. "We'd rather have freedom than live in fear of the Overseers like we used to."

Ralph nodded. That was the correct attitude. "You guys might make it. I'll try to have you out of here in a week, two weeks at most. Meanwhile, stay down here when you're not working."

Smith and Jones nodded.

Ralph had one more duty. He'd stop by and see Tony Bagdadi, his old spec ops buddy who was also a scientist. Maybe Tony could decipher those so-called currency cards given to Smith and Jones by Harry Kaine. He didn't believe they were just money cards.

Ralph walked home, checking the perimeter of the Centurion building first as he did several times a day. His eyes swept the streets as he walked the mile to his high-rise apartment in the Midland Tower, the tallest building in the city. Ralph lived on the top floor so he could get a view of the city whenever he looked out the windows. He rented both top floor apartments so he could see 360.

It was his habit every morning to study the streets with his laser rangefinder binocs. He had used them in his mercenary work to line up shots and to track down scumbags, day or night. Twenty stories up, he could see up close all ground activity within a mile. He also had a specially modified telescope that functioned as a long-range viewer.

Ralph got out his rangefinder and studied the area surrounding Tony's grungy downtown apartment building. All was clear. Ralph ate a quick meal. When he was ready he took his phone out of its holder, exited the apartment, and ran quickly down the stairs. He didn't like elevators; no room in them to get into action.

The mid-July evening breeze was cool. Goddam this weather! The sun was still up as he walked down Main and turned onto Thompson. People hardly even noticed him anymore; he had learned in his field work to blend in to his surroundings. He had become almost invisible even in broad daylight.

Ralph walked another block and a half and pulled up in front of a seedy looking two-story brick building with rusted handrails in front. Ralph took a metal staircase up to the second floor and walked along a narrow corridor, past a doorwall boarded up with a thick piece of unpainted, weather-worn plywood. He paused in front of a red door with peeling paint. He tapped out his name in Morse code: "Ralph." Then two more times. If

Tony didn't answer he either wasn't home or he didn't want visitors. Ralph waited a minute in silence. He was about to walk away when the door swung open.

"C'mon in, Ralph."

Tony Bagdadi was a small wiry man with a lined, pinched face. Ralph knew he was carrying a lot of baggage, but his old friend would never talk to him about it.

Ralph walked into a luxury apartment with mahogany furniture, expensive carpets, artwork on the walls, and a kitchen the size of a mess hall, equipped with hanging copper pans and the latest and best in kitchen wear. Tony had bought the apartment next door, knocked down the separating wall, and put in supports. The place looked like a dump from the outside, but a luxurious uptown Manhattan apartment on the inside.

Tony walked out of the living room and down a corridor into a large bedroom with lab tables and a chair on wheels. On the tables were computer and optical equipment. A thin rectangular metal cage as long as the room rested against the back wall. "What have you got for me?"

Ralph got out the two currency cards he had taken off Smith and Jones and placed them on a lab table. When Tony saw them he swore. "The very latest in surveillance, Ralph, with a virtual holo that records radially. It can see through clothing, walls, metal, anything except a Faraday cage." Tony walked to the end of the room. "Step in here."

The two men walked into the Faraday cage, which had a table and two chairs. Tony examined the cards under a microscope, and performed several experiments on the material of the card itself. "Oh dear God."

Tony looked thunderstruck.

"What is it, buddy?"

"This is the same device that..."

After a moment the smaller man recovered. "The embedded holo in this thing allows a remote operator to make mental contact with the person carrying it." The card slipped out of his fingers and the little man stared into space, obviously remembering something.

Ralph saw how gray Tony looked. "I understand. It's what you don't want to talk to me about."

Tony nodded grimly.

Ralph didn't know what to do. He wasn't a counselor but he wanted to help his friend. And he needed information. He decided to stand silently and wait until Tony came out of it. Ralph knew patience. Out in the field, sometimes the first man to move was the first to die.

Finally Tony's body shuddered. He looked up. "How long was I out this time?"

"Ten minutes maybe."

"OK Ralph, this is something I never told my debriefer or my program psychologist. I have to get this off my mind."

"OK, shoot."

Tony picked up one of the cards. "This…device…allows a remote operator to find the consciousness of the person in closest proximity to it, but it's designed specifically for the person whose holo is on it. It's a receiver that allows the operator to kill the target, but in doing so the operator is also killed. They don't tell the operators that."

"Fuck!"

"Ralph, if it wasn't for Diane – that's not her real name – I think I probably would have jumped off a cliff already. She works with traumatized people like me. Keeps me sane."

"Heard of those people. Women with empathy who are themselves experienced in the programs."

"That's right. For some reason men don't work." Tony was staring off into space again. Ralph sat quiet and unmoving. Tony's body began to shake and he spoke softly. "I bought this place ten years ago, just before I cracked up. We were working on science fiction stuff. There are portals in certain places on, within, and above the earth. If your mind is in the right place you can identify these portals. If you enter them physically they take you off world." Tony looked up. "You think I'm batshit crazy don't you."

Ralph shrugged. "Sounds like the Bermuda Triangle."

Tony nodded. "Long story short, I was trained in remote viewing. I'm sensitive to these portals, which are just areas of space programmed to shoot you off to another location. The jerk-offs in my program are worried that…hostiles can enter through them and come to earth. My job was to identify these portals and try to look through them at the other side before sending anyone through. We didn't want our guys going into dangerous environments." Ralph could see that Tony was getting excited. His face and demeanor had lost their grayness. "You won't believe what's on the other side of those things Ralph. I got to enter one. I wound up in a base on Mars."

"No shit!"

"Yeah, this base has been around since the 1970s. Turns out that Mars, like Antarctica, is divided into areas of control. Anyway, one day our team flew out to the four corners area in Utah. I located one of the portals inside a cave. When our guy went through he came back with a really fucked up alien who looked like the creature from the black lagoon. Our guy was so scared he was literally shitting his pants. He told us later that if we ever went through there again, the beings on the other side wouldn't appreciate it at all. 'We're full-up out here,' he said the creature told him. 'Stay on your own planet.'"

Ralph shook his head. That was almost as fucked up as when he had to guard a bunch of boring civilians at a formal fundraiser and drink bad tasting champagne.

"But that's all preamble," Tony said. "After locating and working with these portals for almost two years I literally went crazy. You wouldn't believe the shit that's out there.

Ralph, the world isn't anything like we think it is. We're living in a Matrix, just like the movie. This planet has hundreds of these portals, some on every continent, and some under the ocean and in space around the planet. God knows who's coming and going."

That reminded Ralph of the time he had to go to a Costco during Christmastime when he was on bodyguard duty. So many people! That was rough.

He could see how bitter Tony was getting.

"Human beings have no understanding of life outside this little fucking rock."

Tony bent over and held his head in his hands. After a few moments he sat up again. "I still haven't gotten to the worst bit. After I fell apart my shepherd in the program told me to get the fuck out, that I was useless to them. So I left right then, in the middle of the night. They gave me something that looked just like the cards you have. I had plenty of money in a bank account but no way to access it. That night I stayed in a festering shithole in a little town outside Huntsville, Alabama. Just before I went to sleep something malevolent entered my mind. Ralph, it was a mental kill shot. I was going, knew I was dead. Then – I found out later it was an operator on a remote consciousness-assisted transmitter – the operator backed off just before I was going to die. He didn't want to kill me. I felt his screams of agony as he died. I got a glimpse into his mind a nanosecond before he was terminated. Ralph, what killed him is evil."

"ETs?"

"No, nothing like that. These are...demons in the human psyche we've created over the last 6,000 years. It's like that movie, Forbidden Planet. We have to overcome the demons in our own minds and discover God before it's too late. Otherwise we aren't going to make it. It's a test, Ralph, and we're failing."

"Is there a God, Tony?"

"You tell me. In the last femtosecond before the operator died I got something amazing, it's the only reason I'm still alive. Beyond the demons and the darkness, at the very end of life, is something I can't explain. A power that created the universe and everything in it. Did you know that there are precisely 235,677,498,003 stellar systems in the Milky Way galaxy?"

Ralph looked at his friend, puzzled. "How can you know that?"

"That's just it Ralph, I know it with 100% certainty. Just before the operator died he contacted something beautiful. Our minds were connected and I saw it, felt it. It's the source of all information and knowledge."

Ralph was beyond surprised to hear his world-weary old buddy talking in this fashion.

"The demons we're facing are like the Guardians of Evil. They are the gatekeepers we have to get through to recognize our higher potential. I think that's what happened to that operator, but don't ask me how or why."

Ralph didn't understand any of it, but he could see Tony was getting depressed again.

"I guess I survived the kill shot by accident, but whoever is behind these" – he held up the two cards – "is evil. Every fucking day I want to thank the guy who spared my life, but I know how he died and I don't ever want to be part of that." Tony looked around the room. "That's why I built this place. I hide in here and keep my head down. I only go out to see Diane."

"Do you know how Harry Kaine fits into this? I heard he just got recruited."

Tony looked up at the big man. "Don't fuck with the people who made these cards, Ralph. As far as Kaine goes, he's just a tool to take humanity off to a different timeline."

Ralph almost fell out of his chair (which was much too small for him) when he heard this. "I heard something about that from a goofy kid who says he was working on a legacy Looking Glass project."

Slowly Ralph reached into his pocket and pulled out a pod. "Look at this, Tony."

Tony put the pod into his mobile and they watched Germaine Robinson's recording of the timeline from 2035 to 3209. Tony shook his head slowly back and forth. "I'm not surprised by anything anymore, Ralph."

"Do you think this is legit? A fucking flash of light and the end of time?"

"I haven't the slightest idea, Ralph. All I know is that the patterns of life that we thought would last are being broken. We're heading off into a new world and no one has a clue what it's going to look like."

Ralph could see that Tony had had enough. "I'm sorry buddy, I should have stayed away."

The little man grabbed his arm. "No Ralph, come as often as you want. You helped me to see behind the blackness today. I've been working with my therapist Diane on this. She said I was almost ready for a breakthrough. I think you pushed me past…the darkness."

"Glad to help." Ralph picked up the two razor-thin cards. "So what we're dealing with here are special projects nutjobs."

"That's right, Ralph. We must destroy these immediately."

"I'm sorry Tony. I've already led them to you."

Tony grimaced. "Don't worry about that Ralph, those boys already know where I live. As long as I keep quiet I'm safe. I haven't violated my oath to the program."

Tony stepped to the end of the table, where two foot-wide solid plastic clamps held two powerful magnets. Tony placed one of the cards between the two clamps, and twirled the right-hand magnet. Instantly, the two magnets came together with a powerful snap, enveloping the material. After a minute, Tony slowly rotated the magnet back to its original position. "Powerful magnetic fields will render these holos inert," Tony remarked. He did the same with the other card. "OK, we're all set."

The two men walked out of the Faraday cage. Tony looked more animated. "Ralph, there are only two groups I know of who use the technology on these cards. One is located in Alexandria, the other in Denver."

"Thanks for that, Tony. Do you know Harry Kaine?"

"The Transhumanist guy? A real dead-head. Aren't most of the clowns in these programs?"

Ralph laughed. He wanted to tell Tony about his two clones Smith and Jones but decided that his friend didn't need any more stimulation. "I'll be in touch Tony, and thank you."

As Ralph walked home he thought about the cards Tony had destroyed. He was pretty sure that Harry Kaine had gotten one of them when he was recruited. Too bad for him! That kind of thing happened when you joined a special access program.

While walking home Ralph called Tanya, the receptionist at the Taubman Building, and asked her a couple of questions. When he got up to his apartment he called Luigi, the guard who worked on the second floor in the restricted area at Taubman.

"Heard you had some action this afternoon," Ralph suggested. Luigi was off-duty now and had his mobile back.

"Yeah."

"I heard Harry Kaine was up there."

Luigi swore. "You aren't supposed to know that, Ralph."

"Got it out of Tanya. Just got off the phone with her. She said that information isn't classified."

Luigi was silent. Ex-sergeant Tofoli's military training taught him never to divulge sensitive information to civilians. Or to anyone.

"C'mon buddy, you know I can keep a secret."

"Around two o'clock some heavy-hitters paid one of our labs a visit. Those guys were black for sure."

"Did they recruit Kaine?"

Ralph heard Luigi laughing. "Yeah, I talked to the SCIF monitor. Guy said he heard noises and turned around to see a gun get pointed at Kaine's head. Kaine literally peed his pants. He saw it when Kaine and the heavy hitter walked out of the SCIF."

"Did you happen to notice if that guy handed Kaine anything?"

"Now that you mention it, I did. He handed Kaine a currency card, I think, as they both walked down the hall."

"Did they sanitize Kaine's lab?"

"Yup. Tighter than a flea's ass. Sealed the door, too."

"Thanks Luigi. I owe you one."

Next Ralph called Danny Radulo, a freelance intelligence officer often used by various intelligence agencies. From Danny he discovered that Harry Kaine had been recruited by an outfit associated with In-Q-Tel, the former CIA's venture capital arm. That sealed it; Kaine had been recruited by the Alexandria group.

He wouldn't last very long, Ralph thought.

When Harry Kaine saw his new house he was excited. He became ecstatic when he was escorted to the basement entrance of a large building with signage that said "Revitagen." His scientists were already at work. Dr. Anders Kopetz gave him the briefing and explained what they had to work with. "Harry, it's a Transhumanist dream. Bioprinters, mind uploads, nano-biotechnology, everything we need to change the world."

Harry was confused. "All that couldn't have come from my two clones!"

Kopetz smiled condescendingly from down his long nose. "Of course not, Harry. This facility is the result of ongoing research in the hidden programs since the early 1960s. Your two clones have, however, enabled us to solve a couple of serious problems in electronic-biological interfaces and nano-biotechnology."

Harry was pleased, even though he saw that his work was going forward without his direction. "Who's running this show, Anders?"

Kopetz grinned, his hawk face assuming the aspect of an eagle about to pounce on his prey. "Why, I am Harry! Didn't you read your instructions?"

"Most of them."

Kopetz laughed. "You missed the organization chart then. I'm in charge of research, you're the PR guy. Public liaison."

Harry digested this for a moment, then smiled. He held out his hand. "Congratulations then, Anders. Science isn't my forte anyway. I like people."

The people you want to replace, Anders thought. He shook hands perfunctorily. "Take a look at that organization chart, Harry. You'll want to know who your shepherd is."

When Harry got home to his very spacious Alexandria home he decided to take Anders's advice and studied the org chart. Kopetz was right. He was head of the Public Liaison Group, and directly under the Big Boss, Lucas Cernovitz. Who was that? The name sounded familiar, it was in the back of his mind somewhere...no matter. He'd find out soon enough.

Harry saw a laptop sitting on his desk when he walked into his study, a room he had chosen on the second floor with a big window. Probably another instruction set. But when he opened the laptop he received a huge shock.

CHAPTER **26**

Seven months later, February 26, 2036

Karen Everard and Kirra Bigbear met up with Melanie Fuscaldo at the Interfaith Center to discuss their strategy at the upcoming debate in New York. Before they got started, Karen got a frantic call from Kjirsten Chastaine. "ABetterHumanity.com. Quick!"

The two women grabbed their tablets.

"It's Harry Kaine!" Kirra said.

"Who's that hunk beside him?" Karen asked.

"Shhhh!" Melanie ordered.

"...And now I would like to introduce Ben Rudovich. Ben is what we call in Transhumanism a post-human, the very first! During the past several years our scientists have made incredible breakthroughs in immunology, neuroscience, artificial intelligence, and bio-nanotechnology. Using a sophisticated bio-3D printer with Ben's DNA, we have been able to create a disease-resistant human body for him that will not get sick. Our scientists estimate the lifespan of these trans-human bodies at approximately 150 years." Kaine turned to Ben. "How are you feeling my friend? Why don't you tell the public your story."

Ben was about six feet tall with olive skin, curly black hair, and brown eyes. He exuded an aura of vitality and good health. "It's simple. I was dying of testicular cancer. I had two months max. A doctor friend of mine heard about a new startup biotech company called Revitagen. Mimi and I talked to Harry here and went to Alexandria, Virginia to inspect his facility. It was state of the art. We were shown the new bioprinter and a way to upload a person's memories and personality using a sophisticated cloud technology that imprints the brain." Ben spread his arms wide and smiled. "You see the results."

"Mimi, please come forward and tell us your story."

A smallish woman walked firmly up to her husband and stared at him adoringly. "It's nothing short of a miracle," she said breathlessly. "My Ben was dying. Now look at him!"

"Mimi, what would you say to those who insist that creating a human body is an abomination, and against the laws of God and Nature?"

"That's exactly how I felt when I first talked to Harry and saw the procedure," Mimi said. "I thought it was outrageous. But Ben was desperate." She turned to Harry Kaine. "I'm a nurse, Harry. I've been around hospitals and doctors my entire life. Naturally I was skeptical. I had no idea these technologies existed, it sounded like something out of an old Star Trek or Twilight Zone episode." She looked up at Ben with widened eyes and a knowing smile, as if the two shared an intimate secret. "Let's just say that I'm very happy, in more ways than one."

"Have you noticed any personality changes?" Harry asked.

"Only for the better. Ben is more loving and attentive than he's ever been."

"Do you have children?"

"A boy and a girl. Ben Jr. and Sarah are both astounded and impressed." She hugged her husband. "We're a happy family again!"

The presentation ended with a camera rolling through the (very impressive) Better Humanity facility in DC, with a link to ABetterHumanity.com in the right hand corner.

Karen Everard was shocked. My God! Harry Kaine had altered the timeline, just as he said he would. How had he made such impossible scientific advances?

The presentation was slick and professional, Melanie thought. Harry must have found a big donor.

Karen, Kirra, and Melanie stared at each other. In the back of the women's minds was the thought that here was a bombshell that could radically alter human society. For the worse.

Ralph Zimring saw the entire presentation. Quickly, he took his tablet and walked over to Tony's place. After Morsing his knock Tony answered, grinning. "I thought you'd be here, Ralph. Come on in."

Ralph saw that Tony was replaying the half-hour vid, which was also available in VR. Both Ralph and Tony put on virtual headsets and watched in 3D. "I'm no scientist Tony, but this technology is literally fifty years ahead of its time."

Tony nodded grimly. "It is."

"I traced Harry Kaine to the Alexandria group. Those boys are running with this, obviously. But what is their intent?" Ralph was thinking about the timeline recording he'd gotten from Germaine Robinson. Germaine's friend, Joe Courvall, claimed that he had actually traveled in time to the future.

Ralph could see that Tony was looking better.

"The word is that Kaine's group intends to make a big splash. They want to reorient human society."

"It won't be human anymore if people are augmented."

Tony shrugged. "What is a human? The flesh and blood and bones, or the self-awareness that inhabits that body?"

Ralph hadn't thought of that, and he was puzzled. His cynical friend was talking like a spiritualist. "You sound like that space cadet Melanie Fuscaldo."

Tony examined the big man critically. "Don't knock it until you've tried it, Ralph. I've talked to Fuscaldo. She's a former microbiologist with the NIH and has scientific training. Besides, these augmented humans will probably only be used for those who have incurable illnesses."

Ralph wasn't so sure. He would have to talk to Germaine and Joe Courvall about it. "Kaine is having his big conference in New York next month."

"That's right," Tony said. "We'll find out then about how big his plans are."

Ralph decided that he didn't like the idea of augmenting humans, no matter what the benefits were. He wasn't going to let human society get fucked up by a bunch of sociopaths in the programs. Harry Kaine was just a jibroni with a big mouth. "Fuck Kaine. I'm going to stop him."

Tony looked up at the imposing giant and thought that Ralph was an army unto himself. "As long as you take me along for the ride."

Ralph grinned. "It's going to be glorious, Tony!"

Revitagen HQ, Alexandria Virginia

Lucas Cernovitz met with his operations officer in a secure conference room on the fourth floor of a building with no signage, three miles south of the Beltway. "How's it working out with Kaine?"

"So far so good Lucas," Paul Wen said. "Kopetz laid it out for him, he doesn't seem to be a problem."

Cernovitz's expression went harsh. "We have nothing on this guy, Paul. That's a problem."

Paul Wen was disgusted. "I left your little present for Kaine on a secured laptop."

Cernovitz grimaced. He had no personal tolerance for pornography, especially child pornography. He distributed the stuff because it was good side money. He looked up at Paul. The guy was solid. Lucas never chose perverts in his inner circle, or those with tainted backgrounds. "Hopefully Kaine is as weak as Bob Saunders said."

"We'll find out, but I'd trust Bob. He can sniff out a scumbag standing in a pile of shit."

"Kaine is going to be good for us, Paul. That New York conference, and how he organized it, will be brilliant. The presentation he did was very convincing. And he doesn't have to be told what to do. I'm preparing a bigger surprise for Harry in New York, if he takes the bait on that laptop."

"We'll use the Transhumanists as a distraction. All we need are a few programmable humans in critical positions. No more rogue elements or operatives who go off the reservation."

Lucas laughed. "These Transhumanists are idealists, Paul. Delusional! Why create an entire race of human prints when you only need a few? It's better to stay underneath the radar."

Paul rubbed his hands together. "Gotta hand it to you Lucas, you're brilliant. We don't have to control everyone. That's what the communists tried to do with their social credit systems, their gulags, their pandemics, and their surveillance state. Let society, the economy, and politics evolve as they may, but under our direction."

"Our subtle direction," Lucas said. "We recruit influential players with terminal diseases and upload them to new bodies. Their lifelong gratitude makes them willing team players. We don't need to blackmail people, we don't need to kill anyone, so there aren't any messes to clean up, and no whistleblowers. We're invisible to society."

Paul frowned. "We'll have to keep an eye on Harry and Kopetz. Harry is all gung-ho on Transhumanism. Wants to save the world."

"So do our two silver-haired friends."

Wen blanched. Cernovitz was cold-blooded, but Wen could see that even he was afraid of them. He changed the subject. "This goddam side business of yours is causing too many problems," Wen complained.

"Let's talk about that regarding Winthrop Sauvage," Lucas said to his OPO. "Any loose ends in that town?"

Paul frowned. "The texts we sent did the job. That fat police chief, Chastaine, is too stupid to find us, especially since his IT guy left." Paul gave Lucas a questioning look. "You think he's worth recruiting?"

"Liqao Chen? Nah. Maybe in the future."

"That *Chronicle* reporter is nosy but the owner of the paper set her and her editor straight. They're both nobodies."

"What about the COSA group? They've been sounding off about child abuse."

"That situation is contained now. We're safe."

Lucas turned his gaze on Paul. "We'd better be. You know the rules. No loose ends."

Paul's breath caught in his throat for a second. Lucas was intimidating. He had failed to dispose of the two clones as ordered, and that was a matter that involved Ralph Zimring. He'd better not hold back on that. "There is one problem though, a guy named Ralph Zimring."

"Zimring!" Paul saw Lucas' face go white for a second. "What's he doing in that piss-ant town?"

Paul looked curiously at Lucas. "He's been there for fifteen years, Lucas. Involved in that little dust-up with the rogue AI last year, which was thankfully destroyed. Had a run-in with that imbecile Lord Byrnes a while back."

"Right, I remember that. The problem is, I don't know whose side he's on." Lucas rubbed his chin thoughtfully. "Zimring killed that psycho Harriman Drake. Got rid of that child trafficker Chaledzne for us."

Paul frowned. "For you, maybe. But it worked out in our favor."

"Same thing." Lucas was staring fixedly out into space, as if trying to recall something. "How much does Zimring know about our operation?"

"Zimring knows Kaine, or knows of Kaine, but nothing more than that. Remember, Kaine was born in Midland. Zimring also knows Danny Radulo, the freelance intel officer." Paul looked on his tablet. "He also knows some broken down scientist who lives like a hermit, name of Tony Bagdadi. Zimring is hard to figure. He's an independent operator, but loyal to Max Berglin."

"The world is too complicated."

"Yeah, but we're all set now, Lucas. Getting Kaine was the last piece of the puzzle." Paul looked tentatively at his boss. "Get out of the child trafficking and child porn business, Lucas. It's going to be our downfall."

Lucas shrugged. "Business is good. The payoff is good."

"I'm not backing you on that one. You're on your own. The world isn't going to tolerate it anymore and the penalties are...excessive."

Lucas gave him a malevolent look. "Fuck you, Paul."

Wen shrugged. They had had this conversation before. "Should we take out Zimring?"

Lucas laughed. "Who's going to do it? You'd need an army and Zimring would probably kill them all. Avoid him at all costs."

Paul nodded.

"When does our facility open to the public?" Lucas asked.

"Kaine wants to call it A Better Humanity instead of Revitagen."

Cernovitz waved his hand. "I couldn't care less what we call it."

"OK then. A Better Humanity opens right here in DC after that stupid debate in New York."

"Unnecessary."

"It's a bone to Kaine. Keep him happy, he'll push forward like an Abrams tank in a flower garden."

"Kaine is just a distraction," Cernovitz said. "He's our cover. He gets the publicity and all the attention, we work behind the scenes."

"The first Life Extension facility has already been open for business! By invitation only, of course, and under the direction of the good Dr. Anders Kopetz. Testing the waters with real people."

"And if they die no one cares."

"They would have died anyway. A valiant effort to save the terminally ill, and all that."

"I hear Senator Blatsky has terminal cancer," Lucas said. "He only has four months to live."

"It's a good start. Blatsky heads the Senate Select Committee on Intelligence. We'll need him." Paul ticked off on his fingers. "There's also Brodsky, the influential writer, who's dying of testicular cancer; Jessica Kassitides, the famous New York talk show host who has Parkinsons; Vish Ramaswami, inventor of the VR healing centers, who's got an inoperable brain tumor, and Omani Hussein, the famous activist, is also on her last legs."

Both men smiled.

Lucas looked significantly at Paul. "Maybe it's time we got ourselves a new body too." Lucas saw a look of fear in Paul's eyes. "Get with the program, Paul. These human bodies will wear down too soon for us to see this through."

Paul Wen shrugged noncommittally and changed the subject. "What do we do if Kaine gets out of line?"

Lucas laughed. "We'll let Bob Saunders take care of him."

After Cernovitz left Paul realized he forgot to tell the boss about the two clones he had failed to dispose of. Oh well. They had already served their purpose.

Harry Kaine walked into what used to be "his" lab, which had expanded into a huge facility almost ready for the public. Anders Kopetz was in the middle of everything, giving orders, answering questions. It was what he used to do. But it was a good tradeoff, for the Transhumanist agenda would now be assured of success by a real scientist.

Something Karen Everard had said about history already being written bothered him. "The conservation law of time says that whatever changes you make have already been accounted for on the timeline. Your future has been written, and it's an epic fail. Nothing you do will change the end result."

Yet when Harry looked around at the sophistication of the new facility in DC, he was astonished. It was twenty times superior to the one in the Taubman Building, and he had unlimited resources now. He was probably far ahead of the events on the timeline because his first Life Extension Center wasn't supposed to open until after 2098. But if his New York conference went well, the entire world would know of the benefits of life extension. He would be over 60 years ahead!

He was cheered by this thought, and dismissed such nonsense as spouted by Karen Everard, Joe Courvall, and their kooky friends. For some reason he felt the impulse to stick it to those nobodies in his hometown, and to prove the superiority of what they dismissively called "scientific materialism."

Harry ignored the dying Guardian society he had seen in 3013 AD. That dead civilization was about to be transformed! A scientific tidal wave was going to hit the world. Harry looked around the facility again, feeling a wave of exultation. He may have sold his soul to the government, but with his promotional and organizing skills he would beat the conservation law of time. Nothing was written in stone! Did not Caesar 13, the chief time technician in 3013, say that past affects future? Harry knew he was on the right team. He decided to throw himself into A Better Humanity with even more zeal than before.

Now it was time to go to his splendid Alexandria home and make some phone calls. There was still a lot of work to do; the New York event was only two weeks away. And then, of course, there were those mesmerizing images on that laptop in his study...

The Gang met in Ger's house the day after Harry Kaine's presentation on ABetterHumanity.com. The mood was glum. Karen and I were angry.

"Harry Kaine is a fake and so is this Ben Rudovitch," Karen said loudly. "He's a fashion model or something. Who does Harry think he's fooling?"

I agreed with Karen.

Just then we heard a door open and footfalls on the stairs. Ralph Zimring entered the room, his head almost touching the ceiling.

"Ralph!" Ger cried. "What are you doing here?"

Everyone stared at the giant, but I could tell Ralph was used to it. He looked at me and Karen. "In case you were wondering, Ben Rudovitch is real. A Better Humanity isn't fake."

Karen wondered how this monster knew about their conversation, but didn't ask. "That's not possible, whoever you are. The technology won't exist to do that until at least 2098. Transhumanism is just a pipe dream." She looked up in awe at the big man.

Ralph stared hard at her. "It exists. My sources in the programs – and they are impeccable – confirm it."

Ger swore.

Ralph explained about Kaine and his association with the Alexandria group. "If I were you I wouldn't bother trying to debate Kaine. He's a master of public relations; Kaine's people will eat you up. You have nothing to gain and everything to lose."

"So you are on our side?" Kjirsten asked. "My father knows about you, a mercenary and a killer. You should be Harry Kaine's best friend."

"Not me! Harry is a jerk; I've been following him ever since I came to Midland with Max Berglin. But debating him is pointless. He'll destroy you."

Kirra laughed. "Wait until you see the surprise we have for that asshole."

Ralph looked interestedly at the petite woman with the bright red lipstick. "You're cute."

Kirra blushed and Liqao rushed up to confront Ralph. "Hey!" he said, sounding like a little kid who had his lollipop taken away by his daddy. "She's my girlfriend!"

The gang looked at each other, trying hard not to laugh. Liqao came up to the middle of Ralph's chest, and looked like a pencil tree trying to stop a charging hippo. "Don't sweat it kid," Ralph drawled. "But she's pretty cute. Have you been treating her all right?"

Liqao gasped and stepped back. "Uh...yeah. I think so."

Ralph pointed a finger at the little man. "You *think* so?"

The pained expression on Ralph's face, and Liqao's attempt to cover his abject terror of the giant, made us all laugh.

"Nice joke Ralph," Ger said, "but you're a busy man. Why did you make a special trip over here?"

"Because, children, a gathering of the clans is going to descend on New York in two weeks at Harry Kaine's conference. A perfect storm, if you will. You don't want to be there."

"Oh, but we do," Kirra said, and described her plan to win the debate with her Native American friends. She showed Ralph some pics of them dressed in full tribal regalia. As she spoke Ralph's jaw dropped further and further. Kirra concluded with, "I call them my indigenous bombs. We're going to bigfoot Harry Kaine and his lab rats at that debate."

At this the giant threw back his head and guffawed so loudly one of Ger's windows almost broke. He began to laugh so hard I thought the walls might fall down. Tears flowed down his cheeks. Finally the big man gasped and said, "I have never laughed so hard in my life, Kirra Bigbear."

Kirra smiled. "It's not about content, it's about who looks the best to the public."

Ralph was doubtful but he said, "OK. Who's going with you?"

"We're all going except Liqao here. He agrees with Harry."

Ralph gave Liqao a dirty look. "You don't deserve her, kid."

"I do too!"

Ralph dismissed this outburst and smiled. "I changed my mind. I wouldn't miss that debate for the world." He looked around the group. "Tell you what. I'll get a team together to look out after you all. I think I can get Max to let me go for three days. Hideki can watch over the Centurion Building while I'm gone. I want to check out Harry's organization."

"Why do we need protection?" Kjirsten asked.

"Because there will be...let's say...military and intel people at the conference who are stage-managing the event. They want nothing to go wrong and are prepared to, ah, make things difficult for anyone they deem to be obstacles to the success of the convention."

Ralph didn't tell these civilians about Lucas Cernovitz, Bob Saunders, and the team behind Harry Kaine's conference. Danny Radulo told him that Cernovitz knew about Karen Everard, Kjirsten Chastaine, and their little group in Midland, and that Cernovitz was responsible for the threatening text messages they'd received. He'd need two guys on the outside and four inside the NYC Center. Tony Bagdadi would run comms. Danny Radulo would want in. Danny knew a couple of retired guys from his old special forces outfit who would be really pissed if they knew he'd had so much fun without them. Including him that was five for his team. There were two or three retired guys from Kharkov, his old PMC, who would want to get involved.

I could see that Ralph was getting motivated. He was rubbing his hands together and looking out the window, deep in thought. "So what do we have to do?" I asked.

"That's the beauty of it, kid. You civilians just do your thing. My guys will see to it that, ah, no harm comes to any of you."

"You're a civilian," Liqao piped.

"Sure, kid."

Kirra laughed. She loved Liqao but he really was like a small child next to Ralph. A cute child though.

"Ok, I'll come," Liqao said.

"Good!" Kirra said. "We'll share a room."

Ralph paused for a moment. "Do you have reservations?"

"Yes," Karen said. "Right in the Convention center itself. We're honored guests, Harry said."

"Good! Give me your mobiles. I'll take it from there." Ralph took our phones and one by one began to clone them on a portable military-grade nethub.

M'basa and Kjirsten exchanged glances. "Are you a star seed?" Kjirsten asked Ralph while he worked.

"I don't know what that is, sweetheart," Ralph said. "But don't worry. Just enjoy yourselves, and jam up Harry Kaine for me and the human race at that debate."

"Sure Ralph," Ger said.

The big man turned on his heels like a ballet dancer and moved quickly and soundlessly down the stairs. Ralph could move quietly when he wanted to, I thought.

On the walk back to his top-floor suite Ralph rubbed his hands, thinking about how much fun it would be to face Lucas Cernovitz and Bob Saunders. Saunders was just a killer who enjoyed inflicting pain on people weaker than he was. Cernovitz was barely human and involved in sex trafficking. Both of them were bugs. He was going to take them both out in New York. The fun part would be to make it look accidental. Oh yes, he was going to have a good time in the Big Apple!

First he would call his friend Joe MIller, the Deputy Commissioner for Intelligence and Counterterrorism at the NYPD. It wouldn't do to invade Joe's turf without a heads up. Then a call to Danny and Tony, who would help him assemble the team. They'd need floor plans for the convention hall, and Danny could hopefully get intelligence on how Cernovitz would situate his team, and who would be there.

Ralph spent a pleasant evening planning his op, which he named Operation Scumbag.

3036 AD

Li Jing and Li Chun continued to monitor the timeline. After their afternoon shift (all of the free humans rotated between the fields, child care, food preparation, and maintenance work) the couple began to prepare the Shifter for their weekly inspection of the timeline. They had two children now, as did almost every other couple. However, for some reason no couple could have more than two children. The Community understood that two births per couple would only allow temporary expansion. It seemed that the free human community would, over the generations, be just as stagnant as Guardian society.

Approximately 10% of the Guardian clones were now androgynous. The Community kept their promise to Christian 172 and stayed on their side of the 42nd Street divide, except for Hannah and Alan Percival (the former Alan 133). The childless couple had devoted their lives to the gathering of intelligence about the Overseers and Guardian society, keeping the Community informed on what the Overseers were doing. They used the hovercraft and also did scouting expeditions on foot.

"Another of the sexed clones came over today," Li Jing said to her husband as they worked to prepare the Shifter. "A female."

"Good. Perhaps she isn't too far gone to breed," Li Chun said.

Li Jing's eyes teared up. "Oh husband, it is so depressing to see them! Apathetic and lifeless, just going through the motions of life."

Li Chun grinned. "We put them to work and give them a purpose. Eating freshly grown food and association with the community of free humans soon increases their life force readings."

"As the years pass Guardian society clones become less and less vital."

"The androgynous clones are hopeless. Not even we can save them."

Li Jing sighed. "Well, we know what happens in the end. To us all."

As the Time Shifter activated, Li Jing and Li Chun anxiously watched the display tank. The key interval was between 2035 and 2098, when the Referendum was passed and the Life Extension Facilities began to be built all over the world.

"Oh my God look at that," Li Chun said to his wife as the timeline snapped into focus on March 24, 2036. The timeline showed a new focal point, a locus of events around a conference given by Harry Kaine. "Run it forward," Li Chun said. "This conference didn't happen before!"

When they tried to run the timeline forward from Harry's conference in 2036, however, there was a lot of fuzziness. "That fuzziness represents temporal uncertainty," Li Chun said. "It always indicates big happenings."

"Yes! Events are in flux back there – and big changes may be coming for us."

"Compare it to the old recording."

Li Jing showed the two timelines side by side. "Kaine's conference definitely didn't happen in 2036 on the old time thread," Li Jing said as they played the old recording back several times.

"On the old timeline the elites mostly kept Transhumanist technologies to themselves until 2098."

"Yes, they used it on themselves and controlled events from behind the scenes with altered humans. They let everyone else suffer. When the Life Extension Centers were finally opened to the public the treatments were casually administered. So many died!"

"It was a bad joke on them, though. Their World Security State was a massive fail. Now they are pathetic Overseers, tending to a dying society."

The two time techs tried to resolve the temporal uncertainty, with no success. "It's too fuzzy to see what Harry Kaine is doing back there, but the energy around it is strong. Our timeline might get even worse!"

Li Chun's face hardened. "I was afraid of that. Remember those two Tower clones? Harry probably made use of their advanced clone biology."

Li Jing was glum. "I suppose it's no worse than getting destroyed in a flash of light."

"What do we tell the others?" Li Chun always deferred to his wife about important social decisions.

"Nothing, husband, because we don't really know anything. Harry Kaine is using the knowledge he gained from the future to affect events on the timeline. He may succeed or fail. I don't know which is worse."

March 23, 2036

The Gang met at Ger's place before they were scheduled to go to New York. Liqao was eager to watch the debate. "I can't imagine how your Indians will beat Harry's scientists."

Kirra snorted. "We've got a big surprise for Harry Kaine, Liqao. You don't want to miss it."

Kjirsten and M'basa didn't understand why Kaine's conference was such a big deal. "Kaine is a clown and his silly Transhumanism won't go anywhere," Kjirsten said. But Harry was footing the bill and they both wanted to see a play and walk around the great city.

No one in the Gang believed that Ben Rudovitch was real. It was impossible to make a Trans-human in so short a time.

"Harry is just hastening his demise with A Better Humanity," M'basa said. "We already know he's a failure."

Karen and I exchanged glances. After what Ralph said, I wasn't so sure of this anymore, but I kept my mouth shut.

Ger and I had had a talk with Ralph Zimring, who briefed us on one of the Bosses. "Cernovitz is his name, the same guy who's in New York with Kaine now," Ralph had told them. "This guy runs child pornography and child abuse content sites under the name of DarkScandals. His admin calls himself Mr. Dark. These guys are totally ruthless." Ralph gave me and Ger a description of Cernovitz and his assistant, Paul Wen. "If you see them just walk the other way."

Earlier, Karen and I had talked about whether we would share a room.

"Do you want to?" she asked.

"Yeah, I do. I've been in that damn lab with Ger all of my spare time." I smiled. "I've been neglecting you."

Karen's eyebrows rose. "Oh? I didn't know we were an item."

I blushed. "That's my fault. I should have...been more assertive."

"You should have."

Karen regarded me with a feminine gaze I couldn't decipher.

"If we live together for a couple of days we'll find out whether we're compatible. You know, whether I fart in bed or whether you leave the sink too dirty."

"Or whether you play music too loud," she offered.

"Or whether you leave your shoes lying around the living room."

"Or whether your feet stink."

We both laughed at that.

"Let's try it," I said.

Karen smiled brilliantly. "I thought you'd never ask."

———————•✳•———————

Instead of flying, which had gotten more expensive over the years as the airlines became less profitable, we took the Amtrack from Chicago to New York. For those of you reading before 2036, private transportation was no longer the preferred method of travel. The railroads had benefited from a huge infrastructure bill passed by the Congress. All of the public passenger trains now had nice sleeping compartments and good food. On the train we discussed what we would do together in the Big Apple, and what Harry Kaine's conference would look like. We had front row seats in the huge NYC Convention Center. I passed around pics of Cernovitz and Paul Wen. "Ralph Zimring says to avoid these guys, or anyone who looks like military," I said. "Ralph has people in place to protect us."

Kjirsten blanched. "From what?"

"It's a private thing between Ralph and Harry's organization," Ger said. "It doesn't concern us. All we have to do is sit in our seats and watch Harry's show."

"No worries, let's just enjoy ourselves," M'basa said.

Kjirsten grumbled, but we were in New York now. Karen and Kirra seemed unconcerned. Liqao was eager to see Harry's technology. Lledren couldn't care less about the event but was using the trip to get away from the soup kitchen for a couple of days. Ger wanted to see a couple of physicists he knew in the area. I was looking forward to getting more intimate with Karen.

When we got to our hotel everyone got excited; Harry had provided us all with very nice rooms with a view of New York. We decided to take a tour of the famous convention center. Ger and I were looking out for military types but we saw no one. Karen and I held hands as we walked; Ger was out front like a scout on the Western frontier. Several passing women and a couple of guys took notice of his tall dark figure with his head of curly hair and his intense blue eyes. I never knew anyone who was less interested in sex to attract so many potential partners.

When we entered the big auditorium where the conference would begin the next morning, we saw media people hanging around. On the stage we saw several people engaged in a debate. I estimated that the place could hold at least 2,500 people.

"Look at that," Kirra said. "They are afraid of us! Already practicing."

"It looks scripted," Kjirsten agreed.

I walked away from the group and saw Ben Rudovitch standing on the stage at the far right, next to a perfectly formed woman who looked like she had come out of a mold.

Just then Harry Kaine entered the auditorium and saw us. He walked up to us, followed by two military types; bodyguards probably. "I see you have arrived at the scene of your demise!" he said to me cheerfully. "I'm glad you came." Harry held out his hand. I took it reluctantly.

"Wait until you see what we've prepared for humanity. I hope you enjoy it."

"Who is that fashion model standing next to your clone?" I asked.

"That's Ben's wife Randi," he said smugly, ignoring my verbal barb. "Isn't she beautiful?"

"I thought Ben was married to Mimi."

Harry waved his hand dismissively. "Mimi couldn't handle Ben's new look. We offered to give her the treatment, but she wasn't having it." He smiled an oily smile. "Progress!"

"Indeed!" I said sarcastically. "Those who can't face a brighter future will be left behind."

Harry grinned, missing my implied criticism. "Precisely, my friend Joe." He shouted at the stage to Ben and Randi. "Come forward! I want you to meet some friends of mine."

I wanted to tell Harry Kaine that he wasn't my friend, but I refrained. After all, he had given us a free ride to New York. As the two "specimens" approached Harry had the look of a magician who was about to spring a surprise on an unsuspecting audience.

When Ben and Randi walked over to us I was blown away. These were not fashion models! Each glowed with sparkling health and energy, as if their bodies had been scrubbed clean of impurities from the inside out. Ben and Randi were way more impressive in person than in a vid.

Harry's face was a study in condescending smugness. "Well? What do you think? Have I done well?"

I looked over at Karen, who was staring at Randi with an awed expression. I had to admit that I was impressed. I noticed that Karen was literally infatuated with Randi, and got irritated. "Would you like to introduce me to your bodyguards?" I asked Harry, looking at two imposing fellows who stood to either side of Kaine.

Harry dismissed my snarky comment. "Mr. Ernst and Mr. Chernkov. They're just part of the woodwork."

Neither of their faces were familiar, so these guys weren't Paul Wen or Lucas Cernovitz. The two bodyguards looked like mountain lions about to pounce on a rabbit. I didn't like the feeling of being a rabbit. Neither did I like feeling that I was a mangy mongrel compared to Ben Rudovich and Randi. Harry gave me a superior look and said, "I'm looking forward to seeing you tomorrow." He glanced significantly at Karen, who blushed. Satisfied, he walked away with his two sentinels and his Better Humans.

"What was that between you and Harry?"

Karen ignored this. She looked like she had just received the shock of her life. "Joe, could Harry actually have created a superior human?"

"Whatever he has done, Karen, is unnatural. Remember the Transformation Centers from 2098? Harry has just accelerated his progress, but the result will be the same."

Karen had a funny look on her face. "Joe, I want to be a part of that."

"What??"

"Harry's procedure. It's for real. I want to help people look like Ben and Randi."

"You're kidding. After what we saw in 3013?"

"Don't look at me like that! You forget that I know more about Transhumanism than almost any layman. I've talked with Harry's scientists and Harry himself, and I've studied the science. I really think Harry has changed the timeline for the better."

I couldn't accept this. "Harry Kaine will fail, Karen. It's inevitable. No matter how good those two fake humans look like."

"Open your eyes, Joe. They're not fake. Ben and Randi are genuine, just like that barbarian Ralph Zimring said. Besides," – she looked at me critically – "You don't know me."

"I thought I knew you." I was seeing a side of Karen I never suspected, but I was still in disbelief. Karen had a faraway look in her eyes. "Are you actually considering going to work for Harry Kaine?"

"Very observant, Joe. I am. I've been reviewing the proposal he sent me last year. I'm up on Transhumanism, and I don't want to be a big fish in a small Midland pond anymore. New York is for me."

I didn't waste time on emotional rants or make-wrongs. I could see that Karen had her mind made up. "Well then. You should get your things from our room and move in with Harry."

Karen looked shocked for a second, then her eyes traveled to the stage, where Ben and Randi were rehearsing their presentation. "All right."

I couldn't believe it. I walked back to the rest of the gang feeling totally gutted as Karen walked down the aisle to the stage. I could tell she was looking for Harry Kaine.

I saw Kjirsten first. "Where is Karen going?" she asked.

"She's joining Harry's team. Apparently Ben and Randi impressed her."

Kjirsten gulped. "They impressed me too. Joe, I had no idea that...a human being could look so good. The clones we saw in 3013 were lifeless, but Ben and Randi...they are incredibly beautiful. And so vibrantly healthy!"

"You're not shocked that Karen left? It was rather abrupt."

"Not really, Joe. Karen has spent an awful lot of time studying Transhumanism, and she told me she wants to get out of Midland. Harry already tried to recruit her."

"She said she was reviewing Harry's proposal. That's news to me."

Kjirsten regarded me critically. "Yes, Harry has been seriously courting Karen for a while. Maybe Ben and Randi clinched the deal for her."

I just shook my head. Clearly, I told myself, I am no judge of human character. At least not women. I wondered whether I should have been more assertive with Karen, and maybe she wouldn't have left me.

"Are you going over to Harry too?" I asked bitterly.

"No, I'm not," Kjirsten replied. "But I understand how someone would."

At that moment Ralph Zimring was beside us. How the giant had snuck up on the group was beyond my understanding. "Don't sweat it kid," Ralph said. "There's plenty of women in New York."

"Fuck you, Ralph. What do you want?"

Zimring grinned. "That's the right attitude." He turned to the Gang. "Look you guys, Lucas Cernovitz and Paul Wen are here, but my team has determined that they don't care about any of you as long as you stay in your seats and don't rock the boat. They're here to keep an eye on Harry and give him a little surprise."

Liqao was intrigued. "When is this surprise supposed to happen?"

Ralph frowned. "Tomorrow night after the presentation." He wagged his finger at Liqao. "If I catch you nosing around Kaine I'll turn you into a pretzel."

Liqao gulped; Kirra laughed. "You've been warned, Liqao."

Ralph walked away.

"Hey, let's spend a night on the town!" Kjirsten said, smiling up at M'basa. "I want to take the subway down to 42nd Street."

"42nd Street?" Kirra said. "What's so special about that?"

"The Tower of the Overseers in Bryant Park!"

M'basa laughed. "That's a thousand years into the future."

"Yes, but I want to see Bryant." She looked at her dataset. "It's only 4 now. We can sit on the lawn and get something to eat. Please?"

"Oh hell, sure. I promised your father I'd keep you out of trouble. Going to a park is pretty safe."

"That sounds good," Kirra said. "I'm starving."

Lledren, Liqao, M'basa and Kjirsten agreed to this plan. Liqao thought: Tomorrow I'm going to find out what Kaine's surprise is, that oaf Ralph Zimring be damned.

"I'll pass," Ger said. "I know a physicist up at Columbia. I'm going out there to talk to him about my magnetic engine."

"Joe?" Kirra asked me.

"I'm bummed about Karen. I'm going to sit and mope and eat room service." I felt that Karen had betrayed me, and that what she had done was totally out of character. Well, I could at least run up a big room service bill on Harry Kaine's dime. That made me feel a little better.

When I got up to our (my) room I saw that Karen had already moved her stuff out. I drowned my sorrows by drinking the complementary bottle of wine from the room's mini-bar, and felt better. I began to feel cooped up and took the elevator down to the convention center floor. The place was huge, with restaurants, a sound stage, little shops where you could get a haircut or a manicure, gift shops, VR emporiums, a large drone player area, and a workout gym with treadmills and weightlifting stations. I had just decided to work off my frustrations in the gym when I saw, to my right, Harry Kaine walking beside Karen in the corridor that led to the auditorium. They were deep in conversation. Mostly drunk, I took off after Kaine. I saw one of the Convention security guards turn toward me, but I didn't care. I caught up with Harry just before they were about to enter a small conference room. Karen turned around, surprised, as I grabbed Harry by the shoulder and spun him around. "You bastard!" I shouted, and planted a facer on his jaw.

"Joe, what are you doing?" Karen shouted as Harry fell to the floor. I didn't give a fuck about Karen anymore. I was about to straddle Harry and punch his lights out when a man wearing a black uni appeared out of nowhere. He took me by the shoulders and shoved me out of the corridor. "My sympathies pal," he said softly to me as I landed hard on the cement floor, hitting my head. "We've got something better planned for that guy."

As I briefly lost consciousness I heard Karen cry out Kaine's name. "Are you all right Harry?" A few seconds later I woke up. I managed to stumble to the elevator and got into my hotel room. I had a splitting headache, and felt like a hockey glove that had been skated over, turned inside out, and thrown in a laundromat washing machine. Fucking Harry Kaine! At that moment Ralph Zimring opened the door and walked in. The big man had a smile on his face, and I relaxed. My head hurt too much to care how he had gotten around hotel security.

"Heard you took out Harry."

I spoke softly. "Tried to. He took my girl."

Ralph grinned. "You did right. Did you get a look at the guy who rolled you up?"

I thought for a second. "He came out of nowhere. When I was lying on the ground he said something to me."

"Good! Then you got a look at his face." Ralph threw his mobile to me. There was a 10 picture camera roll. I rubbed my head, trying to get some life back into it. "The guy 7th from the left. That's him."

"Good work Courvall." Ralph started to walk out.

"I want to see what Harry's got coming to him."

To my surprise Ralph nodded. "We know where it's going down, in a private VIP section of the hotel. I've set up a camera feed for your mobile. You gotta stay here though. I don't want any civilians around if shooting starts."

"Shooting?—"

But Ralph was gone.

The next morning all of us went to the first session of Harry's conference. I had a headache but felt a lot better. There were several venues in different areas of the convention center, but most of the press was in the pit, recording everything on the stage. Every seat in the auditorium was filled. Harry Kaine was Master of Ceremonies, and in his element. I had to admit the guy was an excellent presenter: nothing fake or phony about him on stage. I looked around the room carefully from my seat at the very back of the great auditorium (I was too bummed to sit with the Gang) and saw several men dressed in black standing at the back of the stage. Were these convention security men, or part of Cernovitz's team?

First Harry showed a glitzy VR film about the latest life-extension technologies, a virtual tour of his A Better Humanity facility (which I had to admit was truly awesome; I could almost forgive Karen for succumbing), and then several short but entertaining lectures from several of Harry's scientists as they came to the podium. Damn the man! He was good. He had turned boring scientific lectures into very short but powerful testimonies to gene therapy, mind uploading, bio-nanotechnology, AI, human-electronic interfacing, and even space colonization. The consistent overarching theme of these presentations was, "long, happy lives free of disease with Transhumanist technologies." I was almost convinced myself.

As the day went by I was looking out for Karen, but I didn't see her, much to my annoyance. The denouement of Day One was the presentation of Ben and Randi, living proof of the success of Harry's new biotech. Ben and Randi walked the aisles after their presentation so that everyone could get a close-up look. The audience was awed, and began to madly send messages on their mobiles.

Only one thing happened to mar Harry's bliss. As Ben and Randi finished their stroll down the aisles and walked back to the stage, a third transformed human strode on to the stage. This man looked to be about 65 years old, but in the pink of health.

"Who...who is this?" Harry stuttered, clearly not expecting the new guy.

The crowd gasped. By God, I thought. That's Senator Blatsky of New York, who has terminal cancer! An excited buzz of conversation exploded from the audience. Then people began clapping and cheering the senator. An almost audible relief and euphoria came from the crowd, as if Blatsky was living proof of a cure for cancer for all present.

For a moment, Kaine's entire body showed shock. Then he recovered. "Welcome Senator Blatsky!" Harry said smoothly. "Ladies and gentlemen, the senator needs no introduction. Here is living proof of the success of Transhumanism and A Better Humanity!"

While everyone was celebrating, I recognized the man who brought Blatsky up to the stage from the photos Ralph showed me last night. It was Paul Wen. According to Ralph, Wen was part of the group who were funding Kaine and who had killed Winthrop Sauvage. They had written those threatening text messages to Karen and the Midland police chief and Winnie Sauvage and the COSA group. Despite my anger at Karen's betrayal I was worried for her.

The first night ended on this enthusiastic and celebratory note. Kaine couldn't have planned it better, but I had been with Harry Kaine in 3013. I knew that Senator Blatsky's appearance had upset him for some reason.

I went up to my room in the Convention Center and ordered room service when I should have been in bed with Karen. Unlike Harry, I had not been assertive enough with her. Karen had told me that herself more than once. My pity party was interrupted when my mobile activated. On my screen I saw Harry Kaine and Karen coming out of an elevator, chatting. Next to the elevator was a door that led to the stairwell. Harry and Karen began to walk down a hallway toward a room. Suddenly the viewer rotated and I saw Liqao's face plastered against the glass of the stairwell door! Then the scene shifted back to Harry and Karen. As Harry approached the entrance the double doors slid back into the walls and revealed a luxurious room with plush white carpets, expensive white sofas, and naked young girls and boys walking around. What the fuck! I saw Karen's eyes widen. She turned quickly and ran down the hallway to the elevator, frantically pressing the down button.

The feed showed Harry walking into a bedroom with a girl who couldn't have been more than 13 or 14.

At that moment the feed shifted. It showed Liqao opening the stairwell door at the end of the hallway and approaching the open doors. Suddenly Ralph Zimring appeared, scooped up Liqao, and carried him away.

"Ow! What are you doing you big oaf!"

I had to laugh. It was just like Liqao.

A few minutes later my door burst open and Karen ran into the room. "Joe!" she cried, and threw herself into my arms, almost knocking me down. I dropped my phone on the carpet. "I'm so sorry Joe – Harry...I had no idea..."

She explained that Harry said they should go into one of the VIP suites and have a drink. "I don't think he knew what was in that room, Joe. It was a surprise to him. But when he saw those...children...his eyes lit up. I ran out of there as fast as I could."

Just then I heard a thump and an "Ow! Hey what did you do that for?" We both saw Liqao lying in a heap on the floor. "Keep him out of my hair," Ralph said, and quickly left the room.

Shortly after that M'basa, Kjirsten, Kirra, Lledren, and Ger came into the room. Karen was still in my arms.

Ger grinned. "So that's how the land lies!"

"What are you doing here?" I asked, still holding Karen.

"We're watching the Harry Kaine show, courtesy of Ralph Zimring," Kjirsten said.

Karen and I looked at each other.

"Somebody interface their mobile with the panel on that wall and let's see what happens," M'basa said.

I let go of Karen and we both looked shyly at each other. She felt good, she smelled good, she looked good...I was so glad she was here and not with Harry. We sat next to each other on the carpet in front of the display panel. People were walking around in the luxurious suite, mostly older men. They were being served drinks by naked young girls and young boys. The double doors were open; apparently this was a private area of the hotel and they weren't worried about being seen.

"They're waiting for a turn in the bedroom," Lledren said. "Who are these perverts?"

At that moment a girl opened the door and scrambled quickly out of the bedroom, grabbing her clothing. The other young people dropped their serving trays and all fled naked into the hallway. "Get back in here!" one of the men shouted, his drink spilling onto the carpet as one of the fleeing boys bumped into him. "Goddam kids!" another complained. Harry Kaine got out of bed, holding a sheet around him. The men in the room were running around with outraged looks on their faces, calling for convention security.

The hallway past the suite branched right and left. Two crouching men holding weapons were coming down each of the corridors. Four men entered the hallway from the stairwell, carrying weapons. All of them wore black block; their faces were covered. "It's Ralph Zimring!" I cried. It must be Ralph, the man was a foot taller than everyone else. Several security guards ran into the hallway from the elevator. Ralph spoke to them. "Take everyone out of here, it's gonna get ugly." The young kids dashed down the stairwell, herded by two of the guards. By this time the old guys in the room were panicking as they saw Ralph's armed men, and stumbled to the elevators like over-the-hill Hollywood actors after a bad audition. The elevator opened and they all crowded in with the other two convention security guards. I almost laughed because the scene looked like an old Buster Keaton silent movie.

Harry Kaine was still standing in the bedroom holding his sheet, a look of shock on his face.

273

The four men coming down the two corridors approached Ralph's team, who were standing about forty feet from the open doors to the suite. "You fucked up, Zimring," one of them said. "Just couldn't leave well enough alone, could you?"

The huge man replied, "We can do this the easy way or the hard way, Cernovitz. Drop your weapons." While Ralph was speaking I saw him make a micro movement with his head to his fighters.

"Fuck you, Zimring."

As the man began to speak Ralph moved incredibly quickly, rolling to the floor and taking down the two men standing to the left of the open doors. Cernovitz fired his weapon, but the bullets went into the ceiling. Just as Ralph was making his move his three fighters were already in motion. As the bullets were flying I saw Harry Kaine fall to the plushly carpeted floor. He wasn't moving. In three seconds it was over.

We heard the big man mutter. "This one's for you, Cernovitz." A muffled gunshot was heard.

"Kaine is hit!" I cried.

Liqao, still lying on the floor, spoke up. "See? I was right. The only way to stop Harry is to shoot him!"

Kirra was shocked. "It's one thing to see violence on vid, it's another to see it in reality. One of those men is dying." She turned her head away.

M'basa wasn't impressed. "The genocide between the Hutu and the Tutsi in Rwanda...and the genocide in Darfur...that was real violence. I hope Harry Kaine is dead."

The feed stopped and the monitor went black.

"Who is Cernovitz?" Lledren asked.

"He's the head of Harry Kaine's group," I replied. "Ralph told me last night."

Liqao was satisfied. "That big oaf killed two creeps for the human race."

We sat around for half the night talking about it. Mostly we wondered whether killing Harry Kaine would stop the Transhumanist movement.

"I don't think so," I said. "Remember the time conservation law. Something will happen to negate Harry's death."

Everyone disagreed with me. I cursed myself for bringing up the topic, for it ignited more conversation. I was wishing my friends would leave so I could talk to Karen. I think Kirra and Kjirsten picked up on this because Kirra said, "It's one in the morning. We should all be getting some sleep."

We said our good nights. It was now me and Karen alone in the room. There was an awkward silence.

I broke it by saying something stupid. "I can't believe you went over to Harry," I said, my voice bitter.

Karen's face went white but she looked at me defiantly. "I'm not apologizing, Joe. I made a mistake in judgment. I thought Harry was legitimate."

"Harry Kaine is – or was – a sexual pervert."

"Yes, but look at what he's done. Senator Blatsky was dying. Harry brought him back to life. The same for Ben Rudovitch, and probably that Randi woman."

I didn't want to hear this but I had to admit she was right. "Yes, but for how long?"

"How long do you need, Joe? Even a week in a body like that is better than dying a painful death from cancer."

"Yes...but..." I didn't have a good answer to the personal dilemma of people like Blatsky and Rudovitch. It upset me because I knew that if I were in a similar situation, I might make the same decision Senator Blatsky had. "Yes, but why did you go over to Kaine's side?"

"You saw Ben and Randi up close. Give me your honest assessment of them."

"I...but..."

Karen frowned.

"Both of them are magnificent," I admitted. "Larger than life. It's like they got an injection of magic elixir. "

"Maybe they did."

Karen's face was a study in obstinacy, hope, and...some other emotion I couldn't read. I had no idea what was going through her mind. "Let's get some sleep Joe," she said. "I'm exhausted."

We slept in separate beds that night.

The next morning we heard the six-o'clock news: Harry Kaine had been shot during the skirmish we saw last night, and had been rushed to a hospital in Manhattan, where he was pronounced to be in critical condition. He was then flown to the Better Humanity facility in DC, where an emergency procedure was performed. The cover story was that he had been attacked by a nut job who was against Transhumanism. The news report said that Harry was resting comfortably and that the conference would continue as scheduled. Harry would have a surprise for everyone today.

Karen didn't want to talk. We got dressed silently and had a quick breakfast with the Gang at a little pancake shop near the auditorium.

The morning began with cutting edge scientists giving scripted presentations very cleverly framed, but this time of a more promotional and speculative nature. The people who put this together were obviously master psychologists and persuaders. "What will society look like without Transhumanism?" one of the scientists said. "Pollution, disease, short life spans, pandemics, and conflict. With Transhumanism, all of these morbidity factors are eliminated. Longer life spans, with perfect health, means less competition for scarce resources, less pollution, and a happier population."

It was a good talking point. Another scientist said, "Why spend a lifetime fighting off illness when your immune system can be improved so that you never get sick?" The lab-coated woman said, "Things like the common cold, the flu, childhood illnesses, cancer, and the diseases of old age are all avoidable using the technologies developed by A Better Humanity. Transhumanism is the way forward for the human race."

Right after the lunch break Harry Kaine walked slowly from backstage to the podium with vid and VR recorders tracking his every move. The audience was spellbound. Harry Kaine now looked just like Ben and Randi: a perfected version of himself.

The audience cheered and clapped as Harry Kaine absorbed their adulation and enthusiasm. "Ladies and gentlemen. Last night I had an, er, accident. I was shot by a lunatic who was opposed to the advancement of the human race. So I decided to take the plunge and practice what I preach. Last night I was flown to our headquarters in DC and transformed." Harry paused for effect. "I am a living example of what every human being on the planet can become."

Kaine stepped away from the podium and let everyone see. It was like a before and after image of a bum on the street corner who had been cleaned up, manicured, and fitted for a new tailored suit. "To my friends in the front row," Harry said, speaking to us, "the debate for tomorrow has been canceled. I'll be brutally honest and say that I held a grudge against you and your metaphysical nonsense. I wanted to destroy you. But that is unnecessary now."

Harry stretched his hands to the ceiling in celebration as everyone filmed. He looked like a benevolent god forgiving a bunch of ignorant barbarians. I was impressed. I could see how Karen had been won over.

Harry smiled. "Instead of a debate tomorrow evening there will be a Q and A session with the media." Harry paused for a second to consult his wristband. "Today's schedule has been slightly revised. We're going to show how easy it is to build a life-extension center using our new equipment and technology. As a service to humanity we will donate the first ten centers to lucky cities who can apply online. An online raffle will be held to select the winners."

I could see how upset Kirra was. Her Native Americans were waiting in the wings, but they would not be needed now. I turned to look at Karen. "I apologize for what I said last night. These people aren't fake! They look like perfected human beings."

Karen sighed. "They do. Joe, could Harry be right after all?"

"I hope not. But he's not waiting for 2098, or a Referendum."

"What I want to know is, how did he do it so quickly?"

———— • ✸ • ————

"What do you think about Harry Kaine and Bob and Randi now that you've seen them?" I asked Ralph Zimring just as the Gang was about to leave New York. We were standing in the NYC Convention Center lobby, waiting for taxis.

"I think they're...transitory."

"What do you know about it?" Liqao piped. "You're just a big oaf who kills people."

Ralph smiled tolerantly at what he called 'the little gnat.' "It's a gut feeling I have."

M'basa was enormously impressed by the big man, in a negative way. "You remind me of the assassin K'wame Otoro," he said. "A mass murderer who killed thousands of Hutu."

Ralph shrugged. "I don't know anything about that. The man I killed was Lucas Cernovitz, head of the organization that murdered Winthrop Sauvage and who threatened you guys with those text messages. He runs – ran – a trans-national child trafficking/porn ring that abuses children, and eliminates people who get in their way. And worse."

"It was vigilante justice," M'basa insisted.

Ralph was in no mood for moralizing. "That man needed killing."

We all looked up at Zimring.

"I'm sorry, children. I shouldn't have set up that feed. It was pure hubris; I wanted to show you what you're up against if you try to stop Harry Kaine."

"Thank you," Kjirsten said, thinking of her father, who had received one of the messages.

"Don't mention it," Ralph replied. "It was fun." Ralph said this nonchalantly, as if killing Cernovitz had been a good workout at the gym. "But don't think those boys are through. I decapitated their head, but Cernovitz's assistant Paul Wen and his gang are still alive and well. Be careful."

I gulped, thinking of Karen and her investigative reporting. If she somehow crossed them again...

"Did you really kill that man?" Kirra asked nervously.

"Sure I did," the giant replied. "It was necessary to protect you all. And children around the world."

"Killing never solves anything," Kirra said.

"Of course it does. An evil child murderer has been removed from the earth. Cernovitz's death will prevent more children from being abused."

"Someone else will just take his place."

Ralph grinned. "Setting up a network like that is a lot of work. It takes time and skill. There's a good chance Cernovitz's organization will fall apart now because he was the brains behind it. You see, I researched the situation carefully before I acted."

"Killing blackens your soul," Kirra insisted.

Ralph shook his head. "A warrior knows that life is temporary, and is unafraid to die. Me and Cernovitz understood that; we played a game and he lost. We are all warriors, Kirra, it's just that almost no one understands this."

Kirra's jaw dropped to hear the barbarian talking in this way. "There is more to you than I thought, Ralph Zimring."

Ralph looked at each of us in turn. "My advice is to keep your mouth shut about what really happened here, and watch how events unfold."

"How did you get away with killing that guy?" Lledren asked.

"I have a friend who heads the NYPD's Counterterrorism unit, who has been trying to apprehend Cernovitz for several years. Let's just say that it is being handled as...ah...a national security matter."

"Who shot Kaine?" M'basa asked.

"Not me. Bullets were flying; Kaine got hit. Shit happens."

The big man turned suddenly away from us and walked into the crowd.

On the way back home, on the train, Karen and I agreed that Harry's transformed were for real, a quantum leap in human health. We couldn't understand why perverts like Cernovitz would want to advance the human race. It didn't make sense.

Part IV

CHAPTER **27**

2039 AD, three years later

After Harry's New York conference over 100 cities applied for life extension centers. In just three years, 92 cities around the world had them.

Harry didn't have to do anything. As people saw sparkling Ben Rudovitchs and Randis coming out of the Better Humanity Centers, the picture of perfected human beings, they wanted it too. The Centers were overwhelmed and had to expand to handle the demand.

The Gang didn't understand how Harry was doing this, because the Better Humanity centers were like factory production lines. People walked in at 8 in the morning and were released the next morning, completely transformed. It was impossible! Harry must be getting some very sophisticated help from somewhere, I thought. However, no 'normies,' as the transformed called us, could prove anything amiss. Documentaries and investigations of the Better Humanity centers came up with nothing. All seemed above board. Media and commentators concluded that the transformed really were the next advance in human evolution. Harry Kaine became the most recognized and popular figure on the planet. And he always stuck it to us in his speeches, calling those who believed in spirituality or religion 'useful idiots' and 'scientific illiterates.'

In the largest social metamorphosis in world history, transformed humans began to fill the world's major cities. 'Normal' humans were regarded as inferior (and began to regard themselves as inferior), and for good reason. The transformed, true to Harry's promise, never got sick. Through flu seasons and even a resurgence of a new pandemic in the fall of 2038, the transformed were immune. 'Normies' continued to get sick and die of diseases that did not affect the transformed. This caused hopelessness in those who saw the transformed as an abomination. But everyone had to admit that the enhanced humans, as Harry called them, were biologically superior.

Or were they? As a time traveler I knew   that transformed humans were a biological dead end. I had seen their perfect cloned bodies in 3013, their weakness, and their antisocial tendencies. That was the future of Harry's transformed, I was sure. But of course I had no proof. Any argument against the transformed was easily shot down by the facts, and by reality.

After a while even I began to regard my trip to the future as nothing more than a vivid dream, and the people in 3013 AD as an elaborate fantasy. Under the wave of favorable Transhumanist publicity, morale among our little group – and for most normies who didn't want to be transformed – was at an all-time low. My friends and I were resigned to living in a socially and scientifically engineered society that regarded human beings as mere physical constructs that needed to be "improved."

"Maybe we should all become transformed," Karen said one cold April Sunday, when we found out that one of our favorite VRcasters had died of colon cancer the day before. She was only 48. "It galls me that Harry Kaine and his modified humans strut around like kings and queens."

"It's depressing," I agreed. Karen had changed her tune about the transformed over the past three years as more of the population became altered. We were both on the same page and our relationship was blossoming.

That evening Karen and I went out to dinner and then came back to my place. We watched the new vid programs that had become wildly popular in the big cities, where almost all of the transformed humans lived. The heroes and heroines in these "enhanced" programs were (we thought) completely self-centered and narcissistic. We viewed them as comedies. It reminded us of the old "Real Housewives" vids, and that "Idiocracy" movie.

In the enhanced programs, transformed humans called themselves "Betters," and always referred to normal humans as "normies." It was funny, because we thought the "Betters" were morons – beautiful bodies and pea-sized intellects.

"Karen, these shows are more propaganda than entertainment," I said. "The acting is atrocious, the dialogue is juvenile, and the plots are transparent."

"It's Harry Kaine of course," she said.

I felt sorry for the 'normie' actors in these shows. Normies either bullied transformed characters, or were pathetic losers who constantly got sick. "Becoming transformed is the only way to a successful life, that's Harry's message," I said.

"This is too depressing," Karen said. "I want to watch the returns of the Chicago elections."

We didn't go to bed that night. I hate politics but Karen had to pay attention because she is a journalist. So we stayed up until it was time for work.

"Transformed humans won the mayor and 60% of the seats on the City Council," I summarized at 6 a.m. "I can't believe it."

"99% of transformed humans voted. Only 52% of normies did."

"It's like they are programmed," I said innocently.

Karen glanced sharply at me. "Do you think it's possible? I've studied Transhumanism. Programmable bio nano-bots inserted into the bloodstream to fight disease could also be used to send electrochemical messages to the brain."

I was appalled, and laughed nervously. "Nah...c'mon, that's crazy talk."

Karen just gazed at me silently. We both knew what Guardian technology looked like. "But...but it's outrageous!"

"Is it?" Karen was better than me at confronting uncomfortable things. "We know what Harry is trying to do. He's doing it!"

"I'm glad I'm spending most of my time with Ger developing that magnetic engine," I said nervously.

Karen was decisive. She didn't look like a woman who hadn't slept in 24 hours. "I'm going to Chicago to check out what's going on with these transformed humans. My editor has already given the go-ahead."

I thought swiftly. I didn't want Karen going there without an escort. "I'll come with you."

Karen smiled brightly. "I was hoping you'd say that!"

"You're a minx," I said. "A manipulator."

She batted her eyelashes at me. "But Joe, men are so easy to manipulate! It's not my fault."

I had to admit she was right. At least, I was easy for her to manipulate. "What time do you want to go? I'll call in to work and take a vacation day, then I'll call a popup cab."

"I want to be on the road by 9. My boss has already arranged for me to visit the government offices. I also want to get a look at one of these transformed communities, Joe. I want to see these transformed people up close and personal."

I could see that Karen had the bit between her teeth. I wanted to make sure she didn't do anything too outrageous or impulsive, and to get her out of trouble if necessary.

But I didn't have to worry.

Karen called me at 8:30 the next morning. "Joe, guess what!" she said excitedly. "Ralph Zimring is coming along with me, so you don't have to go."

I was a little miffed. "Don't you want me to?"

I heard some muffled voices. "Ralph says you can come if you want, but you're likely to be in the way."

I had to laugh. The giant was standing next to Karen with his arms folded, an air of resigned acceptance on his face. "OK Mr. Zimring, I'll go to work."

Ralph relaxed. "That's good kid, you're being an adult." He gave me a stern look. "Get that magnetic engine going. Ride herd on Germaine, he's a little too flighty."

Ralph was using Western metaphors. "Yes, Sheriff Zimring. Keep your gun loaded and ready."

Ralph grinned. "Civilians can't carry guns in Chicago, sonny." Ralph pulled out a knife and a wushu chain whip from a pocket tied to his waist. He was wearing his standard Berglin Enterprises uni with a decal that said 'Chief of Security.' "These will do."

God, I thought. This guy is a caricature, somebody from a comic book. Except that he was real, and I knew that with him along I didn't have to worry about Karen.

"What's your interest in this, Ralph? Don't you have duties to perform for Max Berglin?"

"Max saw the election results just like you. He wants intel. He's got three big customers in Chicago and one of them is now transformed."

That was good enough for me. "OK, see ya Karen. Will you be home tonight?"

I could tell she was excited to be going with Zimring. "I'm not sure Joe. See ya!"

I went to work while Karen and Ralph had fun in Chicago. Karen told me about it that night when Ralph escorted her to my place. "Have fun kid," Ralph said to me as he brought her into my apartment.

Karen came in flushed and excited. "You wouldn't believe what we saw, Joe. I should have brought Lledren along to film in VR, but I got most of everything with my mobile."

The first thing on the device was an interview with a transformed City Council member, who looked down condescendingly at Karen. "Mr. Councilperson, are you excited about your group's victory?" Karen asked.

"We felt it was inevitable," the man said in a lofty tone. He looked just like Ben Rudovitch.

I looked over at Karen. "Are these transformed humans clones?"

"I don't think so. Watch."

As the interview proceeded, several persons walked into the council area. Two were transformed humans, three others were "normies." The difference between the transformed and normal people was obvious. All of the transformed were walking advertisements for Transhumanism and perfect health.

"So you've won the city Raymond," a normie woman said after the interview ended. "The mayor of Chicago is now a freak."

"That's Councilwoman Jenks," Karen explained. "Her district is almost all normie."

Miles Raymond smiled his contempt. "You dare call us freaks! Little human, your kind are on their way out, and the sooner the better."

"Humanity 2.0 you say!" Jenks replied. "Raymond, you and yours are nothing more than beauty shop mannequins."

"Spoken like the antique you are," Raymond replied. "Have you gotten sick yet, Congresswoman Jenks? It's flu season."

Jenks flushed. "I...I have just recovered from a slight illness."

The beautiful man nodded. "Quite so, Jenks. And you shall be sick again, and again, and again. And your children will contract illnesses for their entire lives." Raymond smiled brilliantly. "Join us, Jenks. Live happy, healthy lives without care."

I saw a look of longing in the face of the Councilwoman. "Miles...but no. I can't do it."

The mayor shrugged. "Suit yourself; it's your funeral."

As Jenks passed him on the way out Raymond said, "and that funeral will be sooner than you think."

Jenks flushed and hurried out of the room. The other transformed humans in the Council area had smiles on their faces. The two normies were grim.

I looked at Karen, shocked. "Is this what society has come to?"

Karen nodded. "Ralph and I both saw it. Joe, a curious social phenomenon has rapidly developed in the major cities. Transformed humans have formed their own societies and social groups. And they are beginning to gain political power."

Karen and I had seen no transformed humans in Midland. "I hadn't realized how many transformed humans there are now."

"They are mostly confined to the large cities. But there's a small group in Midland now, and talk of building a Better Humanity center at the last meeting of the Midland City Council. Ralph told me about it."

I was irritated. "Is there anything Zimring doesn't know? He's like an ubermensch with superpowers."

"Jealous, Joe?"

"Yes, I am," I said honestly. I had learned to be direct with Karen and not hide my feelings. "Between Ralph Zimring and these transformed humans I feel outdated, just like Miles Raymond said."

Karen smiled. "Cheer up. You'll feel better when I show you this."

Karen's recording continued.

"Would I be able to enter one of your communities and look around?" I saw Karen asking Miles Raymond. "I'm interested in becoming transformed."

I smirked at Karen. "You're a convincing liar."

"Oh, no need for that Joe, as you'll see. These transformed humans are so self-centered that when normies flatter them it's accepted at face value."

Karen's recording showed her and Ralph entering one of the transformed areas of the city in a popup. Ralph was hanging out the end of the thing, unable to fit on the small seats. He looked like a crash dummy being transported to the testing ground.

"OK, watch this," Karen said. The vehicle drove into Riverdale, formerly the most crime-ridden neighborhood on Chicago's south side, past 115th Street. "Driving down 122nd Street, we saw a sign that said 'Entering Transformed Area.' There, do you see it?"

"Yeah." I noticed that several abandoned buildings were in the process of being renovated; a street cleaning machine was scrubbing 122nd when we crossed it. There were several people walking around.

"Watch what happens," Karen said. "Ralph and I get out of the pop-up and start walking around."

The big man must have been trailing Karen as she recorded, because he never appeared in the vid. The neighborhood was in the process of rebuilding. Construction workers were everywhere; a few transformed humans were walking around on the streets. One of them spotted Karen and Ralph; two construction workers walked toward them from a building site. "What are you normies doing in our neighborhood?"

"I gotta tell you, Joe, nobody noticed Ralph until he spoke up," Karen commented. "It's amazing that a seven foot guy can be so invisible."

Karen and Ralph continued to walk and didn't respond. Two more transformed humans came up to them and blocked the sidewalk. "We want to know what you're doing here."

"None of your business," Karen replied. "We're just taking a walk."

"You're in our neighborhood. We don't want normies around here."

"Why not?"

One of the construction workers stepped forward. "Because you're Humanity 1.0. You're dinosaurs. Get lost."

"We just want to look around and see what kind of a society you're building here."

The men took a look at Ralph and then back at Karen. "Well fellows, should we show these normies around?"

"I'm a reporter for the *Midland Chronicle*," Karen said, wanting to be honest.

One of the men frowned. "As long as you write a fair article."

Karen saw that there were two types of transformed humans here: tall and blond, and barrel-chested men of medium height with dark curly hair and dark skin. Karen was struck by their resemblance to a couple of the Guardian clone lines.

One of the other men laughed. "Oh, show them around Brian. Let these normies see the future of the human race."

This met with general agreement among the men, the consensus being that Karen and Ralph, unless they were very stupid persons indeed, would see the superiority of transformed culture. Karen sealed the deal by saying, "I've watched many of the Better Humanity serials. I'm curious, that's all."

The workmen walked off back to their jobs, all looking like they had just come out of a beauty shop even though they had been doing rough carpentry work on the houses they were renovating.

"I'm Brian," the man said, not offering to shake hands. "Come with me. There's a small community center in each of our neighborhoods. We can talk there."

Ralph and Karen followed Brian. "Ralph told me later he was disappointed that the confrontation led to nothing," Karen said. "He wanted to see whether these transformed humans were as strong and agile as they looked."

Karen was amazed at Chicago's lower population density. Of course many people had moved out of the big cities in the aftermath of the 2020 pandemic, but this neighborhood looked like a ghost town that was being renovated.

The community center was a hastily constructed building on a former parking lot. Signs were posted on a big billboard, announcing various events. "122nd Street Beauty Contest This Weekend," one of the signs read. Another said, "Celebrate Liberation Day! Bring something to eat and let's talk."

"What is Liberation Day?" Karen asked Brian.

Brian looked at her as if she was mentally arrested. "Oh that's right, you're normies. Liberation Day is the day when the first Better Humanity Center opened to the public in DC. April 13th, 2036. It's the day we were freed from the tyranny of sickness and the slavery of old age."

Ralph spoke for the first time. "Where are your artistic and religious events?" he asked. "Your book clubs? Your plays and musical events?"

Karen was startled at this question.

"There are no announcements for concerts or cultural events," Ralph added.

Brian shook his head. "We haven't time for that nonsense."

Karen spoke diplomatically. "Transformed humans seem to like to stick together."

Brian nodded his agreement. "We like our own company. We find that Humanity 1.0 culture is...fatiguing."

Karen stopped the recording. "The fatigue of new ideas and artistic insight," she said to me.

She restarted the vid. Ralph was looking at a number of full length mirrors that had been placed on the walls of the structure. People frequently stopped to admire themselves or check their appearance. Beside her Ralph Zimring was yawning. He lowered himself and spoke softly to her. "These people are boring and placid. I'd say half of them are retarded."

Karen had to stifle a laugh. "I wonder whether intelligent people like the famous physicist, Raoul Hernandez, become stupider after they are transformed"? she replied softly.

Karen spoke with a few of the transformed as she and Ralph walked around the center. Then the two walked out and back to their popup, which was waiting for them on the street. "You guys loved it didn't you?" one of the construction workers asked confidently.

"Oh by all means," Ralph replied drolly. "Beautiful people doing beautiful things."

The man beamed and waved to them as they got in the popup.

"Get me out of here Karen," the big man said. "I liken these transformed to a herd of cows."

The recording ended with Karen's laughter.

"Do you feel better about yourself now Joe?" Karen asked me with a smile as we sat at my kitchen table.

"A lot better. Are all transformed humans as shallow as that?"

"I don't know."

"See if you can interview Dr. Hernandez," I suggested. "Maybe Ralph is right. Maybe these transformed humans are slow."

We didn't have to wait long for Karen to interview the famous scientist. We saw the advert on the local news that evening. Harry Kaine was going to bring a Better Humanity event right to Midland a week from Friday. The main attraction: Dr. Raoul Hernandez.

On the day before Harry Kaine's Better Humanity presentation in Midland, Ralph Zimring showed up at the Radisson Hotel downtown, the city's most luxurious and largest convention area. He spoke to one of the Radisson's security guards in the hotel lobby. The sun was streaming through a couple of plate glass windows onto the floor.

"Harry Kaine is here, is that right Jamaal?"

The guard frowned. "I'm not supposed to say anything. You know that, Ralph."

Ralph could tell the man wanted to tell him something so he stood there silently. He had been careful over the years to cultivate relationships with the police and the private security firms. Everyone knew him.

"I saw that transformed freak walk in here with a normie girl," Jamaal said finally. "I wanted to wring his neck but I'm supposed to protect him."

Ralph grinned. "Leave that to me, my friend. What room is he in?"

Jamaal sighed. "Oh fuck it. It's room 1408." The man looked at Ralph curiously. "Kaine's got two guys guarding him at all times. One outside, one inside. They're carrying."

"Thanks for that Jamaal," Ralph replied in a bored voice, striding to the stairwell door. To stay in shape he always walked the stairs. The day he couldn't do 14 floors without stopping was the day he'd retire.

When Ralph got to the fourteenth floor he rested his burning legs for a minute in the stairwell. He opened the stairwell door and saw that a private security guard was standing in front of the door to room 1408, looking at his mobile. Ralph approached the man, walking casually. "Is this Harry Kaine's room?" he asked with the breathless air of a fervent admirer. "I wonder if I could get an autograph?"

The man looked up at him. "You're Ralph Zimring," he said flatly. "Leave or I'll call hotel security."

Ralph yawned and spoke quietly. "Open the door."

Ralph noticed a slight tightening in the man's eyes. He was going to make his move, but Ralph was ready. Ralph moved forward a step, pinning the man against the door with his body. He spoke conversationally. "Why are you protecting that freak anyway?"

A voice was heard inside the room. "Stop it Harry!" a frightened girl shouted, startling both men. The guard's face showed anger.

"What's going on in there?" Ralph asked, stepping back and giving the man some space.

"Let's find out," he said, opening the door with a keycard.

Ralph burst into Harry's room, the guard following. Ralph saw Harry dallying with a young woman. He snapped photos with his mobile as the second guard rushed out from a little alcove at the back.

"Get that guy out of here!" Harry said to his guards, who did not move. "Who is this normie?" Harry demanded, getting up off the sofa.

"My name is unimportant, Harry," Ralph said. "I'm here to tell you that your Better Humanity center is not wanted in Midland. Certain councilpersons you have bribed will be doxxed if you proceed."

"Wait, I know you. You're that normie barbarian, Ralph Zimring."

Ralph executed a passable bow for a seven-footer. "*à votre service*, Kaine." Ralph was feeling expansive. "Really sir, you should take your clown show out of our city. And the feeling is mutual, Kaine. We don't want your 'transformed' humans in our city."

Harry frantically looked around for help, but none was forthcoming. He got out his mobile and called hotel security. His two guards did nothing. He vowed silently never to hire normie security again.

"That will not be fruitful Harry," Ralph said calmly. "The Radisson security staff feel just as I do, as well as the majority in this town." Ralph glanced at the frightened girl, who had wrapped her arms around herself at the end of the sofa. Something would have to be done about her.

Harry snorted. "The majority of normies, you mean." Harry had calmed down and now spoke as one who knows he will be protected no matter what he does. "You can't stop me or the Movement. If you knew the people backing me you'd leave this room immediately."

Ralph smiled. "Oh, Max Berglin and I are well aware of the Alexandria Group. In fact, it was I who, er, eliminated Lucas Cernovitz in New York back in '36."

"*YOU!!*" Harry shouted. Then he smiled. "Your life isn't worth much now, Ralph Zimring. Thank you for outing yourself."

Ralph bowed again. "It was a pleasure. I'm looking forward to, er, engaging with your operatives again. I've been pretty bored lately."

Harry looked the big man over. This normie was impressive. Harry felt a thrill of sexual excitement as he thought how much more impressive the giant would look after his transformation. "Join us, Ralph Zimring. Recognize the inevitable. You would be a formidable asset."

"Thank you, but no." Ralph held up his mobile. "You seem to forget I have been recording this. Placing this vid on the WorldNet will out *you*, Harry Kaine." Ralph pointed the mobile toward the girl. "How old are you?"

"Fourteen. "

"And you were having sex with Harry Kaine?"

"Is that who he is? Yes." She looked hopefully at the giant. "Will you take me home to my family sir? I...I made a big mistake."

Ralph held out his hand and the girl approached and stood behind him. Ralph turned to Kaine, holding his temper. "This is a normie girl, Harry. Why don't you dally with transformed girls?"

Harry smiled deprecatingly. "I would not so demean myself."

Ralph flared and took a step toward Harry. To his credit, Kaine flinched but didn't step back. The two guards had disgusted looks on their faces; they obviously didn't like their client. "Don't hurt him or we'll get in trouble," one of them said to Ralph.

"No problem. I just want to test something." Ralph grabbed Harry by the shoulder and Harry tried to twist away, but Ralph held him easily. These so-called "Betters" didn't have unusual strength then. Satisfied, he stepped back toward the girl, still facing Harry Kaine. "Tell Paul Wen and his gang of perverts that he has 24 hours to get his crew out of Midland. That includes you."

Harry laughed. "Oh you poor, deluded soul. The presentation will go on without me."

"No it won't Harry, for all of you freaks will be gone. Me and my security detail will see to that."

Harry sighed, remembering what had happened in New York. There must be no disruptions here in Midland. "That will not be necessary, Zimring."

Ralph grinned. "You freaks are through here, Harry Kaine."

Harry laughed again. "I think not. Mr. Zimring, this presentation will be done by normies. As will our proposal to the Midland City Council. You see, we are always one step ahead of you."

Ralph was stunned. "You mean...you have found normies to do your dirty work for you?"

"That's right, Ralph." Harry was gloating now. "Enthusiastic participants, I might add."

"I should have guessed," Ralph said with disgust. "You bribed them with a free bio-conversion."

Harry shook his head sadly. "Human 1.0's are so...predictable. It's why your species will go the way of the Neanderthal."

Ralph understood the analogy. When *homo sapiens* appeared on the scene, their superior intellect and understanding eventually displaced the older, more primitive, Neanderthal species. "Over my dead body," Ralph said grimly.

Harry waved his hand dismissively. "Take this...female...away. She is inferior."

Ralph knew he needed to get out quickly before he did something stupid to Harry. He was under strict orders from his boss, Max Berglin. "The game is bigger than you, Ralph," Max had told him during his brief. "Don't get carried away by your emotions."

Ralph grabbed the girl's hand, about to leave. "We haven't seen the last of each other, Harry Kaine."

"I'm looking forward to it. And by the way, enjoy the event." Harry walked to a small cabinet and took out six guest passes. "For you, your boss, and anyone else you'd like to invite. Right here in the Radisson in the convention center, tomorrow, Friday, 9 a.m. sharp." Harry smiled contentedly. "You won't want to miss it, Ralph. You'll be surprised, I'm sure, when you see who we have enlisted."

Disgusted, Ralph took the tickets. As he walked out of the hotel room with the girl he could hear Harry arguing with his security men.

"Where do you live, sweetheart?" he asked as they walked down the hall.

"Chicago. I want to call my mom and tell her I'm coming home."

"Did Harry rape you?"

The girl flushed. "We didn't get that far. Thank God you came in when you did."

Ralph handed her his phone. "Call now. I'll drive you up."

"Oh, thank you!"

Ralph was a sucker for helping the ladies, even though he had never met a woman who could tolerate his lifestyle. The girl had to call twice before the call was picked up. An older woman answered and saw her daughter. "Penny! Oh thank God you're all right!"

"Yes I am mom, thanks to this guy here." She moved the mobile to show Ralph.

"Wow! Who is he?"

"This is Ralph Zimring. He's really tall and really nice. He's going to drive me home, right Ralph?"

"That's right Penny. We'll be there in an hour or so."

In his beat-up car on the way up to Chicago, Penny told him she had run away from home two weeks ago, and that her parents had filed a missing person notification with the Chicago police. "That really made me mad," Penny said. "I just needed a vacation from mom and dad."

When Ralph walked in the door at the house on 33rd Street and Ogden, he was thanked profusely by Penny's parents. "Where did you find her?" the mother asked.

Penny looked up at him anxiously.

"On the streets in Midland, ma'am. She saw me and asked for my help."

Penny's face relaxed. She gave him a grateful smile.

Ralph studied the two parents, both in their forties. "Have you seen any of these, er, transformed humans?"

"Oh we've seen them all right," the father said. "Strutting around like they own the city. Now they've got control of City Council and one of them is the mayor!"

The mother was hugging her daughter, but her face registered anger. "They treat us like bugs," she said.

"Don't worry about the missing person report ma'am. I'm a...security expert and am known to the Chicago police. I'll take care of everything, but you might get a confirmation call from a Sergeant Jackson."

Penny's father shook hands. "Thanks very much Mr. Zimring!"

After he left the house Ralph went downtown and saw Max's transformed client, who assured him that he was satisfied with the software from Berglin Enterprises. Ralph swore mentally as the man tried to get him down to the Chicago East Better Humanity center. "I'll drive you myself," the client said. Ralph declined as politely as he could.

After that Ralph decided to drive around Chicago and see how many transformed neighborhoods there were. Several hours later he concluded that the city was about 35% transformed already. It was fucking amazing. Chicago had two Better Humanity centers and they were cranking out transformed like widget factories. Ralph was stunned at how fast the city was changing.

He had gathered some good intel on this little adventure, and Max would have to be briefed. He would do that as soon as he got back to Berglin Enterprises.

After Ralph briefed Max he went to see Joe Courvall and Germaine Robinson at their lab. Ralph decided to walk, for he couldn't fit into those damn pop-ups and it was only a mile to the Taubman Building on campus. As he walked he thought about his job, which wasn't really a job. Because Max was engaged in some classified research and had contracts with the DoD, Berglin Enterprises had to have a trained security detail at the building. Until the confrontation with the national security establishment several years ago during Max's development of the Cube, he had just been head of security. After that he had turned himself into a sort of roving human intel asset for the city's law enforcement community (and Max). He knew everyone of any importance in the town, and made it his business to know what they were doing. The public knew him only as a very tall gentleman who sometimes appeared in public wearing a Berglin Enterprises uni. But police chief Chastaine knew him well, as did Sheriff Costigan and the various private security firms that guarded public buildings like the Radisson. He was on familiar terms with the local office of the FBI, and sometimes briefed the local (but invisible) NSA operative on events in Midland. He even knew the Midland CIA station chief, although that guy (and his office) was small potatoes now after the cleanout of the Agency. For sure his job at Berglin Enterprises wasn't as good as his old mercenary days, but at least he had his freedom.

When Ralph got to Germaine's lab he saw Joe Courvall and Germaine staring at a rotating shaft coming out of a device surrounded by a circular array of magnets that sat in the middle of the floor. It was a miniature version of the huge device he had seen three years ago, which now sat at the back of the lab.

"What is that thing?" Ralph asked.

Ger spoke like a kid who had just received everything he wanted for Christmas. "The magnetic engine. A prototype."

"It's not plugged in."

"You got it Ralph."

Ralph saw the two nerds laughing. "What can it do?"

"We've got a simple, elegant design, using permanent magnets, that can function as a powerful but portable generator," Germaine said, pointing proudly to the device. "It's a great leap forward, I say without bragging, from the primitive magnetic motors used in electric cars. You see it turning a crankshaft here; it will go on forever. We've modified this prototype demo to generate 60 volt AC for houses and businesses. This thing is going to replace the primitive wires and poles of the electrical grid."

Ralph was amazed. "You finally did it then!"

Germaine smiled proudly. "With Joe's help. It took us four years, and we almost gave up several times because...well, it was a lot harder than we thought. But we did it."

Ger and I exchanged glances.

We had almost killed each other a number of times over the past four years, but we both felt an overwhelming sensation of satisfaction. And relief.

"You guys are never going to be allowed to distribute this. You remember what happened to Max! As soon as people start using this thing Midland Power will notice a drop-off in electrical use. You'll get a Patent Secrecy Order and have the nat sec boys in here."[27]

"Not this time Ralph," I said, understanding the Ralph often spoke in metaphors. "Nat sec" meant national security. "This thing can be built by a competent do-it-yourself-er in his basement for about $3,000. We've worked out all the kinks now. We've got blueprints, a materials list, and an assembly schema."

I could see Ralph's eyes widen.

"That's right," Ger said. "We want to put the specs on the World Net. No charge. Download the PDF's and build your own. Become energy independent."

"Jesus Christ. If you do that you guys are going to blow the lid off the energy industry. Just like Max tried to do years ago."

"That's right Ralph," Ger said. "But this time Joe convinced me to give it away. There will be so many people competing to build them and put them out, not even the government can stop it."

"And we'll build them too," I said. "Maybe I can quit my job!"

"Ralph, I've had preliminary talks with Midland Edison," Ger said. "They might welcome this development. Because of climate change the grid is starting to fall apart. It's outdated, legacy technology that can't be maintained anymore. The public utilities are losing money trying to keep up with the demand for electricity, and maintaining the poles and the wires and the generating plants and the substations. The electric companies can build these mag engines and charge people to set them up, because 99% of people won't want to put them together themselves. I told them that Edison can stay in business because they already have the infrastructure and universal public acceptance in the marketplace. Prices will be rock bottom because of so much competition."

Ralph was skeptical. "Why would Edison pay you to build something they can build for free?"

"Because they don't have time to do it. They're a public utility and have to spend all of their time maintaining the grid."

"Then get to it, Germaine. And you too, Courvall. I can't tell you how pissed I still am that they shut down the Cube."

Germaine smiled. "Don't worry. These mag engines can be kicked out by anyone with an IQ over 100, if they want to spend the time and the money."

"We hope," I said to Ger. "Not everyone is as smart as us."

"It won't be long before the mag engine can be sold as a series of stamped items," Ger insisted. "Anyone will be able to do it!"

To my surprise I saw the big man's eyes tear up. "Goddamit, then put this up. Right now."

"Now?" Germaine said. For the first time he realized that there were legal hurdles to overcome. Tanya told him that Old Man Taubman got a slice of the pie of any private development that came out of his building. The old fart would never agree to give the mag engine away. "Uh, I need to get permission from old man Taubman, and my boss at Carleton University."

Ralph snorted. "Fuck Taubman, we don't have time for that legal bullshit. Get your asses going and put this up on the Net. I'm going to stand here and watch you do it."

Uh-oh. When Ralph talked like that it meant he wasn't messing around. Ger and I looked up at the giant, who had stepped forward toward the mag engine, which was generating 60 volt alternating current, resolutely and silently turning the crankshaft. I felt a slight movement of air as Ralph pointed to a workstation. "Do it."

Ger smiled, trying to deflect the big man. "No can do. There are protocols to this research, Ralph." He pointed to the mag engine. "This thing is worth billions. If old man Taubman found out Joe and I released it for free we'd wind up in jail. Carleton University would sue us too."

Ralph remembered what had happened with the Cube. The national security boys had seized Max's lab and forced them to shut down production. He wasn't an unreasonable man. "All right then, I'll do it. Set it up and I'll push the button."

Ger was frightened, I could see that. My knees were a bit weak also because everyone knew I was working with Ger. I would be an accessory to a crime.

"You can tell them I forced you," Ralph barked in a command voice. "Get going!"

I gulped. Ger backed away. "You do it Joe," he said adamantly. "I will not break my word to my employers, who gave me the lab despite the fringe nature of this project. I'm also a Carleton employee, and the university won't allow me to release this information unless they get a slice of the pie."

"Robinson, your word has to be broken when it means the future of humanity." Ralph turned to me. "Well, Courvall? Are you going to step up and do the right thing?"

"Why can't we wait until Ger gets his permissions?"

Ralph's face was grim. "Because he'll never get them kid, and this thing will be suppressed just like the Cube. Taubman and those bean counters at the university are greedy money grubbers. Their lawyers will be at each other's throats for years, and nothing will happen." Ralph stepped forward to the workstation. "This is on me fellas; I'll fall on my sword. Just say I forced you to release the docs."

Ger looked like a little ducky bathtub toy being confronted by Godzilla.

For some reason I felt...a kind of patriotic fervor. Ralph was right. I was worrying about myself when I should be considering the entire planet. I knew how important this device was to the human race. And here a mercenary was shaming me into doing something I shouldn't have hesitated to do! I stepped up to the workstation, a little ashamed of myself, and began to issue instructions to the machine.

"You're right Ralph," Ger said as he watched me work. "I know Nikola Tesla wouldn't give a shit. I'm over myself now."

Ralph looked at Ger and me like we were three warriors going together into battle. "We're in this together now." The big man stepped back, a pleased look on his face. "Whoo-whoo!" he cried with delight. "We're going to create the biggest shitstorm since the Pandemic of 2020!"

In thirty minutes it was done. We stood there, expecting the men in black to break down the door, but nothing happened. Ralph and I and Ger looked at each other. "That was anti-climactic," Ger said.

Forty-eight hours later it went viral on the WorldNet.

CHAPTER **28**

On Saturday, Karen told me all about Harry Kaine's presentation to the City Council at the Radisson Hotel. I worked Friday until 10 p.m. on the magnetic engine after I got off work at Phoenix, so I couldn't attend. We were still tweaking the design to make it as portable and efficient as possible.

"First Harry had his event for the bigwigs," she told me. "Kaine was there, Ben and Randi, Raoul Hernandez, and even the new Mayor of Chicago, MIles Raymond. Just to show off; you know: 'In person we're even more awesome than on a vid!' Max Berglin and Ralph Zimring were there, and of course the media. The Radisson's auditorium can only seat a couple hundred people, but every seat was filled. And you'll never guess who was on Harry's support staff: Liqao."

"What?" I couldn't believe it at first; then it made sense. Liqao had always favored Transhumanism. The Gang hadn't seen him for months. Kirra had been silent about him and, to spare her feelings, we hadn't asked her about him.

"Yeah. So Kaine does his usual spiel; the transformed humans are brought onstage but they don't say much; they just prance around and tell the assembled multitude how great it is to be transhuman."

"Did you get to speak to the great Raoul Hernandez?" I asked.

"I did have a word with him but I couldn't tell whether he was as dumb as the rest of those bag-puddings. He seemed bright enough."

I laughed. "Bag puddings?"

"Yeah, right out of the box, wrapped and boiled."

"So what happened at the presentation to the City Council?"

"Five normies made the presentation." Karen started to laugh. "One of them was Max Berglin's business rival, Trevor Clarke. Berglin got so mad that Ralph had to restrain him from going onto the stage to confront Clarke."

"What did Liqao do?"

"Nothing much. He was in the background, doing IT support stuff."

Karen showed me the vid of the presentation. I had to admit that Kaine was fantastic at PR. His useful idiots did a credible job of promoting the proposed Better Humanity center.

"The presentation started at 9 and lasted three hours. By the end everyone was hungry. We ate lunch. Then the entire City Council went on stage and they had a public hearing."

"They're supposed to have hearings in the City Council building."

"Yes, but I found out later that in the last Council meeting announcement there was an item about it added to the agenda. No one reads those announcements anyway; but everyone who would be sympathetic had already gotten the word."

I sighed. Kaine's group was so powerful it was steamrolling the human race all over the planet, and literally changing what it meant to be human. "So what happened?"

"It was typical Harry Kaine, everything rigged in his favor. He got that slimy Trevor Clarke with his British accent, Bob Justice, the head of the Planning Commission, County Court Judge Roland Massimino, and Craig Ginzburg, the ethically challenged real estate developer, to make the proposal. The problem for Harry was that Councilperson Statsny, who said she was going to vote yes, didn't show up. So it was a tie vote, 5 to 5. You should have seen the look on Kaine's face! He thought he had the vote locked in. They have scheduled another vote on Monday evening at 7 at the usual meeting in the City Council building."

I smiled. "I suspect that was Ralph Zimring's work. He told me about how Harry bribed a majority on the Council to vote yes. Maybe Ralph did something."

Karen was amazed. "My editor Bob Guza thinks it was just a fortunate coincidence. I didn't know that barbarian was on our side."

"He helped us in New York."

"No, he just killed somebody."

"I don't know for sure what his game is, but Ralph gets around in this town." I told her about how Ralph had forced us to put the mag engine specs online.

Karen saw a story here. "I need to interview Zimring. How does a seven-foot giant remain so anonymous?"

"He's like the delivery person; no one notices him." I changed the subject. "We need to pack the meeting on Monday, Karen."

"Good idea! I'll write a story for the Sunday edition of the *Chronicle* and tell everyone to come down. I'll call every blogger I know, and that rabble-rousing talk show host at WAMF. Harry won't have a chance!"

"That will probably work. Get Lledren to film the meeting and put it on his blog. He's getting more popular."

"I will!"

We left it at that.

———— • ✳ • ————

On Monday at 7 p.m. a crowd gathered at the City Council meeting. Harry had gone back to DC and his transformed humans were nowhere in sight. So much pressure was put on the Council that the Better Humanity Center proposal was defeated 6-5.

I knew this was a temporary victory because Harry Kaine wouldn't give up. His centers were thriving in many of the world's big cities. Now he was going for medium-sized cities like Midland. The centers were so popular they were self-funding, and kicked back a portion of their profits to each city.

In Beijing (and many other cities in China), Kuala Lampoor, Jakarta, Manilla, and Kyoto, transformed humans had either elected mayors or were numerous in the cities' administration. For some reason the transformed human idea had really caught on in Asia and in South America. However, the Sao Paulo center had been attacked by a mass "normie" protest. Harry called this "an illegal and terrifying act of terrorism reminiscent of the twentieth century wars of genocide."

When M'basa heard this he wanted to get on a plane and strangle Harry Kaine. "To compare this to the wars in Sierra Leone and Rwanda is an insult!" he cried at one of the Gang meetings. However, at the African Union meeting in June, a decree banning Better Humanity Centers was unanimously passed. In sub-Saharan Africa not one center existed, to M'basa's great satisfaction.

In Europe Harry was going strong in all countries except Switzerland, Poland, and Hungary. There were Better Humanity centers now in all of Western Europe's major cities.

Here in the US the centers were confined exclusively to the major population centers, with the main HQ in DC. The heartland of the country – like Midland – had so far refused them.

At Karen's apartment she and I were looking at a hardcopy map of the US we had printed out. We put red dots on all cities that had transformed human centers.

"Wow," I said. "It's eerily similar to the Blue-Red divide. Blue states have the trans-formed centers, red states don't for the most part."

"It's just easier for Harry to reach more people in the big cities," Karen said.

Whenever I was unsure of something I consulted my intuition. It was a holdover from my childhood. "Something big is about to happen Karen, I can feel it. Battle lines are being drawn."

We talked about how the timeline was changing from what we saw in the 3013 Shifter and what I had seen in Ger's primitive time viewer. "Everything has shifted 60 years down the timeline. The future is affecting the past because Harry went forward in time and came back with those two clones from 3013," Karen said. "He's using the knowledge he gained to accelerate his progress and change the timeline."

"Harry has help, Karen. There's no way a couple of clones from the future could have resulted in a complete transformation of human biology in just three years."

Karen shrugged. "We don't know anything about that. All we know is that Harry is accelerating events that should have happened gradually."

We both recognized that Harry Kaine was the primary player on the timeline; a human actor around which massive temporal forces were operating.

At the Gang meeting that weekend Ger lobbed a bomb.

"I went to Midland Edison this morning with a proposal. I brought our diagrams and blueprints for the new mag engine. I showed them our prototype."

"What did they say?" I asked.

"Edison is going to give us everything we need to build them, Joe. They'll fund it completely. I get 35 percent, they get 65, after expenses."

"Holy shit Ger! How did you wangle that? We gave the stuff away for free."

"Yeah, but the mag engine isn't as easy to put together as we thought. People have been writing me from all over the world with questions and I don't have the time to answer them. So we're going to set up at Edison and be first on the ground with working portable generators. Edison will market them as the inventor's original device, with a unique stamp. Our machine will be the gold standard brand. Edison will clean up selling and installing them."

"Are you going to cut me in?" I asked.

Ger looked at me speculatively. "You want a piece of the action?"

I knew my cue. "I figure it ought to be a thick percentage."

Ger and I laughed at the original Star Trek dialogue.

"Mostly you hung around and looked pretty, but I cut you in. I told them I also wanted a salary for me and you in case the national security people hassled the project. It's good money."

When Ger told me the salary I almost fainted. "When do we start?"

"It will take them a couple weeks to get up to speed. There's a building on the Edison campus we can use, but they have to set it up and get it ready. I told them exactly what I wanted. You and I discussed it last week."

Ger and I had already discussed what our "dream factory" would look like, and I had drawn up the workstation blueprints in my CAD/CAM software. "Do we get everything we wanted?"

Ger grinned. "We get everything except the two refrigerators filled with expensive ice cream. The Midland Edison people are really fired up, and they should be. Especially after I demo-ed the prototype to Ruth Vandenberghe, the Director of Operations. "

"What if old man Taubman sues us?"

Ger laughed. "He already has, which is why I talked to Ruth. We're Edison employees now, Joe. She said Taubman already contacted the company, and she told him to fuck off."

We both cracked up. I wish I could have seen the look on the old guy's face.

"Ruth says the company will back us and pay all legal fees. As long as the project pans out of course."

"It will."

"Ruth says we'll make so much money we can handle anything Taubman's greed can throw at us."

"It's Ruth now, is it?"

"Don't start Joe!"

Karen and I looked at each other and smiled.

Midland Edison put out a press release that got picked up immediately by the *Chronicle*, who sent it to the *Chicago Tribune*. From there it went across the country. A week later Ger got a call from Lizzy Gross, the famous (former) NPR host, who was still cranking out interviews after all these years.

Ralph Zimring showed up the day after I quit my job at Phoenix and went to work with Ger for Midland Edison. We were standing at a workstation in a building on the Edison campus. Our printer was stamping out parts for the mag engine. I was bent over, checking the parts for accuracy. Ger was watching the process like an anxious mother.

"Max is pissed," Ralph said loudly, striding into the work area at six-foot intervals. He stopped and looked around. "Pretty impressive. I didn't know that a public utility could get into action so fast."

"Money is a big motivator," Ger said, checking the computer code for the 3D printer, which he wrote himself. "We're talking literally about billions when this gets off the ground."

"Yeah," I said, straightening up. "And the fact that the rickety old grid is falling apart. The mayor, I'm told, has been pestering Edison. He's tired of the power outages."

"Why is Max angry?" Ger asked.

"Max just had a call from the Patent Office, and the DHS. They said you won't have any problems with your magnetic engine, but they can't do anything about the Patent Secrecy Order attached to the Cube. The nat sec boys are still pissed about what happened fifteen years ago. That's why Max is upset."

We both nodded. Fifteen years ago Max Berglin fought a local war with the national security establishment when he developed the Cube, the world's first working overunity device. They made him shut down his production facility. Now here we were, waltzing home scot-free. Max's Cube still had a Patent Secrecy Order on it.

"Any scumbags you need taken care of?" Ralph asked. "I'm getting bored again."

"Yes," I said. "I got a call from Vahan Katelian last night. He says he wants in on the mag engine project. The local mafia can make life difficult for us."

The Katelians and the Nalbandians are Midland's version of the mafia, except they are Albanian. They muscle their crews in on all of the city's building projects. Jules Rothman is their formidable fixer.

Ralph's eyes lit up. "Rothman! Oh yeah," he said, rubbing his big hands. "I'll have a word with Jules." Ralph looked around. "This is small potatoes anyway, not in the Katelian line."

"Yeah, but the Katelians and the Nalbandians have a nose for money, especially Vahan. He's smart enough to see that there's a huge payoff here."

"I'd appreciate your help Ralph," Ger said nervously. "We don't want any trouble with those guys."

"Rothman and I don't get along," Ralph said with a grin. "Whoo-hoo! Gonna have some fun, especially if he gives me a hard time."

Ralph walked out, whistling. If Ralph said he would take care of the mafia, it was as good as done. Ger and I got busy. We were almost ready to start cranking out mag engines.

At the end of the day I looked out the window. A light rain was falling. It was the end of July and the low was forecast to be 38 degrees. I cursed the cold but there was nothing I could do about it. Last year we had a summer, but not much of one so far this year. It was colder in the northern hemisphere, but NASA's Vegetation Index continued to show the retreat of the Sahara Desert in Africa, which was becoming much greener due to CO_2 emissions.[28]

More people were moving into that area as the population in Africa grew. The world was going to need our magnetic generators.

———————— • ✳ • ————————

After her work at the library Kirra Bigbear took a popup to a hotel on the outskirts of the city. She was to meet Liqao for a quick dinner and an ultimatum. She was tired of thinking about him and worrying about him. They didn't have a relationship anymore since he quit his job at the Midland Police Department and went to work for Harry Kaine in Chicago over a year ago. She half expected him to be one of those transhuman robots by now. Kirra was very apprehensive as she sat down in the hotel restaurant at a little table for two. She had arrived ten minutes early to gather her thoughts. The restaurant was on the first floor and surrounded on two sides by thick windows, giving a good view of the street.

After fifteen minutes Kirra had finished her coffee and Liqao hadn't shown up. He was early for everything so he probably wasn't coming. She told herself that he wasn't worth her trouble, but she was near tears as she thought about breaking it off. She was just about to walk out when she saw the funny little figure with his thick head of black hair push his way through the revolving door at the hotel entrance. Her heart skipped a little. She knew she loved Liqao but he was going to have to commit to her or it was over.

Liqao walked over to the little table and stood in back of the chair. "Hi Kirra," he said softly.

Kirra felt how she always felt in his presence: a cuddly feeling inside her that made her want to get up and hug him. But she wasn't going to do that today. She wasn't going to tell him he was late, or remind him that he hadn't called her in weeks. It was on him to make the first move. She nodded silently.

Liqao slowly pulled out a chair and sat down. Kirra noticed immediately an emotional somberness, as if he was weighed down by something momentous. As usual he was waiting for her to speak first. Kirra tried to keep her composure as she looked into his eyes.

"I've decided to get the procedure," he said, sighing. Now that he'd made the decision he felt much better. He couldn't have done it without Kirra being here. He saw Kirra's eyes tear up. He recognized a hopeless sadness in them as she brushed away the moisture with the back of her hand. Her eyes hardened and she stood up, straightening, and steeling herself.

"It's over then." She turned to walk away, but abruptly stepped back toward the table. "Are you prepared to become a human robot?"

"They're not robots."

"Oh Liqao, they are. And you know it. Why are you doing this?"

"Because...because I'm curious. I'm fascinated. To live in a body that never gets sick, that looks beautiful all the time...it's something amazing. I've seen the procedure and it's amazing, incredible...." Kirra didn't respond so he continued. "And, after all, it's only one lifetime," he said with one of his cute grins. "We're all reincarnating again aren't we? That's what you always say, that consciousness isn't limited to one human expression. If I don't like it I'll come back as a normie!"

Kirra was shocked, and impressed, at this insight. Her eyes widened. "Liqao...I...I never thought of it that way!" To her surprise she felt elated. "You have given this a lot of thought." She sat back down.

Liqao smiled one of the boyish smiles she loved so much. She thought of the conversation about Kitchi-Manitou they had at the Mason Street coffeehouse a few years ago. Liqao was a physical lightweight, but he was capable of deep insight. Her feeling of sadness evaporated. "So, you're willing to risk an entire lifetime on this – " she tried to find a word – "experiment?"

Liqao was pleased that she held back her criticism. "Yes! It's a life experiment, thank you." Liqao felt a lot better about his decision. "You see, in IT we are constantly upgrading our software, firmware, and hardware. In every iteration we get better. Why not do that with the human body if the technology is available?"

Again Kirra was surprised by his insight. "You've grown, Liqao. I wasn't expecting it." At his inquiring look Kirra explained about the vid Karen and Ralph Zimring had shot of the transformed community in Chicago. "Liqao, these transformed people are...narcissistic and commonplace. They have very little self awareness and no spirituality. Haven't you noticed that?"

Liqao's face fell. "Yes, of course. It's one of the reasons I haven't done it yet. It's why I wanted to see you."

"But Liqao, once you become transformed you have committed yourself for the rest of your life. You can't undo the process."

"It's a challenge, Kirra. The greatest challenge ever. But you convinced me that...there's other lives to come."

Kirra sighed. "I hate that you are even considering becoming transformed. I don't want to lose you."

Liqao saw into her emotional depths. He needed that in his normie life, but becoming transformed was a choice that he would never get again. If what Karen said was true about the timeline, these transformed humans and their society would eventually devolve. But that was far into the future. Liqao understood that he was a gambler, a risk-taker. He would have to gamble with the ultimate chip: his life. Was losing Kirra worth it?

Kirra saw his confusion.

"Let's both do it," he said finally. "Then we can still be together."

Now Kirra was really shocked. "Liqao, you...you really are amazing. I always thought you lacked commitment, but I see now that you are willing to commit totally." Her expression softened as she looked into his eyes. "I really do love you, you know."

Liqao felt himself tearing up. "I think I love you too, Kirra. That's why I want you to come with me."

She had the thought that if Liqao ever committed to her, it would be a deep and lasting bond. Their life together could be special. She looked at him longingly, knowing that she couldn't do it. Becoming transformed was a dead end. The legends of her people told her that connecting with the earth, and with the Great Spirit, was the only true way forward. With a profound sadness she realized that she would have to somehow get over Liqao. He was lost to her now.

Liqao saw the emotions written on her face and understood. He sat back in his chair with a sigh.

Kirra tried to lighten the atmosphere. "When is the procedure?"

"Two weeks from Wednesday at the Better Humanity center in DC. I want to see what the main center looks like."

They both sat silently for a minute. Kirra fiddled with her empty coffee cup; Liqao was staring out the window but his attention was inward.

"Can I call you next Wednesday morning?" Liqao asked her. "The procedure is at 9 a.m." He looked at her hopefully. "I can still back out you know. Harry's centers have got the procedure down to an assembly line almost. If I don't go through with it there are three others waiting for my time slot."

Kirra smiled. "I'd like that; a chance to say goodbye."

Liqao started. "Not goodbye Kirra. We can still see each other!"

Kirra realized that she might have an irrational bias. What Liqao proposed to do was so alien to her and her people...it was unthinkable for her. "Let's leave it for now. Call me on Wednesday."

Both coffee cups were empty now. The waiter came over. "Are you ready to order?"

Kirra got up. "I'm not hungry anymore," she said. She nodded to Liqao. "I'll be waiting for your call."

As Kirra walked out the door she saw Liqao discussing the menu with the waiter. She realized that nothing fazed her boyfriend; he seemed to be able to accept anything. Maybe his irresponsibility was just an eagerness to embrace new situations, and the ability to drastically shift his beliefs to adapt to new developments. She didn't want to lose Liqao, but his impetuous personality was something she would never be able to control. She found herself on the street in front of the restaurant window, but did not look back. She called a popup on her mobile and got in. She wished that the popup was one of the old cabs, where you could actually talk to the human driver. She wanted to ask, does love mean you have to give someone up even when your heart is aching?

Chapter **29**

Melanie Fuscaldo sat in her office at the Interfaith Center. She had had a bellyful of Harry Kaine and his Transhumanism. The Better Humanity center in Midland had been rejected by City Council, but momentum existed for another try. As usual it was the elites in the city, the wealthy and powerful, who were behind the effort. Rumors swirled that certain members of the Council had taken money from Kaine, but nothing could be proven. She did not want one of those abominations in her town.

Karen and Kirra kept telling her about the wonderful program they were going to develop with some powerful Native Americans to counter the influence of Kaine, but nothing had ever come of it. She glanced at the *Courier* news headlines on her mobile:

"**Better Humanity Center to Open in Springfield.**" The state capital! 'A number of state legislators have already booked their appointments in the new center, which is scheduled to open to the public in two weeks. It is estimated that 25% of the state legislature could be transformed by the end of the current legislative session in August."

"**Chicago Mayor Issues Edict: Normies Banned from Transformed Communities.**" Melanie was getting angry now. "Mayor Miles Raymond has issued a city-wide edict banning travel by so-called 'normies' into areas of the city designated for transformed humans. At a news conference today the mayor stated, 'The Chicago police will vigorously enforce this order. Those who do not comply will be arrested.' Studies show that the rift between normal humans and transformed humans is becoming more strident. 'Normies' insist that transformed humans are arrogant, superficial, and narcissistic. Transformed humans say that normies are evolutionary dinosaurs. 'We want nothing to do with those robots anyway,' a spokesperson for the normal community said today after reading the mayor's edict."

Melanie didn't understand what was happening in society. It was worse than the insanity during the 2020 pandemic and its aftermath, and far more dangerous. What made people want to transform? What would happen if transformed humans gained a majority in the Illinois legislature? Society was fracturing, splitting apart, into two different species! It wasn't healthy.

Melanie admitted her bias against the transformed. The transformed communities were like a hive of ants, there were hardly any different opinions on anything. Almost as if the entire population was programmed.... She remembered reading about bio-nano-technology, where programmable molecular entities were inserted into the bloodstream. These entities kept cells healthy and fought off disease, but could also send electrochemical "messages" to the organs and cells. And to the brain. It was a stretch, certainly. But what if it was true? The only good thing about transformed communities is that they were insular and non-aggressive. So far. But what would happen if they became a majority of the population?

Melanie made a decision. It was up to her to do something before the split in society metastasized. No one else was doing anything about it.

At that moment Kirra Bigbear walked into the center. She looked glum. Melanie approached her friend. As Kirra told her about Liqao, Melanie became more and more motivated to do something.

"He says he will call me on Wednesday morning before he undergoes the procedure," Kirra said hopelessly. "I hope he will turn back from it."

Melanie was empathic. "You have to accept that he might be lost to you."

Kirra nodded. "There's always hope."

"Speaking of hope, Harry's movement looks unstoppable now, and we desperately need a program to re-connect people to their spirituality." Melanie told her friend about the nano-bio-entities used in the Better Humanity centers. "I still keep up with the scientific journals," she explained. "A holdover from my former life as a neurobiologist."

Kirra's face assumed determined lines. "In our group meetings we talked about streaming a show exposing Harry; that was Joe's idea. But he got sidetracked with Germaine Robinson and the magnetic engine and nothing ever came of it."

"It's a good idea."

"Who would host it?"

"M'basa and Karen." Suddenly Kirra's face fell. "We could have used Liqao to set it up."

Melanie smiled sympathetically. "Don't worry about that. We have several tech people in my congregation." She had an idea. "We need to take the Bannon approach. Remember that guy during the 2020 pandemic? It started with nothing and just kept getting bigger and bigger, and eventually influenced government policy all over the world. Harry Kaine is an existential threat to humanity and we have to do something."

"I remember. But Bannon was a fire-breather. We don't have any of those."

Melanie smiled. "Oh, but we do. Karen Everard. And M'basa."

"Karen has a full-time job; so does M'basa. Why don't you do it?"

"I would but I'm too mild-mannered. So are you and Joe. You've seen Karen when she gets passionate. And M'basa too, when he talks about the African civil wars. That's what we need. Human beings all over the world feel threatened by the transformed. We already have a ready-made worldwide audience."

"Lots of people are streaming shows about it."

Melanie spoke quietly but passionately. "Yes, but these shows are all political rants. In Bannon's case the subject was inherently political. But Harry represents a materialist assault on the very existence of humanity. An effective message would inherently be a spiritual one. People are ready for it."

"You're right! M'basa as co-host? They could both feed off each other."

"If we put M'basa and Karen together, what will Kjirsten think?"

They both knew the answer. "She'd join the show, of course. Keep her man in line."

Kirra was starting to get excited. She knew Kjirsten wasn't working and was bored. "We'll enlist Lledren to stream it on his platform. He's already popular on his network. If we're successful, the show will get picked up by other networks."

"Where will we broadcast?" Kirra asked.

"Ken Strickland in our congregation is a broadcast engineer. He has a small studio in his basement, and some room for seating. I'll ask him."

"A live show then?"

"Oh yes. As many as we can cram into Ken's basement."

The two women got to work.

When M'basa was asked to be a co-host by Kirra at the Gang meeting that weekend at Ger's place, he agreed with one condition. "People will get bored with spiritual messages. The main thrust of the show must be an expose of Harry Kaine and how his centers are contaminating human biology and destroying the future of humanity. That's going to involve politics and current events. We need facts and evidence to back up our claims. However, we can introduce a spiritual aspect by pointing out how scientific materialism is a dead-end for the human race." He looked at Karen and Kjirsten, his co-hosts, thinking of what they had seen in 3013. "Some of us know this for an absolute fact."

Kjirsten's eyes widened. When M'basa spoke like that, everybody listened. She looked around Ger's living room. Everyone was at attention, leaning forward in their seats.

M'basa turned to Kjirsten. "That's where you come in, Kjirsten. You have to dig. Watch all the important news shows, scour the WorldNet for info on Harry and what he's up to. Try to find out who's behind him; talk to whistleblowers if there are any. If we're going to do an hour every day that's a full-time job."

Kjirsten was starting to get motivated and excited. Her father's police work had imbued her with a sense of justice, and an understanding that solving crimes was always just a lot of hard work uncovering facts. She liked the idea of doing something important for the good of all. "I'm ready!"

In two weeks it was all set up. Melanie announced the show to everyone in the Interfaith congregation every day. The show would run for an hour between 7 and 8 in the evening after work. Karen, M'basa, and Kjirsten made a topic list for the first seven shows. The first show would be titled, "Who Is Harry Kaine?"

Melanie was a little frustrated after the first week of show preparation and rehearsals. "We're losing the spiritual message in the politics," she told Kirra.

"That's because M'basa and Karen are such forceful personalities. Karen is a journalist after all."

"That's what I'm afraid of. The show is going to be too political."

Kirra shrugged. "Kitchi-Manitou tells us that the world evolves. So do undertakings like this."

Kjirsten spent the entire two weeks researching show topics and writing down sources and links. She was into it now. She had worked for her father at the police station for a while after graduating from Carleton, but had never found her niche. She discovered that she was good at research and organizing the topic lists.

On Sunday, the day before the first show, Kjirsten got an email from Ralph Zimring with the subject line, "Here's a vid you'll want to see." It showed Councilwoman Statsny and Harry Kaine meeting at a back table in the Mason Street coffeehouse. The voices could barely be heard over a lot of chatter, but Harry was ogling the councilwoman and saying, "I could make it worth your while to vote yes on the Better Humanity center." After further conversation Kaine handed Statsny a currency card.

Kjirsten was on the phone instantly. "Mr. Zimring—"

"Ralph."

"OK, Ralph. How did you get this vid?"

"Let's just say that in my fifteen years at Berglin Enterprises I have developed a large network of trusted...correspondents." The big man smiled. "I like what you are doing, and so does Max. I'm your ace in the hole. Just don't mention me or Max on air."

Kjirsten was baffled. "We haven't even had a show yet! How do you know what we're doing?"

Ralph ignored this. "Don't call me, I'll call you if I get anything you might need. I'll be listening to the show when I can."

The connection went dead. Kjirsten was surprised and pleased. Ralph Zimring was a nosy parker, but in a good way.

They were ready for their first broadcast the next day, a warm mid-August evening. M'basa and Karen had rehearsed the first three shows for the past week. At 6:55 on Monday everyone was ready. Ken Strickland sat in a small control room off the main bench, which just had three mics on a table. Two swivel web cams showed the scene up front, operated by Ken from his booth. Karen and M'basa sat next to each other, with Kjirsten slightly apart. As many from the congregation as could fit in Ken's basement showed up.

Kirra was excited as she sat in the audience, but a little depressed as she thought of the worldwide behemoth that was A Better Humanity. The idea was to attack Kaine with the truth, and hammer their message over and over. Each show had five bullet points or takeaways that they would repeat over and over. The show would hopefully appeal to anyone with self-awareness who was turned off by Harry's degraded scientific materialism and the self-centeredness of the transformed.

Ken gave the signal from the control booth. They were on the air! Kirra could see how excited Karen and M'basa were. Kirra felt a little thrill go through her.

Karen began. "Welcome to the Humanity 3.0 show! Streaming on the Vox Populi network, channel 3027. I'm Karen Everard with my partner M'basa Ogunfatidime. To my right is Kjirsten Chastaine, in the Lightroom." Karen leaned forward. "We are here to tell you about the scam called A Better Humanity, and to tell the truth about Harry Kaine and his wealthy elite backers. Every weekday evening at 7 we'll have the latest updates, and a positive message about human nature and unaltered human biology. There's a live chat you can comment on at our site, lightroom.org. Let's get started. M'basa, would you play the first vid?"

"This is humanity's hero, Harry Kaine," M'basa said.

The vid (from Ralph Zimring) showed Harry Kaine in the cafe with Councilperson Statsny. The audience, who had no clue about the content of the show, began to buzz.

"Is that Statsny?" someone asked.

"Looks like her, doesn't it," another said.

Harry was turning on his charm. "Councilperson, we all know the importance of my centers. You would only be doing your public duty by accepting this small, and entirely appropriate, gift."

"Is that a currency card he's handing her?"

People were on their mobiles, texting and messaging their friends.

Kjirsten spoke up. "Here are three links to text articles from open sources detailing similar activities by Kaine's group. From Kyoto, Japan; Berlin, Germany; and right here in Chicago. Who is Harry Kaine, really? And who is behind him?"

"That's a good question Kjirsten," M'basa said. "We all know Harry; he lived here for most of his life. Where is he getting the squillions in currency to set up his centers across the world? Somebody really big is backing him. Who is it?"

"Harry Kaine portrays himself as a benefactor to mankind but he also promotes a philosophy of scientific materialism," Kjirsten said. "Harry thinks human beings are meat, and to 'improve' the race, his idea is to transform our biology. What is this process? It is a closely guarded secret. I have posted links to open source articles where you can find out what little we know about the Better Humanity medical procedures. They require investigation."

"That's a call to all scientists to come forward," M'basa said.

"Ladies and gentlemen, transformed humans call us 'normies,'" Karen said. "They refer to us as evolutionary dinosaurs, and refer to themselves as Humanity 2.0. We want to show you what a transformed community looks like. Recently I was able to film a documentary in Chicago of a transformed community. I was on assignment from the *Midland Chronicle*, so it's official. This was filmed last spring, before the recent edict of Mayor Miles Raymond forbidding 'normies' to enter transformed communities."

The vid played. The audience saw the incident in the Council chambers between Councilperson Jenks and the Mayor, Miles Raymond. They saw Karen taking a pop-up to the transformed community around 122nd street, the confrontation on the sidewalk, and the activity at the 122nd community center.

"There you have it, ladies and gentlemen," Karen said. She turned to M'basa. "I was struck by the condescending attitude displayed by the Chicago mayor, and the 122nd street community's focus on physical appearance. Did you notice the mirrors on the walls?"

"Yes," M'basa said. "There are no religious or cultural activities among the transformed, only beauty contests and discussions about soap operas."

Karen laughed. "Bad soap operas!"

"I'm more concerned about why the transformed community is banning our people," Kjirsten said. "There must be a reason."

M'basa spoke up. "What you have seen is a very small sample of a transformed community. We want our viewers to send us film of other transformed communities, if you can get in. Is the 122nd neighborhood in Chicago the rule or the exception? We have to find out whether the transformed are inclusive or exclusive. Inclusive societies are based on cooperation. Exclusive societies, like the Nazis, the Soviets, and the CCP, are always exclusive. Exclusive societies are aggressive toward their own people, and other groups of people."

"We also need to know what procedures are used in making transformed humans," Kjirsten said. "Harry Kaine has refused to release the methods he uses. We think that the biology of transformed humans is programmable. If so, how are human beings being programmed?"

"We're going to have a lot more on that this week," Karen said. "Nano-biotechnology has great benefits but it can also be used to influence a person's behavior. See the show notes for open-source articles about that."

M'basa spoke. "Guys, we're out of time and we haven't got to one-tenth of what we had planned for you tonight. Send in your comments, articles, and vids to lightroom.org! Stay tuned tomorrow for another episode of the Lightroom, where we shine the light of truth on the darkness of Transhumanism and scientific materialism—-"

The show ended. The audience broke into excited chatter. At that moment Karen's mobile rang. It was Councilperson Statsny.

"Ken, make sure you're recording this!" Karen shouted at the control room.

Ken put everything on speaker.

"Where did you get that recording?" Statsny demanded. "It is an unwarranted intrusion into my private life!"

Karen was polite. She had dealt with Council before in her duties as a reporter. "Councilperson, this is a private vid, recorded by a private citizen. I am not at liberty to reveal my sources."

"You call yourself the Lightroom? You're just a bunch of sneak thieves, spying on people doing their public duty!"

Karen was a little pained by this description of the vid, for it had occurred to her as well. But corruption is corruption. She remained calm. "Did you or did you not take a bribe from Harry Kaine, Ms. Statsny?"

"I...you're in trouble, Everard! I'll be speaking to your editor right away."

The connection ended.

The audience was still hanging around.

"Did you get that Ken?" M'basa asked.

"I certainly did! It will make a good lead-in to the show tomorrow."

Karen was happy. She, like Liqao, had once gone over to Harry's side, lured by fame and success. She was glad now that she had turned away. This show might be exactly what she had been wanting: to speak truth to power. It was exciting and a little dangerous. The same people who had killed Winthrop Sauvage and sent them threatening messages were behind Harry Kaine. Ralph Zimring had essentially told them that three years ago at Harry's conference in New York.

Ralph Zimring's vid on Statsny, and the Councilperson calling in, had given the show instant cred on their very first broadcast. Later, Ralph told Karen that he'd called Statsny and given her a heads-up. Ralph Zimring was devious!

And so it went throughout the week. On Tuesday the three talked about the split of society into normies and transformed, on Wednesday what little was known about the shadowy group that was funding Harry Kaine, on Thursday about the opposition to the Better Humanity centers in cities around the world. On Friday they ended the week discussing the bio-nano technology that they suspected was being used in the transformation process. A whistleblower from the Beijing center (using a nom de guerre) – a scientist who worked in one of the transformation laboratories – told the three hosts a rather frightening story about how it worked. "After the procedure, each transformed human becomes a walking mobile communication device," she said. "Vital signs can be monitored with a remote scan, and your altered biology is like a transponder that responds to that signal. Your location is known at all times."

"Can a transformed human's biology be hacked, like a server or a computing device?" M'basa asked.

"That is something requiring investigation," the scientist said. "My guess is that it can."

Melanie and Kirra sat in the audience. Although both women were excited for their friends, they were dissatisfied with the show's political and current events focus. As the two women walked to their vehicles Kirra said, "The show has accumulated a huge following in one short week, but something is missing."

"It's time to bring your Native Americans to the show," Melanie replied. "We've been talking about this for years, now we have a platform."

Kirra was excited. "Great idea! We'll finally have that debate with Harry Kaine, but this time on our ground not his."

"In person, if you can arrange it."

Kirra nodded. "I'll get on it."

CHAPTER **30**

Kirra waited all day at the library for Liqao's call. This was the day he was to undergo the transformation process in DC. He had *promised* to call her. Out of habit, when she got home she made dinner, but today she could only eat a few bites and refrigerated the rest. She had lost her appetite. After taking a walk and nervously watching her mobile, she decided to go to Ken Strickland's, sit in the audience, and listen to the show. She told herself Liqao wasn't worth all of her worry, but this thought made no difference to her emotional turmoil. At 8 p.m., after the show, she went home feeling depressed. By the time she went to bed Liqao still hadn't called.

It was over then. He hadn't thought enough of her to say goodbye, but it didn't matter anyway. He was one of *them* now. A sudden fury poured through her body. Damn Harry Kaine! Kirra was shaking in reaction to the hatred pouring through her. Normally she would meditate and release the bad energy, but she felt like smashing something. Kirra realized with a start that this sort of anger was probably behind every crime of passion that had ever been committed. She wanted to kill Harry Kaine!

Kirra threw her mobile on the end table and left her apartment. She started walking quickly. It didn't matter where she went; she would walk until she felt better

A half-hour later she was still walking, oblivious to her environment. It was getting dark now. She realized she had headed in the direction of downtown and had no idea how she had crossed the busy downtown streets. Kirra shook her head to clear it and stopped. She was passing the Berglin Enterprises building on First Street. Her unconscious mind had sent her here for some reason. She stared up at the modern steel and glass building with its large "Centurion Building" gold-colored letters plastered over the front door. Just as she was about to walk back home she saw a very large figure dressed in some kind of black outfit come around the side of the building.

It was the famous Ralph Zimring.

The figure noticed her immediately and stopped about ten feet away. "Kirra Bigbear, isn't it?"

"That's right." She looked him over. "What are you doing in that costume?"

"Just playing around," Ralph replied. "I patrol the streets when I get bored. It's kind of like hide and seek; I try to not be noticed. Sometimes I rescue damsels in distress." Ralph looked thoughtfully at her. "Are you a damsel in distress?"

Kirra laughed; her depression lifted a little. "Well yes, but not something a knight in shining armor...or in night pajamas...could do anything about."

"Ah then, it's boyfriend trouble."

Kirra was pleased that the giant was so perceptive. "Yes."

"Liqao Chen I believe," Ralph remarked. "I remember him. He's a little Dennis the Menace."

Kirra laughed even though the reference escaped her. "Yes, he's gone over. He broke my heart."

Ralph grinned. "I'll scare the shit out of the little gnat if it will help."

Kirra laughed again. Her depression was gone. "How did you get to be so good at relating to people?" She was curious.

Now it was Ralph's turn to laugh. "I used to relate to people through a gunsight or with my fists. But I've been around, seen a lot of things and met a lot of people in my work with Max Berglin."

During this exchange the two had turned and walked the way Kirra had come. "I'll walk you home if you'd like," Ralph said. "But if I see any bad guys I might have to desert you."

"You're no rescuer of damsels," Kirra remarked. "What kind of knight abandons a lady after the first sign of trouble?"

Ralph smiled. "A busy one."

"And what makes your life so busy? You're the head of security for Berglin Enterprises. Seems like a cushy job."

Ralph laughed again. He couldn't remember having laughed twice in one week, much less twice in five minutes. "It is a cushy job, so I have to keep busy doing, er, other things."

"You're a mysterious fellow, Ralph Zimring. Are you married?"

Ralph guffawed. Did that count as a laugh, he wondered? He didn't miss his game of good guys and bad guys tonight. "Never have been able to find a woman that could stand my lifestyle. So, no."

Kirra had never met a man who was so...*male*. She liked him.

"You're too young for me," he said, picking up on this. "But you're godawful gorgeous."

Kirra blushed. She was glad night had almost fallen so he couldn't see her face.

Ralph must have read her mind. "You just blushed."

"I did not!"

"Oh yeah you did."

Kirra couldn't hold on to her anger and she smiled. "Liqao never told me I was beautiful."

"Liqao is a little shit. He knows nothing about women."

"And you do? Mr. No-woman-can-stand-my-lifestyle?"

Ralph grinned. "I didn't say I was good with women. But I do understand women."

"What's the difference?"

"I know what women want. I'm just not constitutionally able to provide it."

They had to stop and wait for a red light.

"You mean you're a selfish pig that doesn't want to change to keep a woman happy. Most men are like that. Liqao is like that."

"If a man has to change to accommodate a woman, then there isn't enough love there to sustain the relationship. Love comes when two independent people accept and like each other."

The light turned green and Ralph began to stride across the street, but Kirra stood rooted to the curbside. "I...you...why, you're right! That's something Kitchi-Manitou might say."

Ralph came back and gently took her arm, escorting her across the street. Kirra liked his touch; it was gentle but firm, the touch a man should have.

As they walked to Kirra's apartment she had to explain Kitchi-Manitou, the Great Mystery of the Ojibwa, to Ralph. He understood immediately.

"How can a military guy who kills people understand these metaphysical concepts?"

"Kirra, some of the most intelligent and wise people in the world serve in the military. We understand camaraderie and love because in combat, it's you and your team against all odds and everyone's lives are at stake. You cooperate or you die. That breeds love. Moreover, both men and women join the military because they want to protect and serve. If you have a good unit those feelings become stronger and stronger. When you see your buddy die in a mine explosion or get his head shot off, you understand grief, and how precious life is. It makes you think about the value of people. Military men sometimes make the best husbands because they have the most experience. Most successful military careers involve a lot of travel; you get to see a lot of different cultures and you get to really understand the world, and people. You experience people on a much deeper level than most civilians."

They had reached Kirra's apartment complex now.

"So why haven't you married if you understand people so well?"

"Because I'm a bastard," Ralph said simply.

Kirra laughed so hard she almost fell down. "Ralph Zimring, you are a treasure. I thank you for this talk; you've made me feel so much better." She gazed up at him curiously. "Have you ever considered becoming a counselor?"

"In my declining years, perhaps."

Ralph saw something out of the corner of his eye and moved his head. Kirra turned to look. When she turned back around the big man had vanished. He hadn't made a sound.

CHAPTER 31

Liqao Chen decided not to go through with the procedure on Wednesday. He knew he should call Kirra, but he was too ashamed. He had fucked up again with her; but he still hadn't made up his mind about the Transhuman program. He had been hanging out at the DC Better Humanity center for the past week, asking questions about the procedures, observing those who came in and were processed, and in general displaying unbounded enthusiasm for the amazing new bio-technology. The center provided a bed and food for everyone undergoing the procedure (in a large open room with curtained partitions), so he had been taken care of. His boyish enthusiasm was welcome both to staff and incoming patients. He even took vids with his mobile. No one seemed to care that he had backed out; lots of people did and then reconsidered later.

Liqao got a call the day after he backed out from the great Harry Kaine himself.

"Liqao Chen," Harry said, turning on the charm. "I've heard good things about you."

Liqao was suspicious of people who knew things about him. "Oh, really?"

Harry laughed. "My spies are everywhere."

Liqao believed him. "To what do I owe the honor of this call?"

"You seem to have a level head on your shoulders, my friend. You've been helping in our Chicago operation, and I wanted to ask if you would like to join our organization permanently. We have a spot for you. An important one."

"I'm all ears."

"I understand that you don't want to go all the way, and that's fortunate because we need normies, er, regular humans, to speak for the transformed community. You've been working with us now for over a year and would, I think, be a great asset to the community."

Liqao was intrigued. "My main interest is in the technology. From what I've observed it's fascinating, and I'd like to know more."

"That might be arranged," Harry said.

"I'm nobody. Why are you interested in me? You can find more famous personalities."

"That's just it, Liqao. I have already enlisted plenty of celebrities. What we need are average citizens."

Liqao saw the logic in this.

"I'm also interested in you for personal reasons. One, you are friends with Karen Everard, M'basa, Joe Courvall, and Kjirsten Chastaine. You already know I went to the future with them. And two, you're from Midland, my hometown. That might not be a big deal to you, but I have a fondness for the place and want to see a center there. And third, your four friends don't like me and I don't like them. They've started a VRcast and have been ripping me every day."

This was news to Liqao. "Really?" Liqao could see that Harry Kaine was affronted. "Why would you care what they think?"

Harry looked petulant. "That damn podcast is inciting opposition to the program all over the place. It's not right and it isn't fair."

Liqao saw no sense in this. "How isn't it fair for my friends to exercise their right to voice their opinions? Society went through all that after the Great Reset."

Harry looked shocked. "Whose side are you on, Liqao?"

Liqao dismissed this as nonsensical. "Are you saying that going to the future was real? I thought it was all bullshit."

"It's true," Kaine said belligerently.

Liqao changed the subject; the idea of time travel was ridiculous. "I really would like to get a closer look at the technology you're using here. I'm impressed with what has happened to me so far. I'd like to learn more about the procedures."

Kaine's irritation magically disappeared. "Tell you what I'll do. I'll arrange a short interview with our chief scientist, Dr. Anders Kopetz. Anders happens to be in the facility doing an inspection."

Liqao's boyish face lit up with eagerness. "I'd like that very much!"

Liqao had his own kind of charm. Harry beamed and said, "Consider it done."

An hour later it was arranged. During the interview with Dr. Kopetz Liqao discovered two amazing facts. The first was that the primary increase in knowledge that led to the Transhumanist breakthrough came from two bio-enhanced humans Harry Kaine had picked up from a soup kitchen in Midland. He got Kopetz talking. "These altered humans were street bums, but they were wearing an amazing film that we later reverse-engineered. It's a remarkable and durable thermofilm that has enormous adaptability to temperature."

Liqao immediately correlated this fact with a comment made to him by Lledren Cadwallader.

"Was this uniform blue and black?"

Kopetz was astonished. He leaned his tall frame into Liqao's face. "Where did you hear that?"

"From a friend of mine who works at the soup kitchen where Harry found the two, er, derelicts," Liqao said innocently. "He said the two men looked lost, and were wearing some kind of funny costume. He thought they were hard-up performance artists."

Kopetz relaxed. That was the official shore story.

"But how could a couple of street bums be enhanced?" Liqao asked.

"That is classified information," Kopetz said. "If you join us we'll tell you."

Liqao groaned. "It's not the time-traveling crap, is it?"

Kopetz jumped. He was very angry. "Where did you get that information you little rat?"

Liqao almost laughed. "Harry told it to me; but it's bullshit of course."

Kopetz frowned. Kaine was much too loose with sensitive data; he would have to inform Bob Saunders. He studied Liqao Chen. The little man wasn't hiding his contempt for the idea. Kopetz shrugged and let it pass.

To get rid of Liqao he agreed to let him have a guest pass into the main lab, where new arrivals were processed. "Stay away from the doctors and the patients," the hawk-faced man said, his face hardening into stern planes. "If you get in the way security will throw you out. Understood?"

Liqao gulped. "Yes sir."

It always worked, Liqao thought as he made his way to the main lab, his lab guest pass around his neck. For his entire life he had used his boyish looks and innocent demeanor to get his way. As he walked he filmed with his mobile. The place was as big as one of the old big box stores, with a large waiting area filled with people. The corridor floors were made of brown-lacquered concrete with high ceilings and cameras everywhere. The walls were painted in pastels and decorated with art. Brilliant, full-length holos of transformed humans also appeared on the walls. The full-spectrum lights were bright and cheerful. Doctors in blue gowns were everywhere. Liqao was amazed. Each of these centers probably employed dozens of medical doctors! It was said that the centers trained their own medical personnel and had themselves created a boom in the private and university medical schools.

Liqao passed a door with a sign that said "1 Designer Genetics." Another door was labeled "2 Bio-Printing Lab," and another said "3 Uploading Area." Liqao was bursting with excitement to see these labs, but the Main Lab was the big attraction. According to Kopetz, this was where incoming patients were examined and prepped for the transformation procedure.

The first step for new normies was to watch a video presentation by Harry Kaine. Liqao walked slowly along the walls and got in position to see it. After watching the vid Liqao was convinced again that he should be transformed. Kaine was a mesmerizing presenter, exuding confidence and certainty that the future was a transformed one. The images of transformed humans as they came out of the labs were truly impressive. It was said that the conversion rate for Kaine's presentations, once a person arrived at the center, was somewhere between 90% and 95%. Although he had backed out, it was mainly because of Kirra's influence. For a normal person off the street Harry's "Introduction to Transhumanism" would be totally convincing.

The next step was the Main Lab, a large open area that could hold a hundred patients. This was what he really wanted to see, for he had been denied access to it after he backed out. Liqao watched as each normie was led to a bed with a portable medical station. They could see the other patients and observe what was happening. Liqao was impressed, for this setup inspired trust. It was as if Harry was telling the converts, "What is there to hide? Our operation is totally transparent." Each patient was hooked up to an IV, which contained a blue fluid. Liqao saw a look of serenity on the faces of each patient as the drip began to flow. Liqao listened as one of the medtechs (who were dressed in yellow gowns, doctors in blue) explained. "Nothing to worry about, Robert. You'll feel a pleasant sensation as the serum prepares your biology for enhancement. A mild sedative is administered, but you'll be aware the entire time. In about twenty minutes I'll come get you."

Liqao saw that each medtech handled about ten patients. After the serum was administered, a tech would remove the blood tap and ask the patient to sit up. "How do you feel, Robert?"

"I feel great, doc! What is that blue stuff? It's better than coke!"

The tech laughed. "You'll feel even better after the next procedure. Come with me to Lab 1. You're undergoing the genetic enhancement, right?"

"That's right."

Liqao saw an amazing transformation. The man, who had come in looking rundown, now looked vital and healthy and was practically glowing. What the fuck? What is that blue stuff?

As he observed, Liqao saw that almost all patients were directed to Lab 1, the genetic enhancement lab. Occasionally a patient was directed to Lab 2. Liqao remembered that Lab 2 was labeled "Bio-Printing." During the time he watched, every patient who received the serum looked happier and glowed with health. Liqao noticed that no patient was ever brought to Lab 3. As the hours passed, Liqao understood that there were two types of transformed: normies who were genetically modified to be disease resistant, and those who received what Harry called "comprehensive enhancement." Liqao guessed that Lab

1, with its "Designer Genetics" sign, was for the former. Lab 2, the Bio-Printing lab, was for the latter. But what was Lab 3?

As it got dark outside the place was still humming. He remembered that the centers operated 24 hours a day. He was getting really hungry and left the main lab and went into the cafeteria off the main hallway for a couple of sandwiches and some coffee.

Liqao resumed his vigil in the main lab. During the night hours traffic fell considerably, but a few people were still walking in.

Liqao was determined to get into Lab 3. He would make his try in the early hours after midnight. Many of the staff had left after 8 p.m. Liqao didn't change his seat in the main lab in order not to attract attention, but no one noticed him. No one ever noticed him, he thought. He was invisible, like the janitor.

At 2:15 a.m. he decided to make the attempt. The main lab only had three people in it. All the rest of the walk-ins had apparently been processed. Liqao walked out into the hallway and casually strolled past labs 1 and 2. Lab 3 was at the end of the long corridor. He stood at a vending machine, looking for something to munch on. Just as he put his currency card into the machine and his orange pops came out, he saw the door to Lab 3 open slightly. He heard muffled conversation. Liqao saw a sleepy tech lead an elderly patient dressed in a white gown down the dimly-lit corridor. No one noticed him as he began to munch on his orange pops. "Welcome Jane!" someone said from inside the lab. The door was opened and the tech walked back down the corridor, his back to Liqao. The patient walked in and was greeted by two blue-frocked doctors. Liqao saw his chance. Just before the lab door closed Liqao slithered into the almost completely dark lab. He saw several beds that were surrounded by medical stations that gave off a pleasant bluish-green light. What looked like a perfectly formed human body lie on one of the beds. The two doctors led the old woman to another bed alongside the body. Two glowing shells were slowly moved into position over the bodies. The two doctors checked their instruments and went through a long checklist, checking vital signs. After all was ready, the shells began to glow brighter and brighter. Liqao wondered what was going on. He had positioned himself along one of the walls in a corner, behind some storage lockers.

What Liqao saw next blew his mind. After the glowing shells had been in place for about twenty minutes, they began to dim. Each doctor raised one of the shells. The elderly woman's body was now limp and lifeless. But the other body (which Liqao suspected had come from the Bio-Printing lab) began to stir. One of the doctors leaned over the bed. "How do you feel, Jane?"

Jane slowly sat up on the bed. "Wow. I haven't felt this grand in years. Can I see what I look like?"

"Sure!"

The two doctors led the woman over to a full length mirror, which had small lights positioned around the frame. Liqao saw the beautiful body preen itself in the mirror for a bit. "Why, I'm beautiful!" the woman cried.

"Yes you are."

"Oh, thank you so much! I have a new life!"

Liqao saw the look of pleased satisfaction on the faces of the two doctors. "Now Jane, we just have to do some tests to make sure everything went perfectly," one of the doctors said. "Will you please lie back down on the bed?"

At this point Liqao decided to leave. Very carefully he moved toward the door and opened it ever so slightly, looking down the corridor. It was empty. Liqao silently slipped into the corridor, walking down it and back into the main lab. He walked unobtrusively along the walls until he found the front door (there was only one entrance into the DC center), and walked out into the parking lot. His vehicle was still there. A couple of security guards were patrolling the parking lot.

Liqao looked at his chronometer. It was now past 4 in the morning. Liqao got in his vehicle and began driving north. He was stunned. The technology he'd seen was far in advance of anything he had imagined. Where did Harry get it? He dismissed the story of the two clones from the future. Even if true, it wouldn't account for the blue serum or the incredible consciousness transfer tech he'd seen in Lab 3. Karen had called it mind uploading. No, this stuff must have already been in existence. But where? Moreover, the amount of money needed to set up the DC Better Humanity center must be in the hundreds of millions at least, and there were over 100 of these centers around the world. Where was the money coming from?

Liqao pondered these things and many others as he continued to drive. He wasn't paying attention to where he was going. After three hours he became aware of his surroundings because the landscape was getting brighter. He checked the roadside sign. It said, "I-70 East." Another sign said "Pittsburgh 77 miles."

Liqao realized he was on his way back to Midland.

CHAPTER **32**

Harry Kaine was sitting in the Berlin center after giving a live presentation. It had gone well; they always went well. The movement was a great success. He was certain he had beaten the time conservation law. The two clones from the future had been the difference! That's what Anders Kopetz told him and he believed it.

Harry had never scrutinized the entire Trans-Post-Human suite of technologies. He simply accepted them. He was enjoying his life enormously. A Better Humanity had taken him all around the world, and he had helped so many people!

There was only one irritation in his life: Midland, Illinois, his hometown. Harry ticked off his grievances mentally. 1) The City Council had rejected his center. Twice. 2) Those time-traveling friends of Karen Everard and that irritating Melanie Fuscaldo had combined to produce a critical podcast that now had many millions of views every week. 3) Those fools were backed by a mysterious outfit that was rumored to include Ralph Zimring. Zimring was an almost super-human fighter who had defeated the program's best field operatives in New York. He had killed the great Lucas Cernovitz, the program's founder and leader.

What really bothered Harry was the podcast, which he listened to every day even though it made his head explode. Those space cadets were unrelenting critics of himself and Trans-Post-Humanism. It wasn't fair! What did they care whether people became transformed? It was none of their business. The arrogance of those people! The most irritating thing was that he listened to the Lightroom in order to discover things about the program Kopetz and Paul Wen wouldn't tell him. Harry ground his teeth whenever he heard one of the podcast's so-called "whistleblowers," who were always techs or doctors associated with one of the centers. He didn't know whether the "whistleblowers" were genuine or fake, because he had never met them.

Kopetz wouldn't discuss it. "Harry, you're doing a fantastic job," the chief scientist would say. "Stay in your lane, enjoy yourself."

Harry admitted to himself that he was obsessed with the Midland "spiritual" group, and always had been. They called themselves star seeds and old souls. Star seeds my ass! They couldn't even define what "spiritual" meant! There were no metrics for it. Yet week after week the podcast message was the same: Trans-Post-Humanism was Godless scientific materialism that ignored the "divine nature" and "higher purpose" of humanity. Higher purpose! What the hell does that mean? "Divine" is a made-up word for ignorant persons who couldn't understand science.

It was a source of profound mystery to Harry that so many people identified with that nonsense. He consoled himself with the idea that over 1.2 billion humans had now become transformed. Yet there was something about his glorious new body that he didn't understand. Sometimes, when he found himself contemplating leaving the program and enjoying a more simple life, he would feel a sort of compulsion to "stay in his lane," as Dr. Kopetz always said. This had happened a dozen times now. He began to wonder about these Post-Human bodies. It was almost as if they were directed along a certain path. M'basa Ogunfatidime, on last Friday's broadcast, had even suggested that all Trans- and Post-Human bodies had a "kill switch."

When he had asked Kopetz about that, the doctor's eyes had flamed for a second. Then he had said in his measured way, "Don't believe those silly conspiracy theories Harry! Low-information persons are irrationally trying to drive a wedge between the normie population and the transformed."

"Why haven't you become transformed, doctor?" he had asked.

Kopetz had responded dismissively. "Oh, I'm too busy for that now. Besides, I'm in great health."

Harry, who could smell bullshit a mile away, was convinced the doctor was hiding something. Harry was not a profound thinker, but he sat in his luxurious suite at the Berlin center and reasoned it out. He could get no answers from his own people. The most informed persons about the program, outside the program itself, were the Lightroom folks, who interviewed people who worked in the labs. And Ralph Zimring. All of them were in Midland. Therefore he would have to go to Midland.

Harry smiled broadly. He would call them and ask to be a guest on their moronic show. How could they refuse? The great Harry Kaine himself! He would ask them the questions Paul and Anders had refused to answer. And while he was at it, he'd charm that beautiful bitch Karen Everard into his bed. He had almost had her before; he wouldn't fail this time.

Alexandria, Virginia. A Better Humanity HQ, 6 a.m.

Paul Wen met with Bob Saunders, the program's second-in-command, and their Ops director, Manuel Stoller. They were in the Better Humanity HQ in Alexandria, a nondescript four-story building without signage just south of the WIlson Memorial bridge by the Potomac River.

"Stoller, what's going on in that piss-ant Illinois town?" Wen asked. He stared at the blond-haired, blue-eyed man with the lantern jaw. Stoller was basically a blunt weapon with no subtlety. Wen knew he was wrong for the job, but Lucas was gone now. When Cernovitz ran the show he knew just when to use Stoller. Wen admitted to himself that Cernovitz had been the brains behind the program, although he had muddled along fairly well for the past three years. Thank God for Harry Kaine and Anders Kopetz, who kept the Centers running as smoothly as possible.

"It's just a podcast, Paul," Stoller said. "Nothing to worry about."

Wen had learned from Lucas how to handle subordinates. He checked his fact sheet. "A podcast now with 40 million listens every day, and growing. I thought we already took care of those Midland people?"

Wen saw Stoller wave his hand dismissively. The Ops director's face showed incredulity. "They're a bunch of nobodies, Paul! Post-millennials with day jobs, for Christsakes. The show is part-time. It's a nothing-burger, just like all the rest of the opposition to our centers."

Wen sighed. "That show was responsible for a riot at our Sao Paulo center in Brazil, and protests at the London, Berlin, Chicago East, and Shanghai centers."

"No way," Stoller said. "I investigated those personally. All of them were locally organized."

Wen activated an action chart on his desktop. A hologram sprang up from the desktop. "Look here, Stoller. You too, Saunders. This is a 3D diagram, with time the primary variable." The two men saw a 3D image of activity at all of the centers. Superimposed on that was a radial chart of worldwide broadcasts opposing the centers.

"Shit," Stoller said. "That show in Midland is the hub for the whole world!"

"That's right gentlemen," Paul said. "The Lightroom show is the hub connecting all of this rabble rousing. Everyone is taking their cue from those three whiners in Midland, Illinois."

"So, boss, you're saying that we should take them out," Stoller suggested.

Wen smiled. "Cut off the head and the body dies."

Bob Saunders was pleased. He hadn't fucked anyone up in a long time. Besides, he reasoned, normies are becoming like Neanderthals; a species on its way out. Killing a normie was no big deal. Saunders had already decided to become transformed.

"Listen up gentlemen. Harry Kaine is determined to go to Midland. He's going to engage those Lightroom people in a debate next Monday, which I have reluctantly sanctioned. We can use that to our advantage. Get to Midland now with a team. Scout the area, look for your opportunity."

"We cut out the cancer before it metastasizes," Stoller offered.

Wen sat back in his chair. The meeting was proceeding satisfactorily. "The primary goal is to stop the show," Wen suggested. "If accidents can be avoided..."

Stoller and Saunders glanced at each other. Both men shrugged. "Sure boss," Stoller replied. "We'll take care of it." The two men rose from their seats.

Wen sighed. Here is where they missed Lucas. He would have come up with an elegant plan to shut the show down without creating a big mess.

"One last thing, gentlemen. I see the guiding hand of Ralph Zimring behind these nerds. Be very careful. We don't want bad publicity."

"Zimring!" Saunders shouted. "He can have nothing to do with these...children. Mercs don't give a shit about civilians."

Wen frowned. "Zimring does. He's gotten himself religion in his old age. Remember, it was Zimring who killed Lucas in New York." He saw Bob Saunders frown. Wen knew Bob Saunders despised Ralph Zimring. Zimring was the man he wanted to be, but could never be. This fueled his anger. That, and Zimring's reputation among those in the business.

Paul Wen regarded Bob Saunders, with his blocky body and his thick face and curly blond hair. Saunders was a valuable man certainly, but not a nuanced operative. "Do you want Duchene Comstock involved in this?"

Bob Saunders' eyes lit up. "That fat sneaky bastard!" Saunders thought for a moment. "He's a conniver all right. And he hates Zimring and Berglin. You'll recall he was also involved in the fight against Max Berglin."[]

"'Yes, by all means call Duchene," Stoller said. "If those are your orders, we'll liaise with him."

Paul nodded. "You guys take care of it."

Stoller and Saunders nodded, and walked out of the room. On the way out Stoller said, "I recorded everything. Wen isn't too bright."

Saunders grinned. "If anything goes wrong we'll pin it on him."

"Fuck Duchene Comstock," Manny suggested. "Let's handle this ourselves."

Saunders nodded his agreement.

After the two operatives left, Paul Wen sat in his chair watching the sun slowly set. He thanked the heavens for Harry Kaine, who had been brilliantly holding together the company's public image with his enthusiasm and marketing ability. But Harry had erred badly in Midland with the bribery of the city council. It was his first major mistake, and completely unnecessary. Wen had a soft spot for Kaine: the man had essentially been responsible for dozens of Better Humanity centers, and over one billion transformed humans.

More than likely Harry's error in Midland had resulted from a personal attachment to his home town. Nevertheless, he would have to monitor Kaine more closely from now on. That was trivial, for all transformed humans were like a mobile phone: you could read their vital signs and know their location at any time. The bio-nanobots that ensured a disease-free life doubled as little electrochemical transponders, and could be scanned using the proper equipment. All of that data went into a central database right here in Alexandria.

If Harry did become difficult, disposing of him would be easy. They also had vid of Harry's sexual liaisons with underage girls. On the other hand, Harry was irreplaceable. Lucas had scoured the world for someone to replace Kaine, if it became necessary, but he had never found anyone with Harry's ability. The man was a once-in-a-generation marketer.

Lucas' goal had always been to subtly control "normie" society with a few influential and grateful influencers, but Paul understood that Harry Kaine had a bigger vision: to completely replace normal biological humans with the transformed. Harry's vision of the future – a future of transformed humans with superior biology – was so real it was like he had actually seen it.

Paul asked himself whether he should undergo the process. The transformed bodies literally oozed vitality and health. The only thing holding him back was, if the transformed are programmable, who is doing the programming? Yet he had never seen any evidence of this programming, and he was around transformed people all the time. They all seemed overjoyed to the point of arrogance with their new lives. Was the nano-bio programming automatic to the transformation process? He would have to consult with the program's chief scientist, Anders Kopetz, who was responsible for the administration of the protocols all over the world. Kopetz could give him the answers he needed.

Paul remembered the time he and Lucas had been read in to the program, in New York. The two Better Humanity directors had been briefed by two odd-looking fellows with silver hair, who were dressed in form-fitting, silver-colored unis. The men spoke with a funny accent, as if their words were being translated through some kind of speech filter. When he and Lucas were briefed about the new biotech, they were blown away.

That had been the beginning. He and Lucas had received periodic orders about changes to the program on a secure channel. These were always sensible orders. Otherwise they had been left to themselves, except for the monthly briefing with the two silver-haired men. He wondered why a replacement for Lucas had never been assigned. Paul knew himself as a good number two, but he wasn't a leader. He shrugged.

He must be doing a decent job or he would have been replaced a long time ago.

Paul realized he had been daydreaming. He had a lot to do and not a lot of time to do it. His main job was to keep an eye on the centers all over the world from the global CCC in the basement of the building. This was possible because every room in every center was monitored and feeds sent back to Alexandria via a network of dedicated satellites.

Paul exited the fourth floor conference room and took the stairs two at a time down to the basement. Dozens of techs occupied data stations, taking the feeds and processing and storing the information. Paul entered the main control booth, a circular soundproof room surrounded by ten-foot high bullet-proof transparencies. As he studied the feeds Paul became more and more angry. Another riot in Sao Paulo (which Anders was dealing with), protests around the London and Shanghai centers, and political opposition rising in Europe and India. Yet this past week over 1,357,000 humans had been transformed!

Paul Wen used his comm board to contact Kopetz. The tall, bearded scientist with his salt-and-pepper Van Dyke and great hawk nose answered immediately. "What is it?" he said sharply. "I'm very busy."

"I can see that." Kopetz stood in a white lab coat, monitoring the "production line" labs in Sao Paolo. These Brazilians! The very poor in the barrios flocked to the center, seeing an opportunity for the "good life." This caused conflict between them and the wealthier citizens, who saw the center as an abomination. Sao Paolo was one of the busiest centers, but also the one with the slackest procedures. It was absolutely imperative that the procedures were flawlessly executed for every human. "Failures" (deaths) always created public outcries, despite the legal disclaimers everyone had to sign before being admitted. Despite Harry's genius, Paul knew that Kopetz was even more important, for word-of-mouth was A Better Humanity's best marketing tool.

"Anders, when will you be back in Virginia?"

Kopetz frowned. He hated it when people called him by his first name; it showed a lack of respect. "I'm leaving tonight. We're back on track in Sao Paolo."

"Good! Will you stop by HQ tomorrow morning after you get in?"

Kopetz hesitated. He had a liaison planned for noon in New York...but he could catch a red-eye to DC tonight. "Sure. I'll try to be there by 6."

Anders arrived punctually (as usual) at 6 a.m. and the two men went up to the fourth floor conference room. Paul shoved a large coffee toward him and they both gulped down the hot liquid. The goddam cold! Paul thought. Not yet the end of September and the temp was in the low 40s.

"I'll be brief Anders," Paul said after they had settled in. "What is the nature of the nano-programming in the bio-enhancement procedures?"

Kopetz was clearly startled by this.

"Uh, in order to ensure cellular health and to support the immune system, our molecular bio-bots are able to respond to certain, ah, electrochemical signals."

Wen looked hard at the hawk-faced man. "Cut the bullshit. In plain English, how susceptible are transformed humans behaviorally to outside control? And how sophisticated is that control?"

"You've been listening to that podcast, Paul," Kopetz accused.

"Yes I have, because I'm thinking of undergoing the procedure myself. I want to know what I'm getting into."

"I wouldn't recommend it," Kopetz said firmly.

"Why not?"

Kopetz smiled with his mouth but the smile didn't reach his eyes. "Our silver-haired friends won't allow it."

"What are you talking about?"

Kopetz sighed. "You and Lucas are – were – too important to the organization." The chief scientist's smile broadened. It wasn't a nice smile.

Paul Wen began to feel sick to his stomach.

"You've done a good job so far, as has Harry Kaine. And it's too bad about Lucas; he was a good man. Here's a friendly warning Paul: stay in your lane and all will be well."

For a split instant Paul Wen looked up into the eyes of a monster; a monster who talked casually and who was outwardly friendly. Then the doctor's eyes assumed their normal blandness and Kopetz began to speak expansively.

"Really old boy, it's not as bad as that. Humanity is getting an upgrade that's all. Our silver-haired friends, simply put, are putting a stop to humanity's destructive and warlike tendencies, which could shortly, er, reach out beyond the boundaries of our little solar system. The human race will eventually become a good citizen of the Greater Community when enough transformed find positions of power. The plan is almost complete, Paul, and you've been an integral part of it. Congratulations!"

Paul felt like a little child who had just discovered that grownups weren't all nice people. He felt lost.

"There there, don't worry. Stay human. I myself would never undergo the procedure."

"So all of Harry Kaine's presentations are so much bullshit."

Kopetz was shocked. "Not at all, Paul! Harry is sincere. He genuinely believes that the human race will be better off as Trans- and Post-humans."

"But you don't."

Kopetz stared at him cynically. "Do I really have to explain it to you? Review your recruitment into the program and what you've been doing here for the past several years. All will become clear to you."

Paul was numb now. "And what is your role in this, Anders?"

Kopetz shrugged. "Like you, I am a cog in the machine. Once my work is done in the program – and it is almost done now – I shall retire and live out the rest of my life in luxury."

"That's insane. Why not upload your consciousness into a perfect body and live for a hundred years?"

Kopetz looked at him in disbelief. "You still don't understand, do you Paul?"

Paul Wen felt himself falling down a deep, black hole. He rose quickly from his chair, his hands out before him, his head shaking in denial. "No...no...Anders, it's not true."

Kopetz smiled broadly, as a man will who is perfectly comfortable with his life's work. "The Trans- and Post-human bodies will eventually fail, of course. You've surely heard Harry's ridiculous stories of the future."

Paul was beyond disbelief now. "You mean...they're true?"

Anders shook his head sadly. "They're just Harry's fantasies of course! But my dear fellow, we've rushed this program of Transhumanism forward so fast that the procedures must be constantly tweaked. The biotechnology is fantastic, of course, and duplicatable. But I and my staff – with the, er, assistance of our two friends – have to monitor every center very closely."

Paul put his hands on his face. A terrible thought occurred to him. "Our friends...who are they?"

Kopetz was watching him, an amused smile on his face. "Do you really want to know? I think it would be better if you didn't."

"How long do the transformed have?" Paul said faintly.

"Who knows? It could very well be a hundred years...or it could be much less than that." Kopetz smiled avuncularly. "But however many years there are, they'll be good years. Disease free, happy years." Kopetz rose. "Now, if you'll excuse me Paul, I have a liaison in New York with a remarkable woman. I don't want to be late!" Anders Kopetz walked out of the room, whistling.

After Kopetz left Paul Wen stood where he was, unmoving, looking blankly out the east window, watching the sun slowly rise.

CHAPTER 33

Liqao Chen drove all the way to Midland, only stopping once for coffee and sandwiches at a roadside diner outside Pittsburgh. He had done the trip in a little over nine hours, which had given him plenty of time to think. His expertise was not in the life sciences, but what he had seen in Lab 3 was other-worldly. Consciousness transfer to a 3D printed body! It was impossible. He was fascinated and abhorred.

His experience at the DC center had matured him. Talking to Kopetz, he had suddenly realized that Kirra had been right all along. He was living his life being a little kid. A cute little kid; and Kirra liked that. But it was time to get serious. What did he really want from life? He had no idea. He had quit his job at the MPD and abandoned his deepfake recognition software for lack of commitment. He was 25 now, with no direction. He must talk to Kirra; she would have great advice.

Liqao almost drove off the road as he realized that he had forgotten to call her last Wednesday. That might be it for him, and he didn't like the thought. He felt a pang in his heart.

As he arrived in Midland he looked at his chronometer; it was now 3 30 in the afternoon. Kirra would still be at the library. Liqao put his foot on the accelerator and hurried over.

When he got out of his car he almost ran through the library's front door. The first thing he saw was Kirra at the front counter, flirting with a tall, well-dressed man. Liqao stopped and stood in the entrance, staring. Kirra gave the man a big smile and the two completed their conversation. The tall man began to walk toward him and the front entrance. At that moment Kirra turned her head to look at him. When she saw Liqao her face blanched and she took a step backward. Kirra raised her hands to her face.

Liqao's face fell. She hated him, obviously. He wouldn't bother her anymore.

Liqao turned and walked back to the parking lot, his emotions (for the first time in his life) in complete turmoil. He got in his car and drove to his apartment. When he arrived he checked his mobile: there was no message from Kirra.

He waited around his apartment, trying not to panic. Maybe she was busy at work. He decided to wait around until 5, when Kirra got off her shift.

As 5 p.m. came and went, Liqao was about to go to the Mason Street coffeehouse to get something to eat when he heard his mobile ping. It was a text from Kirra!

So you didn't get the procedure.
No. I couldn't do it.
You didn't call.
I'm sorry. I was too absorbed in myself. As usual.
That's refreshing honesty at least.

Liqao didn't know what to say next. He found courage.

Can I come over? I want to tell you all about my adventure.

There was no response. Liqao got more and more nervous. He wanted her to say yes. After ten minutes his mobile pinged again.

All right. For a little while.
I'll be there in ten minutes.

Liqao felt a great sense of relief, and something tight loosened around his heart. This was a new feeling. He almost felt a sense of desperation as he parked his vehicle and walked up the stairs to Kirra's apartment. What could he say to her? He had basically left her and hadn't called or messaged even once. He realized what an idiot he'd been.

Liqao knocked on the door. When it opened Kirra was standing there in jeans and a burgundy top. Her raven black hair fell over her shoulders, her intense blue eyes gazed into his. Her ruby-red lips were full and inviting. She was gorgeous, in spirit and in body. Why hadn't he noticed before?

Kirra stepped back to let him in. Embarrassed, he walked gingerly into the living room and stood there, unmoving.

"Well, that's a change!" Kirra spoke lightly, as if she was also embarrassed.

Liqao wanted to say something to break the tension but he couldn't think of anything witty or appropriate. "How so?"

"Because you usually rush over to the couch and start talking about yourself."

Liqao grimaced. "That bad, eh?"

Kirra saw at once that Liqao was different. "Yeah." She spoke with some bitterness.

Liqao looked into her beautiful blue eyes. "I'm sorry Kirra. I've been a fool."

Kirra's eyes widened. "What does that mean?"

Liqao's face assumed an almost desperate air. "It means...that I've always liked you. Loved you even. But I was too stupid to know it."

Liqao saw Kirra's eyes tear up. "I'm sorry Kirra...I'm doing it again aren't I?"

"Do you mean it?"

"Mean what?" he said with his boyish smile.

"That's the Liqao I know," she said.

"Oh, right. The thing about loving you." He gazed into her eyes. "I don't know what love is, Kirra. But you're the closest thing I've ever come to it."

He saw her smile; it lit up her dark face; her blue eyes sparkled. The little earrings she wore shook gently. "God you're beautiful."

"That's better Liqao."

"Can I kiss you?"

Kirra flushed. "OK."

Liqao stepped slowly toward her. The moment was incredibly intimate and he hadn't even done anything yet. He took her gently in his arms. They were almost the same height; he was but a half-inch taller. Liqao bent his head and placed his lips gently on hers. They were so soft! She tasted incredibly good. Suddenly he felt a jolt of electricity go through his body and he tightened his grip and placed his lips firmly against hers Her body felt fantastic against his. She responded to his movement and pressed herself against him. He was breathing hard; she was too.

They found their way to the bedroom.

Soon after getting approval from Paul Wen, Harry Kaine called the Lightroom podcast's guest line, right in the middle of the show. "Hello, this is Harry Kaine."

The voice on the other end spluttered. "H...Harry Kaine?? Is this a joke?"

Harry had learned a lot from Anders Kopetz about how to phrase sentences for maximum impact. It had to do with making important statements, but speaking very casually. "Not at all. This is Harry Kaine. I want to be a guest on your show."

Harry heard the famous phrase. "Breaking News!" someone shouted. "Harry Kaine has called into the show!" He recognized the voice as that of the bearded Ken Strickland, who was the show's producer.

"Put him on live!" Karen Everard said excitedly. Harry heard the audience in the background buzzing. He was pleased with the impact he was making. His mobile showed Karen Everard sitting at the table in front of her mic, which was connected with a red extension cord. The others had different colored cheap cords that you could buy at a hardware store. What a slapdash outfit, he thought. But Karen herself was even more beautiful than in New York. For some reason normie women were more attractive than transformed women. At least for a while.

"Hello Karen!" Harry said warmly. He saw the women in the audience sit up a little straighter when he turned on his charm, and the men tense up. Harry smiled. "I was telling your producer that I'd like to come on your show," Harry said smoothly. He almost laughed as the other woman, the redhead, tossed her head and smoothed her hair. Her boyfriend, M'basa, was glaring at him. Good! He already he had them emotionally. Years of practice had made this ability almost unconscious.

"We'd love to have you Harry Kaine," Karen said. "But why in the world would you want to grace our rabble-rousing little podcast?"

Harry detected the note of sarcasm and was pleased. She was trying to play him, but he was an old hand at that game. "Well Karen, I have to say I'm a little upset by the constant attacks against me and my centers. So I thought we'd have a little debate. After all, we never got to do that when you were in New York back in '36."

Harry saw Karen's face tighten. That was good, he had gotten to her.

"If I recall, Harry, it was you who cancelled that debate," she said.

"I think he was afraid back then and he's afraid now," M'basa interjected.

Oh this is good, Harry thought. He loved it when he made people angry. Harry smiled expansively. "I'll concede the point, er, M'basa." He pronounced the name with just the slightest tinge of condescension. He saw M'basa's eyes flare. Oh, this one has a temper! "During our little talk you'll be able to explain to me, I'm sure, your vague and unsupportable concepts about spirituality – whatever that is – which you claim without evidence is somehow superior to a worldwide organization that has helped over 1.3 billion persons."

The redhead's eyes hardened. "Oh, we'll do that! Game on, Harry. You're welcome anytime."

Harry smiled. "Good! It's settled then. I also have a few questions for you about your so-called whistleblowers. Or are they frauds? Charlatans, perhaps? I've heard some deplorable stories about shows like yours that pass off imposters as genuine."

Karen had dealt with Harry before. She knew his game. "Oh Harry, we'll bring on the chief scientist at the Sao Paulo center. I think you'll recognize him."

Harry coolly dismissed this. "Oh yes. Dr. Mario Caralho. I'm afraid we had to dismiss the man for incompetence." Harry sighed with disappointment. "With so many centers and so many being transformed, it's hard to maintain the quality of our personnel." Harry pointed a finger. "But I assure you, Karen, that when we find malefactors, we terminate them immediately."

Karen smiled as a combatant might smile to her opponent. "You're good Harry, very good. You've become much more sophisticated in your old age."

Despite himself Harry felt that one hit home. He was thirteen years older than Karen, but his perfect bio-enhanced body didn't show it very much. Nevertheless, it was a sore point with him. Harry felt himself warming to the confrontation. "And you, Karen. You're even more beautiful than I remember. You see, I have a fondness for normie women." Harry sighed with genuine regret. "Despite the perfection of transformed women, there is a sort of...crudeness, or perhaps a lack of polish, in normie women that I find attractive."

Karen heard the slightly disparaging tone Harry used when saying "normie women," and almost let the cat out of the bag in her irritation. She remembered the feed Zimring had given them in New York, when Harry had been in bed with an underage girl. This was information Ralph had said should never be divulged. Karen found her composure but acknowledged the hit. "Touche, Harry. You won't mind if I bring a few guests?"

Harry spread his arms out. "The more the merrier, Karen." He said her name in tones of intimacy and saw her quick intake of breath, and then her realization that he had reached her.

Kjirsten spoke. "Why don't we have a full debate? As long as you like, Harry. We'll stream the first hour live and record the rest for streaming on later shows."

Harry almost rubbed his hands in satisfaction. "That will be marvelous, er, Kjirsten? Do I have that right?"

M'basa frowned and the redhead flared a bit. Oh, this is going to be easy, Harry thought. He had collected several tells from each of the hosts now. It would be child's play to move them about the debate chessboard, and make fools of them. "How about next Monday for the debate?" Harry suggested amicably. "Precisely at 7 p.m.?"

"Book it, Ken," M'basa snapped. The two women nodded.

"Then ta-ta for now. I'll come with my two assistants." Harry broke the connection.

CHAPTER **34**

I got home from work at 8 p.m., wanting to fool around with Karen. Tonight was her night to come over to my apartment. I wandered around the place looking for her, then remembered she was doing that damn show. Normally easygoing, I was suddenly very angry. It was bad enough that I got home late every day, but Karen often didn't show up until past 11! Then she was so tired she wasn't in the mood for anything except a shower and sleep. It had led to several arguments.

"I work a 16-hour day Joe," she snapped one night. "Sex is the last thing on my mind."

"We don't have much of a relationship," I said in a complaining voice.

"Who's fault is that?" she said, her ever-ready temper coming to the surface. "Are you trying to blame it on me?"

I sighed. Getting angry with Karen was never a good idea. She never backed down from me, or anyone. It was hard to reason with her sometimes. "Forget it. We're both tired, let's hit the sack."

After we got in bed Karen was out immediately. I lay beside her, wondering what I was doing. I was as busy as she was so it was just as much my fault as hers. But I was tired of always giving in to her stronger personality. I admitted to myself that she and I weren't working anymore.

I lay there wondering if there was anything I could do to salvage things. After an hour of thinking, trying not to move and disturb her, I decided that we somehow had to connect during the day. She went to her job at the *Courier* and then to Ken Strickland's house for the show, eating out without coming home in between. Ger and I were both up before 6. We spent 16-hour days on the phone with suppliers, with Edison procurement staff, with the engineers, and on the workstations tweaking the designs and testing the finished products. We were way behind schedule, only producing a couple of the mag engines each day. It wasn't good enough, and we were both frustrated.

It had been another long day at the Edison center for me. The assembly process was hitting continual snags. Mostly the problem was a shortage of the powerful permanent magnets required by the mag engine. Edison had compounded the problem by starting a local publicity campaign with a slick vid, demonstrating the device which ran our little factory. It had caused a sensation. We already had orders for over 2,000 of the things at $3,000 a pop. But we couldn't keep up with the demand. I was making money but I didn't have time to spend it or enjoy it.

I knew what would happen tomorrow. I'd get up at 5:30, try to slip out of bed and not wake Karen, then take a shower and be off. We wouldn't see each other all day and then when we did interact at night we would be both too stressed to even talk. I had to break up with her or find some way to repair the relationship. It wasn't working for me anymore.

When Germaine Robinson got home he felt a vague feeling of dissatisfaction. This discomfort had nothing to do with his work. Midland Edison had come through on their promise, and was almost as excited as he was about making and distributing the mag engines. Old man Taubman had settled with his company (a private corporation that included Joe, himself, and Edison) for a percentage of sales. Work wasn't the problem.

It was a cold evening at the end of September. Ger was used to the cold now, everybody was, so it wasn't that. Something was missing from his life. He thought about Ruth at Edison, an older woman. He liked older women. No, she wasn't it.

Ger made a pot of coffee, turned up the heat, and sat down at the kitchen table. He thought about the women in his life. Women were attracted to him but he never had time for them. They had always distracted his attention away from his work. In the past he had always had a single-minded, driving urge to invent something that would change the world. He had done it! He had developed the time viewing device, he had discovered and perfected the mag engine. In the depths of his soul he felt he had achieved his life purpose. He felt content about his work for the first time in his life.

His dissatisfaction couldn't be something as trivial as a woman, Ger thought as he sipped the scalding hot coffee at the kitchen table. Although, there was Tanya at the Taubman Building. His thoughts had been straying to her more and more lately. He used to see her every day and took her interest in him for granted. But now he worked full-time at Edison and hadn't seen her in months. She had made no secret of her liking for him, but he had always blown her off.

A troubling thought entered his consciousness: Had Tanya found someone? Why was this possibility so disturbing? Another thought entered his brain: Who but Tanya would understand how important the mag engine was to him? He used to bounce ideas off her all the time. And she was smart. She would be genuinely happy for his success.

Ger was almost in a panic now and rose sharply from his chair, almost spilling his coffee on the table. The thought of Tanya being with some dumbass rankled him. She was his! Or was she?

Ger looked out the darkened window at the maple tree, its colorful leaves drowned by the descending blackness. He'd stop by the Taubman Building before going to Edison. He'd do it tomorrow morning before Tanya got on shift.

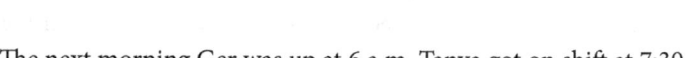

The next morning Ger was up at 6 a.m. Tanya got on shift at 7:30, but she usually picked up a coffee at the Mason Street coffeehouse right at 7, just before walking to the Taubman Building. He'd be there at 6:30, sitting right at the front by the register with a cappuccino. When he got there he became uncharacteristically nervous. What would Tanya say to him? Would she even notice him at all?

Ger began to remember his interactions with her. He had usually been brusque with her, and completely self-centered. He began to feel embarrassed about his conduct. He'd have to go very easy.

As 7 approached he started to feel nauseous. What was wrong with him? He turned his head to look out the window and saw her approaching the door with her blonde head down, covered by a cute little cap. A wind was blowing and it was cold. She walked by his table and to the front counter. She hadn't even seen him.

Ger watched as Tanya ordered and had a word for the server, asking her about her young daughter. Tanya turned with her coffee and saw him. She yelped and dropped her coffee on the floor. "Germaine!" she cried. "What are you doing here?"

Everyone in the place turned to look at them. Ger quickly got up and picked up the coffee, which had fortunately not burst. All of the Mason Street take-out containers were sturdy.

"Can I walk with you to Taubman?"

"Uh, yeah." She spoke as if mystified that he would notice her at all.

So this is how she expects me to act! Ger felt embarrassed. He handed her the drink and looked into her eyes. She had a wide Slavic face, high cheekbones, delicate nose, and full lips. He had never noticed before.

Tanya recovered her composure and smiled. "Why this morning?"

Implied in her question was, "Why at all?" Ger felt a sense of despair at his lack of consideration.

They were walking toward the door and Ger opened it for her. He moved to the left to break the wind for her as they started to walk down Mason to the traffic light.

"I deserved that," Ger said. "Are you...are you seeing anyone?" he asked nervously.

"Not presently," she said, startled at the question. "I just got rid of an academic who could only think about his work."

There was a subtle message there, he thought. "Uh, yeah. I know the type."

She laughed, then spoke seriously. "What is this Ger? You've never been interested in me before."

Should he be brutally honest? He felt he could with Tanya; they were like buds.

"It's OK," she said, encouraging him. She was either reading his thoughts or was super-sensitive. He liked it.

Ger took a deep breath and plunged ahead. He wasn't good in social situations, and especially with women. "I blew you off before because...because I'm self-centered," he got out. "My work was the driving force in my life I didn't want to do anything else."

"I know," she said. "You have a strong life purpose. Everyone can see that."

"Can they?" Ger was mystified.

"You really are quite silly you know," she said, like a big sister to a little brother. "Of course! It's why I...I haven't given up."

They were past the light now and into the intersection. Ger suddenly grabbed Tanya, dropping the coffee into the intersection, and kissed her on the lips.

Horns were honking now. "What are you doing?" somebody shouted, his car about two feet away from the kissing couple. Ger noticed that Tanya had responded eagerly to his embrace. Somebody laughed, another cheered. "That's right honey!" a woman shouted who had reached the other side. "Don't let him go!"

Ger was totally absorbed but came back to reality. The dropped coffee was forgotten. He hustled Tanya across the street and waved to the vehicles in acknowledgment.

Tanya's face was flushed with excitement. He couldn't tell whether it was because of his kiss or because of the danger involved in his impulsive action.

"Wow! Where did that come from?"

"Was it any good? Or did I just scare you to death?"

Tanya laughed. "It was good, Germaine. Very good."

Ger realized that he had been moving very fast. He was inexperienced in relationships and didn't want to give Tanya the wrong idea.

She picked up on this. "Don't worry Germaine. This is good enough for now."

Ger relaxed, but he wasn't sure what to do next.

Tanya put her hand on her hip. "Are you going to walk me to Taubman like you promised?" she said, smiling.

"Oh yeah! Yeah, Taubman. Right!"

"For such a smart guy you're pretty forgetful."

Ger grinned. "You're a lot to forget about."

Ger noticed that Tanya walked with a skip in her step all the way.

CHAPTER 35

At an Undisclosed Location

Two silver-haired men sat across from each other at a small table. On the table a curious sculpture was between them. A very thin wire held up a startlingly real carving of an animal about to leap; its hind legs tensed for a spring. Two small transparencies, one each on the north and west walls, showed black.

The two men were both tall and thin, with long, fine, silvery hair. Other than a thin face and very large eyes, each could have passed for an unusual-looking human being. Each of the eyes had a double eyelid. Both men looked remarkably similar, but there were subtle differences in their facial features. One of the men was almost one-quarter of an inch taller than the other.

"The experiment has so far been a success," the taller man said.

"We have exceeded our brief," the other said. "We have been interfering in their development."

"Not at all! We have done nothing untoward. The humans asked for the biotechnology. We merely acquiesced to an independent request. It is permitted."

The smaller man squirmed, but reluctantly agreed. He was more concerned about what would happen to them after their mission review.

The taller man sensed this. "Amongst trading nations there are certain ethics and standards that are generally followed and honored, as you well know. Should a race spoil its own natural environment, other races have the right to come and displace it. That is our justification for being here."

"This is a controversial clause and is rarely acted upon."

"Certainly. But it is still generally held amongst most races that are involved in trade and commerce in the Greater Community."

"These humans are remarkably clever," the smaller man offered nervously. "They have advanced far more rapidly than we thought."

The taller man shrugged. "So be it. They are very dangerous as well; more violent even than the Mazarek."

"But nothing compared to the Maitre."

"One less violent species is all to the good. Violent races interfere with commerce and supply chains."

This was agreed to.

"The human doctor Kopetz has done marvelously. If present trends continue, a majority of the planet will become altered."

"And thus, more harmless."

"Yet these humans are so unpredictable," the smaller man said. "And there are so many of them!"

The taller man waved his hand. "It is well. We have slowed them down, even if the normal humans act in time to shut down the centers. The battle between the altered and normal humans will cause conflict for centuries."

"A united humanity is dangerous."

"Unless all become transformed, of course. The bio-programming will see to a uniformity in thinking, and a certain passivity that will temper this race's emotional outbursts."

Both men were pleased with the discussion so far.

"There is still the temporal anomaly our superiors wish handled," the smaller man said. "The humans have constructed a primitive temporal device and have been altering their own timeline! The future humans have one, and could adversely influence the present situation."

"Primitives should not experiment with temporal phenomena."

The smaller man brushed back his silvery hair. "I was just thinking that an insertion from the future, before the year 2035, could completely destroy our work."

The taller man thought for several minutes. "Monitor the timeline and report on the activities of the future humans."

"What do we do if the future humans interfere again with the past?"

"That is beyond our brief," the taller man replied. "We cannot ourselves create more temporal stress on the earthian timeline."

The smaller man was satisfied. "Very well. What we have done is satisfactory; we are blameless."

The two men nodded to each other.

"The Better Humanity program will continue, but it is out of our hands now. The humans have, as they say, opened Pandora's Box."

CHAPTER **36**

Ralph Zimring kept close tabs on events and people in Midland. It was part of his duties for Max Berglin, who was paranoid about the classified programs. The CIA had been neutered, but the special access programs were beyond the reach of any reforms from the civilian sector. Those guys operated without any government oversight and controlled technology that had not been released to the public sector. It was his job to watch the activity of those furtive bastards in Midland. He was basically Max Berglin's intel officer now; Hideki ran the security operation at Berglin Enterprises.

Ralph had formed an on-call team of former special forces operatives, and mercs he knew from his former employer, Kharkov PMC. He knew a lot of guys who used to work the Circuit or were bored with civilian life. Among them were Danny Radulo and his old buddy Tony Bagdadi, who ran comms for the team from his isolated Midland apartment.

One night while on patrol Ralph was cruising in his vehicle at the airport. He always checked incoming flights, and especially those from private planes. His buddy, a former Flight Control Officer at NASA, gave him the schedule every day. It was illegal of course; but fuck that. DeShawn Crawford had got bored at NASA and liked the action at Midland International Airport.

Tonight a private plane was landing; an old N288KA Gulfstream III. Those planes had been used by the now defunct CIA for rendition flights: the covert transport of terrorist suspects. They masqueraded as private business flights but had essentially been secret prisons in which individuals were held and tortured by the CIA. Ralph knew that the Alexandria group had one. There could be as many as a dozen guys on that flight, or just a couple.

Ralph decided not to interdict at the airport. He would watch from a safe distance with his infrared night vision glasses. DeShawn had provided a feed (again illegally) so he could hear what DeShawn did. "N288KA, you are clear to access private runway S22," DeShawn said. The plane landed and taxied beside a small, unlighted outbuilding. Four men exited the plane. All were carrying war bags and a combat backpack. Ralph immediately recognized Bob Saunders and Manuel Stoller, two scumbags recruited by the late Lucas Cernovitz. The other two were Chernkov and Ernst. Both of them were inside the wire guys.

The four men went into the outbuilding for several minutes. They came out and got into a vehicle. Ralph followed the car to the Radisson downtown, only a block away from the Centurion building. So they were going to keep watch over him as well! The Radisson was only a block from the Midland Tower, where he occupied the two apartments on the top floor.

Why were these scumbags in Midland? A detail to protect Harry Kaine? Or was there some other nefarious reason? Ralph wasn't a person to ignore unpleasant possibilities. He knew about the tension between Kaine and the kids doing that podcast. Paul Wen's guys might be in Midland to shut down the show, which had generated huge negative publicity for the Better Humanity centers. Something would have to be done about that.

Ralph parked his car at the Berglin Enterprises front parking lot with a view of the Radisson entrance. Like a cat, Ralph liked stakeouts. In the field he had learned to catnap and stay fresh. He dozed from time to time, pretty sure that Wen's men wouldn't do anything tonight. And so it proved. Ralph went home at 6 a.m. and ate six eggs, a dozen small breakfast sausages, and had three cups of coffee laced with whiskey. He was ready for the day.

At 8:30 Ralph saw the four men exit the Radisson and call two pop-ups. Ralph followed in his vehicle. The men went to the house of Ken Strickland and drove around the place, scouting exits and entrances. Then they drove around the neighborhood. Ralph guessed that their next stop would be the Midland Tower apartment complex. Grinning, he raced over to the Tower and walked up the stairs to the top floor. He then stood in front of the elevator waiting for Wen's team to show up.

Ralph didn't have long to wait. Ten minutes later the elevator door opened. Four men confronted the giant.

Ralph spoke casually. "It appears that I have you at a disadvantage, gentlemen. You are confined to close quarters, while I am free to attack from the open. May I ask what you are doing here?"

Saunders and Stoller flared. "Zimring!" Saunders shouted. "I'll kill you!'

Ralph smiled. "You tried that one cold desert night at the Kirkuk oil fields, Saunders. How did that work out?"

The four men in the elevator stood down. They were curious, and no one really liked the abrasive Saunders anyway.

Bob Saunders remembered that confrontation. He had been a very young mercenary fighting with the Peshmerga and the KRG, the Kurdistan Regional Government, against the Iraqi government. The Peshmerga had lost the Kirkuk oil fields in a battle with the Iraqi government in 2017. One cold night he had been sent by the KRG commander as a saboteur to disable the Kirkuk-Ceyhan Oil Pipeline, which transported oil from the north of Iraq to the Turkish port of Ceyhan. At that time it was the Iraqi government's largest crude oil export pipeline. He had just rigged his explosives when a sixth sense told him that someone was behind him. Saunders whirled around with his Glock 17 drawn but a hand cracked his wrist and his weapon fell to the ground, bouncing off a rock. He was a tall man, almost 6' 4" and 235, but the giant lifted him into the air as if he was a toy, staring into his face.

"Merc, in the employ of the Peshmerga," the man stated flatly. He was released and fell awkwardly to the desert floor. "Your Glock is over there," the giant said calmly, as if they had just finished watching a video together.

"Fuck you, whoever you are. You broke my wrist."

"Meant to. You'll live."

"You're Kharkov aren't you?" Saunders said through his pain. His anger at having been so easily defeated helped him to ignore the pain.

"That's right. Kharkov PMC. We're working with the Iraqi government to protect the oilfields from saboteurs like you."

While he talked Saunders had gotten up and silently palmed his SP2 combat knife from his boot sheath with his left hand. It was pitch black dark. He moved slightly toward his opponent, who was turned slightly to the side. His opponent was a big target. Saunders decided to go for the kidney. Just as he struck, the dark figure moved like lightning. His knife was knocked out of his hand.

"Didn't break your wrist that time," the voice said. "You can still pull with the left?"

"Yeah, I'm ambidextrous. Who the fuck are you?"

"Zimring's the name. Ralph Zimring. You're good. If it was anybody else you would have gotten me."

"You're fuckin' irritating, Zimring."

"Pick up your C4 and get that wrist attended to."

"Sod off. Peshmerga don't worry about broken wrists."

"I like your attitude. A man should fight enthusiastically for the brand. You're from the Midlands aren't you?"

"Sod the fuck off." Saunders was really pissed now. What was this, a pool party? He walked back to the pipe and retrieved his explosives. He didn't think the wrist was broken because he could still use it a little. He decided to put the wrist in a cast and come back the next day. But he never got the chance. The Iraqis and the Americans launched an attack that night and drove the Kurdish forces back several more miles.

"How's the wrist?"

Saunders started. They were in the elevator and Zimring was standing there in the hallway, maddeningly casual. "God I hate you, Zimring."

"Your hate is why you're not the fighter you think you are."

Ralph saw the other three grin. Saunders' face turned red with anger "Gentlemen, if you're here to detail Harry Kaine I have no beef with you. But if you're here to fuck with that podcast, or anyone associated with it, I'll end you just like I ended Lucas. Now get the fuck out of here."

Ralph pushed the Down button and the elevator door closed.

Well that was fun, he thought. He hoped Saunders would start something.

As the elevator went down, Manny Stoller laughed. "So that's Ralph Zimring. He's got a sense of humor, doesn't he Bob?"

"Shut up Manny." Saunders turned to face the other three. "I faced him in hand-to-hand combat. He's a living legend and for good reason." Despite his hatred of the man, Saunders knew Zimring was right about his temper. A top-notch fighter should never lose his head.

Stoller sobered. "Fuck Wen. I say we blow town and have a good time in Chicago."

Saunders rubbed his chin and considered this. The other two men perked up. "Not a bad idea, Manny. Who gives a fuck about that nerdy podcast?"

"You guys go ahead," Ernst said. "Somebody's got to watch over Harry and make sure he doesn't do anything stupid." Chernkov, the fourth man, nodded his agreement.

Stoller swore and remembered his duty. He also remembered the silver-haired man, who scared the living shit out of him. "You're right, Ernst." He looked at Saunders. "Bob and I will fuck up that studio, and Zimring too. You two watch Harry and hustle him off to the airport after the show."

"Fuck Zimring!" Saunders shouted. "If Zimring shows up he's mine. Then to Chicago for a little fun!"

This was enthusiastically agreed to by all four men.

"Where will you two be?" Ernst asked.

"Somewhere Zimring can't find us."

Chernkov was skeptical. "Zimring will be watching the place. He might have help."

"I hope he is," Saunders replied with heat. "I want another chance at him."

"We'll fuck him up like he did to Lucas," Stoller promised.

Chapter **37**

Kirra called Chief Chad Standing Bear and John Washakie. She asked them both the same question. "Do you want to debate Harry Kaine? He's coming to Midland on Monday at 7 p.m."

Standing Bear agreed immediately. "It's for real this time? We've had a few postponements."

"I apologize Chief, but this time it's on. Harry has already sent an advance team to the city."

"Good. The tribal council is very concerned about the whites, Kirra. They are taking humanity down a very, very dark road."

"Don't I know it!"

"There has even been talk of eliminating Kaine among the United Tribes, Kirra. Already, I hear, over a billion humans have undergone this...this evil transformation."

"It is true, sir."

"I'll be there."

"We've already reserved rooms for you and John at the Radisson downtown. Rooms 113 and 114."

"Ah, Washakie is coming. Splendid! There's no one I'd rather go to battle with."

Kirra smiled. Throughout their history, the vast majority of the North American plains tribes were first and foremost warriors, not chanting spiritualists or victims, as they were portrayed by revisionist new-agers and privileged white liberals. Washakie and Standing Bear considered themselves to be true men, not effeminate weaklings.

She called John Washakie and received an enthusiastic response. "Are there any talking points?"

"No sir. The debate will be entirely fluid. But we've already prepared for New York a long time ago. The conceptual battlefield is the same."

"Standing Bear and I, and the United Tribes, are agreed. The Transhumanists must be stopped, and stopped now. No more appeasement. More than a third of the state and city legislatures and executives nationwide are now altered humans. Even some of our own people have succumbed. It is grotesque."

"I agree sir. The tribes must again stand up for humanity, for Gaia, against the whites, who so often travel down the wrong paths. It is not racist to say this."

"Not at all! Look what happened in the 20th century, the century of darkness. First the takeover of Russia in 1917, which led to the gulags and over 70 years of Soviet mass murder of the Russian people. Then the fascists in Nazi Germany who were responsible for the deaths of 20 million people. In 1949 the rise of the Chinese communists, who killed over 80 million of their own people and were not defeated until the appeasers were shoved aside and the darkness was confronted. All of this trouble caused by the whites, who have lost their connection to the Great Spirit and to the foundational energy of the earth."

Kirra was thrilled at this passionate speech. "Get him, John Washakie. Show the world how true men and women live, aligned with the beautiful consciousness of the earth."

After her conversation Kirra felt inspired by the two men. The whites often went down the dark road of totalitarianism and control. They compensated for this by taking an equally irrational position in their spirituality, insisting that purely passive "peace and love" are the only options. Why, their own Yeshua, the Christ, physically threw the moneylenders out of the temple! "My house is a house of prayer, you have made it a den of thieves." Then he underwent a flogging and the carrying of his own cross to the hill at Golgotha, where he was nailed to it for hours. Like Washakie and Standing Bear, Yeshua was a true man. Although he was a healer, he was also strong and brave, not a timid, chanting weakling. Kirra was feeling confident about the upcoming debate.

I decided to attend the Kaine debate after work. Maybe if I went, Karen would see it as a sign of my support for her. Normally, after working a 13-hour day, I was too stressed to watch the podcast. By all accounts Karen was leading the charge against Harry Kaine, but I found politics and current events too irritating. Besides, I already knew what a scumbag Harry Kaine was. My job was to help Ger develop the mag engine and give humanity clean, unlimited energy that wasn't dependent on polluting and inefficient fossil fuels. That itself would resolve many political and economic problems.

———— • ✻ • ————

On Monday evening Ken Strickland's house was packed. His basement could hold ten rows of eight tightly packed chairs. There were at least twenty more people standing squashed along the walls, but no one minded. The great Harry Kaine was to appear tonight! Midland's own, and probably the most influential person in the world. Even those who disliked Kaine were excited to see him in person and to attend the debate. Everyone knew that Harry got his start in Midland and still had a fondness for the city. In the audience were twenty transformed persons from Chicago. This was a condition of the debate demanded by Kaine. They sat together in specially marked chairs. The regular humans looked at them curiously, for there were few of the transformed in Midland. The atmosphere was electric.

Ralph Zimring placed himself just outside the room at the bottom of the stairwell and checked everyone coming in. Shortly before 7 p.m. he saw Ernst and Chernkov come down the stairs and questioned them. "We're just here to protect Harry," Ernst said. "We're unarmed."

"Where are Stoller and Saunders?"

"Some place you won't find them," Chernkov said with a sneer.

Ralph swore. He had made a tactical error not calling in some of the team. It was just him against four. He was stuck here for now, checking the door traffic, but he'd have to leave the house during the debate and go man-hunting. He would have to trust that Ernst and Chernkov wouldn't do anything stupid in a packed house. There was a long line of people coming in and it was backing up. Ralph swore silently again and let Ernst and Chernkov through.

Ralph didn't know that Kjirsten had called her father, Police Commissioner Chastaine, to keep an eye on the place. Jack Chastaine had sent two squad cars to patrol the area, mainly to protect his daughter.

Karen, M'basa, and Kjirsten sat to the audience's right at the long show table. Ken Strickland had rigged three more mics to the left at the table. The debaters would be right on top of each other, and the audience.

As 7 p.m. approached M'basa began to get irritated. "It would be just like Kaine to cancel at the last minute with a text," he grumbled.

At that moment a stir was heard outside and three people entered the crowded basement. It was Harry! The audience watched as Kaine made his entrance along the side wall, followed by two transformed wearing white lab coats. There was almost no space to move, but Harry made his entrance look like a king who greets his grateful courtiers. He had a word for everyone who acknowledged him, and shook hands with several on the aisle. He exuded an aura of self-satisfied confidence and happiness.

Ralph, looking on, saw that the man had charm and presence. Both men and women sat up straighter when he passed. Harry seemed to absorb the adulation from the crowd. He must have been gawping because Ernst and Chernkov were grinning broadly at him as if to say, "You're toast. Our guy is the star of the show."

Ralph snorted. The debate was meaningless for him. The real action would occur when he tracked down Stoller and Saunders.

Harry and his two scientists crowded into their seats and the debate began. Ralph listened with half an ear, wanting to make sure the crowd was settled in before he left the building. The redhead started it off by saying, "Harry, you and your 'new humans' are biologically programmed robots. It's tragic that you don't have enough self-awareness to even know it."

"If being sick and slowly dying an uncomfortable death from old age is what you want," Harry said with a smile of confidence in his own long-term health, "then by all means stay a normie." This was supported loudly by his transformed cheering section.

"Your lack of self awareness makes you think that human life is just an episode between two periods of oblivion," Karen countered. "That's why you embrace failed philosophies like Transhumanism."

Harry felt this barb hit home, but he waved his hand casually in dismissal. He knew his face showed nothing but casual unconcern. "My dear, Transhumanism goes beyond the delusional subjectivity of an unprovable spirituality."

M'basa stepped in. "Absent the understanding that there is something greater than the physical body, a person will believe the most attractive or seductive lie. That lie is Transhumanism."

Harry laughed lightly. "Poor M'basa. Ask my friends in the audience" – Harry pointed to his transformed cheering section – "whether perfect physical health and a long life are lies."

The twenty transformed stood up and cheered this, the picture of robust health and confident happiness.

Ralph scowled. Harry had definitely won that round, but he didn't have the time to stick around. It was time to locate Stoller and Saunders.

As Ralph walked up the stairs he saw two men appear at the basement landing. One was tall and thin, and impeccably dressed in a nicely tailored suit and tie. The other was a very large, bulky man dressed in full tribal costume.

Ralph stopped halfway up the stairs and stared at the big man.

"Chief Standing Bear," Ralph said. He was impressive, Ralph thought, with his dark skin, bright tribal colors, and regal bearing. Ralph grinned. "You're a man fit to counter Harry Kaine," he said.

The big man smiled grimly and recognized a fellow warrior. Standing Bear held out a pair of huge hands. "Harry Kaine must be stopped. Even if I have to strangle him with these."

"Now that's what I like to hear," Ralph replied.

Standing Bear's eyes shone brightly. "The time for talk is over. Harry Kaine will not leave this building alive."

Ralph heard John Washakie's intake of breath. "Has it come to that my friend?" he asked Standing Bear.

"The Council of Nations has decided. I am their emissary. Harry Kaine is too dangerous; he must not be allowed to continue. The future of the human race is at stake. The transformed are an abomination."

Ralph sighed. He would dearly love to see the confrontation. "I'm going to miss the fun, but my job is to deal with Manuel Stoller and Bob Saunders, Kaine's paid thugs."

Washakie nodded. "Perhaps I should come with you. Even up the odds."

Ralph studied Washakie. Whipcord lean, the man looked like a fighter underneath his suit. "Why not? Are you carrying?"

Washakie grinned and pulled out a wicked looking knife from a hidden sheath in his right boot. "I'll use this."

Ralph nodded. "Come on then. We've got some hunting to do."

When Standing Bear entered the room he was too big to make it to the microphones in the crowded room, so he stood in the back. He raised his arm and pointed his finger at Harry Kaine and the twenty transformed in the audience. Harry was in the middle of a speech. "White man," he interrupted, "you have no understanding of your higher potential. You have no understanding of consciousness and your relationship to the earth. You have severed your ties with all that makes a human being divine and sacred."

For once in his life Harry Kaine was speechless. The personality of the huge man in tribal dress dominated the room.

Standing Bear looked around the audience. "Ladies and gentlemen, the Better Humanity movement's purpose is to program the human race and to suppress human creativity and potential."

"That's not true!" one of the transformed jumped up to say. "That's propaganda!"

"Don't believe me, see for yourself. I have a vid here that will explain everything."

A buzz went around the room as Standing Bear pulled out a data pod. "Have your producer play this."

The vid showed Anders Kopetz in a white lab coat talking to a thin man with long silver hair, big eyes, and a large head. There was something other-worldly about him. The two were standing in one of the DC center labs.

"These two forgot that everything in the DC center is recorded," Sleeping Bear explained. "This was forwarded to the Council of Nations by one of the tribal whistleblower techs who works in the DC center. You all recognize Dr. Anders Kopetz, chief scientist for the Better Humanity operation."

Harry Kaine's face went white.

"...and so I say, if you want to preserve your life, do not undergo the procedure," the silver man said.

"Who is that guy?" someone in the audience said. "He looks weird."

In the audience, Liqao immediately recognized the room. He stood up excitedly. "That's Lab 3 at the Washington DC center! I was just in there!"

This caused a buzz of conversation as Liqao had to explain. In the booth, Ken Strickland stopped the vid until everyone settled down. Then he resumed.

In the vid, Kopetz looked confused. "Why should I avoid the procedure? To my understanding the new biotech is flawless."

"It is, so far as we can make it."

"Who's we?" Kopetz asked.

The silver man smiled thinly. He had thin lips and an ascetic face. His large eyes were expressionless. "You do not have a need to know, but I think you have a suspicion."

Kopetz's eyes widened. "So it's true then."

"Of course it is up to you. I can guarantee that as long as you live you will experience perfect health."

Kopetz caught the nuanced statement. "Ah, I see. Thank you for the warning." He turned away, and then turned forward again to face the silver man. "Is the procedure common...where you live?"

There was a faint smile. "In modified form, of course, for our biology. It is not permitted to release anything unless...it is asked for."

Kopetz's eyes widened in understanding. "Yes, I see. What is given is precisely what is requested."

The silver man's expression softened somewhat. "That is correct. Non-interference and free will must always be maintained here. It is a pleasure to work with you. Your understanding is superior."

At that moment two doctors and a tech walked into the lab. When they saw the silver-haired man they blanched. "All is well," Kopetz said. "Our friend was just leaving."

The vid stopped.

Harry Kaine smiled benignly to the crowd. "You see, all is well. The transformed will all experience perfect health."

The transformed in the audience nodded to each other, well satisfied.

"As long as you live," Liqao said to himself. He was beginning to understand something, and he didn't like it.

Harry Kaine frowned. "Liqao! Will you join me in Delhi? My next engagement is in two days. We'll need someone to supervise the technical infrastructure."

"Thank you Harry, but no. I'll be in Midland for a while." He looked down at Kirra and smiled. Liqao spoke to Standing Bear. "What did that guy mean by, 'what is given is precisely what is requested?'"

"Yeah," someone in the audience said. "How long is, 'as long as you live'?"

"And what is non-interference?" Karen asked.

Standing Bear spoke firmly. "I have no understanding of these hidden meanings. However, the Native peoples of North America have a warning for all human beings throughout the world, which I have been authorized to give you: Do not go down the road of transformation. It is a dead-end."

Harry Kaine scoffed and smiled indulgently at the crowd. "My friends," he said expansively, "join us in a long life without disease or illness. Do not listen to those who, through ignorance and superstition, fail to grasp the wonders of scientific advancement for the human race." Harry looked over fondly at the twenty transformed in the audience and spread his arms wide. "The transformed community will welcome you."

Standing Bear stood straighter. He spoke with the authority and wisdom of all indigenous people. It was as if he were standing with the spirits of tens of thousands of years of native ancestors. "Our teachings foretell of false prophets at the end of the world." He pointed to Harry Kaine. "I say this with no animosity, only sadness. I beg you, dear human beings, not to undergo the transformation and alter the biology given to you by the Divine Power, the Great Spirit. If you do so you will be lost."

Standing Bear looked at Harry Kaine. "Is the transformation process reversible? Can a transformed human return to his or her natural, God-given state?"

Harry looked shocked that the question could even be asked. He appeared hesitant at the powerful presence and speech of Standing Bear. "Well, uh, no, as far as I understand, the transformation process is not reversible." He regained his composure and smiled. "But why would anyone want to!" Harry smiled cheerfully and turned around, displaying his perfect body. "Look at me! Who would *not* want to be like us?" The transformed in the audience cheered this.

Liqao saw that many in the audience who had been swayed by Standing Bear's speech were now doubtful again. He had experienced the same thing when he first saw Harry Kaine after Kaine had been transformed. Liqao observed again that the transformed humans were, without question, glowing examples of perfect health.

In the glass-walled audio room, Ken Strickland put up one finger, for one minute left in the hour and in the live show.

Standing Bear had the last word for the live stream. "Look, my friends," he said, pointing to the transformed, all admiring themselves and each other. "Is this a healthy admiration or a self-centered one?" He paused for a moment. "If you decide to transform yourself – look at your future."

"Play that vid again Ken," M'basa said.

The vid was played, and a long discussion followed. I could see that Karen was leading the podcast crew. She had the debaters and the audience well in hand.

At this point I caught Karen's eye and she gave me a surprised smile, noticing me for the first time among those squeezed into the chairs in the crowded basement. Yup, I should have come before. The show had ended at 8 but the discussion continued until 11, and would be broadcast in follow-up shows. Most of the audience stayed until the end.

Karen introduced me to Standing Bear as people stood around and talked, still excited about the debate. Harry and his two scientists hung around, answering questions. Harry circulated between the normies and the transformed, glad-handing and generally being the charming, self-absorbed narcissist he was.

"This is Joe Courvall, the guy I was telling you about," Karen said to Standing Bear.

The man was enormous, I thought, as I stood in front of him and gave a little bow of acknowledgment. This was, as Melanie would say, a very large-souled person. He had black pupils and pitch black hair.

He looked sharply at me with eyes that showed a fiery intelligence. I got the impression that he was holding himself back so as not to overwhelm me. "This is the man who is developing the new energy invention?"

I bowed again. "My friend Germaine and I are working to mass produce them."

Standing Bear nodded "This is well. There are too many poor in our society. Access to a clean energy source will surely be a boon."

"My thoughts exactly."

Standing Bear frowned slightly. "Your motivation is not money?"

He said this as if implying that all inventors were greedy. "It doesn't hurt, sir, but I want to see one of these devices in every home. Our next goal is to develop a larger device that can connect to all houses in a neighborhood or subdivision. Eventually we can power entire cities with them, and remove the wires, poles, and electrical generating plants."

Standing Bear moved his heavy, bulky body a step back, startled. "You think big! It is well."

I smiled. "The electrical company wants to do this because they can make more money. I don't care how it's done or what the motivation is, as long as it gets done."

"You are not part of this broadcast?"

I looked at Karen and felt guilty. "Uh, I am so busy that I don't have time."

Standing Bear smiled knowingly. "Take care, young man. Your woman, I think, would like to see more of you."

I blinked, a bit taken aback. Apparently Standing Bear saw a lot more than I did. I turned to Karen. "I thought you were mad at me."

"I am!"

Standing Bear smiled. "You are too timid, Joe. This woman loves you, but you must show her your devotion."

Karen looked into my eyes. "Thank you Standing Bear. I have told him this before."

"A man must know himself and what he wants."

I heard Standing Bear say this as from a distance. I kept telling myself I wanted a relationship, but realized that I had not played a man's part. I smiled. "You're right Karen, you've always been right. I've allowed my enthusiasm for my project to interfere with our relationship. It won't happen again."

"You've said that before."

Out of the corner of my eye I saw Standing Bear with a smile on his face. I took this as a sign that I was doing OK. "What can I say? I'll leave work every day at 7 and be there for you during the podcast, even if I have to stand in the corner."

Karen smiled. "That will work. You're forgiven."

Standing Bear gave us both a big hug. I felt like a child's teddy bear within his huge embrace.

After we were released I acknowledged the big man. "Thank you sir. I feel much better now."

He looked at me intensely, then to Karen. "That is unusual," he said softly, as if he had reached a new understanding. "Both of you have the energy of the puma." He whirled and saw M'basa and Kjirsten, sitting at the table and talking to Harry's two scientists. "And those two." He looked around the room, his gaze settling on Kirra and Liqao. "And those two as well!" Standing Bear was clearly mystified.

"Is there something wrong, sir?" I asked.

"No, not at all." Standing Bear spoke softly, almost to himself. "It is just that...six of you with the same totem...it is remarkable. There must be a soul connection."

"What are you talking about?" Karen asked. She was genuinely curious.

"Native peoples are connected to the earth and its creatures, something you whites lost long ago. We believe that every human being is connected to the Great Spirit, and that each person has a guiding animal spirit throughout their lives. With this animal a connection is shared, sometimes through the dream state. All six of you have the energy of the puma."

"What is so special about the puma?" Karen asked.

"The puma represents the companion on journeys to other worlds."

Karen and I stared at each other. "The eight members of our group have a strong bond, but we are totally different people," I said. "Maybe we do all have a soul connection."

"Yes!" Karen said. "We sometimes call ourselves star seeds."

Standing Bear's eyes widened. "It's true then!" The big man was animated.

"What?" I asked.

"The legends of our people say that when the whites become imbued with the spiritual force, when they begin to wake up, the world will finally change for the better."

At that moment a crash was heard, breaking glass, and wood splintering. "I'll kill you now Zimring!" someone shouted. A police siren sounded harshly.

CHAPTER **38**

Liqao thought quickly. "Come on, Kirra. I'll take you out of here." Liqao put his arm around Kirra and led her up the stairs and outside. They saw two men smashing a basement window at the end of the driveway, and throwing themselves into the house. "I thought something like this would happen," he said to his girlfriend as they walked across the front lawn away from the action, and down the street to his car.

Kirra was thrilled. Her boyfriend had thought of her first! "I thought you were going to rush in and make a nuisance of yourself."

Liqao grinned. "That was the old Liqao. I'm a new man now!"

In the basement, people began to run to the stairwell, clogging the entrance. In the control booth Ken Strickland was shouting. Stoller and Saunders had smashed one of the basement windows and hurled themselves into the control booth. They both began smashing equipment. Ken tried to stop them and got a bashed arm in the process, and was roughly thrown out the door of the control room by Saunders, splintering it and breaking the window glass.

Ralph couldn't get down the stairs because of the congestion. He knew he was too big to fit through the basement window. Shit! Ralph forced himself down the stairs, throwing people around, and apologizing profusely. "Sorry ma'am, gotta get through. Sorry sir, gotta catch those thugs down there." Eventually the stairway cleared as almost everyone got up the stairs and out the basement door. Ralph saw Stoller and Saunders, who had completely smashed everything in the booth. Harry Kaine and his twenty transformed humans were milling around in the corner. "If you ordered this Kaine, I'll take care of you later." Ernst and Chernkov were guarding Kaine. They both stepped in front of Harry and his huddled group.

"I didn't!" Harry shouted. "I swear!" Ralph watched with contempt as the altered humans, concerned only with their own safety, backed off, inspecting themselves for injuries. As he approached the control room he saw the M'basa kid on the ground, holding his head. Kjirsten and Karen were tending to him. A few of the braver in the audience still hung around the back. Chairs were spilled and turned over on top of each other.

Ernst and Chernkov burst forward, surrounding Kaine, and ran along the wall to the exit, followed by the other transformed. Ralph let them go. He saw Standing Bear lunge at Harry Kaine, but the group went by too quickly for the bulky Arapaho.

"Zimring!"

Ralph saw Bob Saunders emerging from the wrecked control room with a wrench in his hands. He was followed by Stoller. The bearded producer stumbled back into the control booth.

Saunders dropped the wrench and grinned. "You want a rematch?"

Stoller backed off. "I got no beef with you, Zimring."

"But I do with you, Manny. I'll take care of you after I dispose of your friend here."

Stoller backed off and went back into the booth. Ken Strickland tried to stop him and was shoved aside. Stoller stood on a smashed desk and scrambled out of the window. There he met John Washakie. Both men engaged in hand-to-hand fighting.

Ralph looked his disdain as Stoller disappeared out the window. "Your friend is a chickenshit."

"He aint my friend, sod you. You ready?"

Karen, looking up from her tending of M'basa, offered her help.

"Stay out of this sweetheart," Ralph said grimly. "Keep M'basa and his girlfriend in the control booth and shut what's left of the door." He cracked his knuckles and flexed his arms. "This is gonna get ugly."

Karen and Kjirsten dragged the groaning M'basa into the booth, clearing away the broken glass and smashed equipment off a part of the floor. Karen left Kjirsten to tend to her boyfriend and stood in back of the broken door to the control room. She was morbidly fascinated with what these two barbarians would do to each other.

"Pick up that wrench," Ralph taunted. "You're going to need it."

He saw Saunders flare and knew he had triggered the man's temper.

Ralph had no illusions about the fight. Saunders was good, and was fifteen years younger. He saw the bulge in Saunders' boot, on the left foot. So – his combat knife was in the same place it had been that night at the Kirkuk oil field in the Iraqi desert.

Ralph wasn't carrying but had his own battle knife in a shoulder sheath. "Hand-to-hand, or do we use our weapons?" he asked.

"Hand-to-hand until one of us is at a disadvantage. Then we pull out the knives."

"To the death?"

"I'm so pissed now I'll kill you if I can."

"So be it."

Ralph waited for the smaller man to attack. He kicked a few chairs out of the way to give himself some space. Saunders did likewise. The room was clear now except for a few stragglers, who pinned themselves against the back wall by the stairs.

Now the two fighters had an arena about 20 feet square to work in. Ralph felt better about his chances.

Standing Bear stood against the back wall, his eyes glinting with excitement. His Arapaho blood sang with ancestral memories of the warriors of old and their combat. This would be to the death!

Saunders moved in quickly, feinting a lunge to the left and kicking with his right foot, catching Ralph in the stomach. Saunders came closer, working with elbows and fists, pounding Ralph's ribs and trying to take his wind. He smashed Ralph under the jaw with his head. Ralph caught an elbow on the nose, breaking it. Blood was flowing down his face and it hurt like hell. I'm getting too old for this shit, he thought.

"First blood!" Saunders cried, stepping back a little.

The man was lightning quick and Ralph was out of training for actual combat.

"You aint so tough old man," Saunders gloated. "Today is when you buy it."

Ralph was hurting but he laughed, showing his opponent that his attack had been harmless. He could tell the younger man wasn't buying it.

"Fuck this," Ralph said. He was the stronger but slower of the two men, and decided to bull-rush Saunders. Ralph paused for an instant, leaning in very slightly, inviting an attack. As Saunders moved in Ralph dove for the legs, knocking his opponent off balance. But Saunders was too quick and squirmed out of his grasp, regaining his feet quickly. Both men were hot now with battle lust. All of their training was mostly forgotten. They stood toe-to-toe, smashing each other. Because of his height Ralph went for the head, Saunders went again for Ralph's ribs and stomach, trying to take his wind. Like animals they slugged each other until Saunders stepped back. "Fuck you Zimring." In one lightning quick move Saunders retrieved his SP2 and went for Ralph's gut, cutting edge up. Ralph only had a split second to turn slightly away from the lethal blade. He felt Saunders' combat knife go deep into his side. For a second Saunders was immobilized as he tried to withdraw his knife so Ralph would bleed out. Ralph had just enough time to grab his Tanto from his shoulder sheath with his right hand. With his left hand he grabbed Saunders' wrist in an iron grip, preventing him from retrieving his blade. With his right he stabbed Saunders in the back. Ralph's blade went deep. Saunders used his left hand to hold Ralph's knife hand, but he was at a disadvantage because he was trying to hold Ralph's stronger right hand with his weaker left one behind his back, and Ralph was the stronger man anyway. Saunders knew Ralph could withdraw his knife and fuck him up good.

The two men were now in a death embrace.

Two police officers came down the stairs and stood by the entrance. Both men stared fixedly at the two fighters, but didn't approach them.

Ralph grinned through the pain. "It's a standoff, Saunders! I gotta admit, you're good."

"You gonna withdraw that knife?"

"You're too good a fighter to bleed out like a pig. I figure we're even now, you got me good. I'll take my hand off my knife if you take your hand off yours."

"For fuck's sake Zimring, you're a strange bastard. But OK." His battle lust was gone now. He could feel the pain of Ralph's steel and it was beginning to hurt bad.

I watched, awed, against the back wall as the two men stood head-to-head, the other's knife buried deep into their opponent. I saw the two police officers standing next to me, amazed. I could see that Karen was shocked at the brutality of the battle. I was too.

"You first," Ralph said.

"Fuck you Zimring." But Ralph felt a slight release of Saunders' grip on his weapon, so Ralph did a slight release. Saunders removed his hand completely and Ralph responded.

The two combatants stepped away from each other. Blood from both fighters was dripping onto the floor. "I'll call 911," one of the cops said.

The other officer spoke. "Lie down you two idiots, and don't remove the blades until the ER gets here. Somebody get pillows and put them under their heads."

Ralph looked at Saunders. Both men were holding their knives in place. "It was a good fight," Ralph said through his pain. "We gotta get on the ground to slow the bleeding."

Saunders was in too much pain to argue. Both men lay down on their sides about ten feet from each other.

There were a few pillows on the remaining upright chairs. I grabbed two of them and propped up Ralph's head. "Thanks kid," Ralph said. "You'll do."

"Gimme one of those," Saunders demanded. I tossed another pillow in his direction, but Saunders couldn't twist enough to get it behind his head. "Get the fuck over here and support my head. I can't move or I'll dislodge this fuckin' knife."

I did so. I wondered what the police were doing, just standing around. They looked like they were watching a vid.

"This was just a little brotherly spat between two friends," Ralph said, grimacing as he felt the cold steel of Saunders' blade inside his ripped flesh.

Despite himself Saunders managed a painful laugh. "You're fuckin' funny Zimring." He looked at the cops. "You got your BWC's on?"

"No."

"Good! Keep 'em off."

A few minutes later an ambulance arrived. The EMTs were incredibly efficient. They quickly wrapped roller gauze around the knives for bleeding control, and secured the knives in place. "We will transport with the knife still in," one of them said. "There's no way for us to know what blood vessels may have been severed by the blade. The very blade that cut the vessel may also be stopping blood loss from the cut."

"You're not talkin' to a couple of pilgrims," Saunders growled. "Just get on with it."

The EMTs looked at each other and shrugged. Two gurneys were unloaded and the two fighters were loaded. Ralph was so large I had to hold his legs to stabilize the body on the stretcher as we went up the stairs. It took three of us to lift the gurney over the stairs.

After the wounded men were loaded I directed the EMTs back to the booth where M'basa was groaning and holding his head. He was on his feet now, although a little unsteady. "Take a look at him," I said.

"What happened here?" they asked, looking around at the smashed audio engineering room.

The two cops walked back and were recording the damage. "None of your business," one of them snapped back to the techs. "Do your jobs and shutup."

M'basa was quickly examined. "He's got a concussion. He should go to the hospital, but we only have two gurneys."

"I'll drive him," Kjirsten said.

The two EMTs left and the ambulance took off. Karen helped Kjirsten get M'basa into the car and the two left for the hospital. Karen, with true reporter instincts, had been recording everything.

The remnants of the audience who had stuck around were gawking, and recording on their mobiles. "The Lightroom is over the target," a man said, looking around at the destruction. "Taking a lot of flak."

Karen interviewed everyone present, including Ken Strickland, who looked shocked. There was a cut on his arm and blood was dripping from it. "There goes our show," he grumbled. "And my studio."

Karen apologized. "We're really sorry, Ken. We had no idea this would happen."

"There's nothing left. They destroyed everything."

"Are you hurt badly?"

"Nah. A bruised arm and a scratch from some broken glass." Ken looked around the control room. "Maybe I can salvage something from this mess." He began to examine the extent of the damage.

"What happened to Harry Kaine?" an officer asked.

"He and his pals ran out before the fight," Ken replied.

At that moment John Washakie walked in, his suit a torn mess. "Stoller got away," he told Standing Bear, grimacing in pain. "I'm in no shape for fighting these days."

Standing Bear grinned and put his big arm around his friend. "You'll do, John." The two men walked out of the house and called a popup.

That night a dozen vids went viral on MyVR, the popular streaming site. Harry's frantic exit with his transformed was pilloried by normies and defended by the transformed, although it was generally acknowledged that the debate was a standoff. Ken Strickland lost his recording of the four-hour show, but downloaded the mobile recordings of Karen and the stragglers to his personal workstation upstairs. He vowed to keep the show going. "I'll call the insurance company tomorrow and tell them they have to have my recording studio up and running by the end of the week."

"That's the spirit!" I said, my arm around Karen. "If you need material you can interview me and Ger about the magnetic engine. It's almost as important as this fight against the transformed."

"Good idea," Ken said. "I'll cobble together three programs from the footage I collected and put them on the Lightroom site. Tomorrow I'll upload the break-in and the fight. That will give us even more cred," Ken said, grinning. "Wednesday, Thursday and Friday I'll show the debate from the mobile recordings. You guys show up next Monday. We'll stream live from my bedroom if we have to."

"I have a better idea. We can use Lledren's studio. He works at the soup kitchen on Mondays."

Karen kissed me. "You're a genius," she said.

It looked like things were OK with us now. I put my hands on her shoulders. "I won't let you down again."

"It was my fault too. I'm just as busy as you."

CHAPTER **39**

At the hospital Ralph and Saunders went into the ER. Both knives had to be surgically removed and both men had to stay in hospital. They were both placed in the same room in separate beds.

"You guys were lucky," the ER doctor told them. "You'll both be back on your feet in a couple of weeks."

"Fuck that...er, sorry doc, screw that," Saunders said. "I got things to do."

"If you get on your feet before then you'll bleed internally. If you want to kill yourself, that's up to you."

He paused and looked curiously at the two men. These guys weren't the usual gang members shooting each other up. They looked like professional fighters. "You say you stabbed each other? And then lay down together like good buddies?"

"Yeah," Saunders replied. "But we're not good buddies."

The ER doctor thought he'd seen it all, but these two were right out of a John Woo movie.

Both men were exhausted and felt like punctured voodoo dolls. "You got anything to help us sleep?" Saunders asked.

"I'll send in a nurse." A minute later a pretty nurse walked in and distributed three sleeping pills. "These will keep you out for a while." Saunders and Ralph took the pills and fell into a deep sleep.

The next morning both men woke up and had breakfast. Ralph looked disgustedly at the plate. "Hospital food," he grumbled.

"And not enough of it," Saunders echoed.

Both men were in a lot of pain but had refused pain killers. "Don't want to get addicted to that stuff," Ralph explained.

The ER doctor shrugged. "Suit yourself."

Within three hours Saunders and Ralph were bored out of their minds. "I'm not going to make it for two days, much less two weeks," Saunders grumbled.

"What are you going to do after we get out of here?" Ralph asked. "Go back to work for Paul Wen?"

"Nah. Stoller is a coward, and Wen isn't much better." Saunders told Ralph about his meeting with one of the silver-haired men the audience had seen in Standing Bear's vid. "Me and Wen and Cernovitz and Stoller met him one day at our HQ in Alexandria, right at the beginning. He shows up in the middle of the room with the door closed. Nobody saw him enter. He tells us that Kopetz is basically running the centers, and that we have to run Kaine and keep him out of trouble."

Saunders looked at Ralph, who could see the fear in his eyes. "The fuckin' creepiest motherfucker I ever met, Zimring. Almost like an alien. No emotions, he's talking to us like we're slaves, like we'll obey everything he says or else. And you know what? We did. We didn't question anything he said."

"Did he threaten you?"

"Didn't have to. Never even raised his voice. But I felt like my mind was sorta in a groove to obey after that. Never told us his name either."

"Maybe he didn't have one."

"What do you mean?"

"I mean, maybe he *is* an alien."

Saunders guffawed, then winced in pain. "Sod that."

"How did he get out?"

Saunders frowned. "Now that's a funny thing. We four got to talking. When we looked up he was gone, but nobody heard the door open or close."

"Projection."

"What?"

"Maybe he wasn't physically present. I know about that stuff from Max Berglin. There's some crazy shit in the hidden programs. One of the things Max told me is that they've learned how to project a person's consciousness remotely. I saw it demonstrated once. Mind blowing and scary."

The two men were silent for several minutes.

"I'll keep working for Max," Ralph said. "I'm getting older now, slowing down a little."

Saunders was thinking. Ralph could see that his suggestion that the silver man was an alien was disturbing to Saunders and decided to change the subject. "It was a good fight."

"Yeah. Never woulda thought it would end up like it did."

365

The two former mercenaries went eagerly over the battle in detail, each describing their thoughts as the fight unfolded, and their moves. Then they talked about how they got into the military, and their careers. By the time they were done the day was gone.

"You're not near the asshole I thought you were, sod you," Saunders said. "How old are you anyway?"

"57 this year."

Saunders expressed grudging admiration. "You fight well for an old man."

"I have to get back in training."

The pretty nurse came in with dinner on two trays. Ralph perked up considerably. Saunders, more reticent, just stared.

"Haven't you ever seen a woman before, cowboy?" she asked Saunders.

Saunders' face turned red. "Uh, sorry. Thanks for the dinner."

The nurse plopped the two trays on little tables beside each bed. Ralph could see that she had no interest in him, but that was normal. He was just too big. But he could see her interest in Saunders.

"I'm Sheila. I get off shift in two hours but you aren't going anywhere," she said to Saunders. "Can I stop in at 9?"

"Uh, sure. I'd like that."

She turned to Ralph. "You don't mind?"

"Not at all. Be warned though, this guy is a mercenary."

The nurse's eyes lit up and she turned to Saunders. "Really?"

Ralph tuned out the boring chit-chat and closed his eyes. It was a trick of his he had learned in the field to block out everything but what he wanted to concentrate on. He began to think about Germaine Robinson and his time viewer, the two clones (supposedly from the future) who were now fully human, and Saunders' silver-haired man. According to Joe Courvall, four of the podcasters, and Harry Kaine, had actually traveled to the future New York of 3013 AD. That was a bunch of bullshit. Yet he had seen the images from Robinson's time viewer. Was the silver-haired guy from the future? Ralph thought about that for a minute, and rejected it. According to Courvall, the future was filled with clones who looked like mannequins and who were dying off. Smith and Jones, his two clones, were also like that when he first met them. Ralph was startled by the thought that Harry Kaine, and the transformed, also looked like that. But could two clones, even if they were from the future, cause such a rapid advance in bio-technology? Maybe, but maybe not. If not, where did the new biotech come from?

"Hey Saunders!"

"Aw shut up Zimring! Me and Sheila here are having a nice talk."

Sheila looked at her chronometer. "I'm late! I'll see you at 9." She rushed out of the room.

"Now look what you've done," Saunders grumbled.

Ralph ignored this. "What did that silver-haired guy look like?"

Saunders was clearly reluctant to talk, but after some prodding Ralph got it out of him. "Tall, thin, a big head, big eyes, thin lips, and thin hair that went down to his shoulders. He looked like an old man would look, but I could tell he was younger, spryer. Middle-aged, I'd say."

"So he didn't look like Harry Kaine."

"Nah, nothing like."

"Was he involved in funding your project? It would need billions to build centers all over the world."

"Nope. All that was handled by Cernovitz. One time Lucas said we were being funded by InQTel."

"InQTel was the CIA's venture capital group, but they're not around anymore."

"A lot you know, Zimring. That outfit just privatized; went totally black. Nobody knows who's in charge of Better Humanity, but whenever Lucas needed currency he got it."

So the silver-haired guy probably wasn't from the future, and he didn't fund Harry. "I want to meet this guy," Ralph said.

Saunders' eyes got big. "You're crazy. Nobody *wants* to talk to him. He's scary."

"Look, Saunders. Harry Kaine is taking over the world with this altered human shit. Have you seen those people? They're narcissistic robots. They're not...human! Somebody has to take them down. I have a great team, and you have access."

"I don't know, Ralph. You're a crazy fucker."

"If we don't do it, who will? There's over a billion of these freaks now and they're all over the world." Ralph grinned. "It will be my swan song. One last great mission. A mission to save humanity. You and me against the world."

Despite his fear of the silver-haired man Saunders was impressed. Ralph was speaking his language. He was a loner, but he had a savior complex. It's just that he had never found anything worth saving before. That was why he had been a mercenary for hire.

"I think the silver guy gave Harry this technology to transform humans," Ralph said. "We have to find a way to fuck them up."

Saunders scoffed. "No one knows anything about them. There are two of them, and they come and go and no one can stop them. If they're aliens like you say, they probably have technology that's way beyond us."

"Yeah, but that's the point! Maybe Harry needs them to supervise the biotech. And maybe the silver guys have somebody in the organization who does that. If that's the case maybe we can stop them. Then what would happen?"

Saunders was getting interested. "The key science guy in Better Humanity is Dr. Anders Kopetz. He's almost as scary as the silver guys. He supervises all the centers. Word is that he meets with them every month."

Ralph exploded. "Now you're talking Bob! You see, it's not so hard. We find Kopetz, we can track down the aliens. Or whoever they are. And then we fuck them up and save the human race."

Saunders was intrigued now. His roommate was talking just like Stoller, except Zimring wasn't a coward. And it was for a good cause. Since Lucas died the outfit was leaderless. Or maybe Kopetz was the leader now.

"You're a smart fucker, Ralph. You simplified the problem. All we gotta do is find one guy and track his movements."

"That's right, Bob. Are you with me?"

The two warriors turned in their beds and stared at each other. Unconsciously they had begun to refer to each other by their first names. "Fuck yeah! What's the plan?"

"That's where you come in. You have to be the inside man, get Kopetz's schedule. I'll get my team together, you try to find out when Kopetz meets with Bob and Ray."

"Bob and Ray?"

"An old twentieth century comedy team. That will be our codename for the two, er, aliens."

Bob pondered this for a moment. "Kopetz's schedule is heavily guarded. It is released to Kopetz from HQ in Alexandria every Sunday night by the head honcho, Paul Wen, on a secure comm line."

Ralph thought for a moment. He decided to postpone the problem that the aliens might only be projections. Get Kopetz, take it from there. "OK, you try to get Kopetz's schedule by keeping your ears open. I'll send a couple of my guys to scope out the place. How's security there?"

Bob's eyes widened when he thought about it. "It's lax, Ralph, really lax, at least around the building. Better Humanity is so big no one ever thinks to fuck with them. The building itself has no signage anyway. No one would suspect that building is the HQ for Harry's entire worldwide operation."

Ralph was really fired up now. "OK, give me the building location. Where are the servers?"

Bob was amazed again. "Why, they're all downstairs in the basement, in one room!" Then he frowned. "But the entrance to the basement is guarded by a biometric scanner. If your DNA doesn't match—" Saunders made a slashing motion at his neck. "Pffft."

"OK, I'll talk to Tony about the servers." He looked at the time. "I feel like starting right now. How do we get out of this place?"

At that moment the pretty nurse entered the room. "You two aren't going anywhere. In case you forgot, you each have a four-inch stab wound and there's some internal bleeding."

Ralph swore. "Damn." He looked at Sheila and Saunders, ogling each other. He was the unwanted roommate at the dorm. "Wheel me out of here," Ralph suggested. "Then you two can fool around."

Sheila was intrigued and Bob Saunders blushed.

"You don't mind?" she said. If they were careful it could work. "Our ICU is only 20% full. There's a spare room next door."

"It'll only be for a few minutes," Saunders pleaded.

"It better be longer than that," Ralph suggested.

Sheila laughed and wheeled Ralph into the single.

"Don't make too much noise," Ralph cautioned her. "I'm a light sleeper."

The two mercenaries were in the ICU for ten days. The ER doctor was using some kind of healing laser Ralph had never seen before on them twice a day. The time passed pretty quickly for both men as they told each other about their entire lives, and fleshed out their plan to shut down the Better Humanity centers. "Cut off the head in Alexandria and the body dies," Ralph summarized. "Neutralize Kopetz, talk to Bob and Ray if we can."

"All right," Saunders agreed. "I'll go back to HQ and do some digging. You get your team together and be ready on my signal."

The two men left it at that. On the eleventh day they were released after the ER doctor pronounced that both men were fit enough to walk out of the hospital. "No redness or swelling in or around the wounds, no yellowing of the skin, wound depth rapidly decreasing...I'm pretty good." The doctor was insistent that the pair were to undergo no strenuous exercise for another week at least, "or you might hemorrhage inside."

The two combatants looked at each other as if to say, "Fuck that."

"Thank you doctor," Ralph said politely. The doctor shook his head and walked out

As they walked out of the hospital Ralph stopped at the front desk and asked for a hospital cab to take them to his high-rise apartment. He wasn't going to try to squeeze into one of those popups! There, Bob Saunders was able to book a flight back to Virginia.

CHAPTER **40**

Paul Wen sat with Manuel Stoller in the CCC of the Better Humanity HQ in Arlington. "What just happened in Midland, Manny?"

"We destroyed that podcast, just like we said we would."

"I hear that the studio is going back up. In two weeks they'll be broadcasting again."

"Yeah, but they'll be looking over their shoulders when they talk about us."

"What happened to Saunders?"

Manny shrugged. "In hospital. Zimring and Bob stabbed each other. Hopefully Zimring is dead."

"I wouldn't count on it."

The two men heard a beep from the biometric scanner in the hallway. Bob Saunders walked in. "Speaking of the devil," Paul Wen said. "Is that all you guys could think of, smashing that studio?"

Saunders shrugged and looked contemptuously at Stoller. "At least I stayed to fight. You climbed out a window."

"Yeah, and that's the problem," Stoller said angrily. "While you were indulging in a personal grudge, me, Ernst, and Chernkov got Harry out of there and avoided the police. You and Zimring were questioned by the local authorities."

Saunders didn't give a shit anymore about A Better Humanity, but this from Stoller was just stupid. "There's nothing that dipshit police chief, or the local sheriff, can do. Do you seriously think those yokels can mess with us when we have half the Congress in our pocket, and control many of the city councils?"

Paul Wen nodded. "It was messy, that's all. Duchene Comstock would have come up with something clever."

"Fuck Comstock. We handled it ourselves."

"You should have killed Zimring," Stoller asserted.

"I almost did. With no help from you, you chickenshit."

"I took care of Washakie, you fuckin' moron!"

Paul Wen put out his hands to left and right. "Enough, gentlemen. What's done is done. It was a bad idea to try to stop those stupid kids in Midland, I see that now." Wen lowered his head to the table. He looked up at Stoller and Saunders. "I have to deal with our two friends now, who aren't going to be pleased about your little escapade."

Bob and Ray! Saunders thought. "When do those guys talk to you?" he asked casually.

"Every first Monday of the month, precisely at noon. Right here in the basement."

"Kopetz talks to them too," Saunders said.

Paul Wen jumped out of his chair. "What??"

"You mean you didn't know?"

Wen's face clearly showed shock. "Where did you get that information?"

Saunders was mystified. "It was on that podcast debate! Some whistleblower recorded them. Didn't you see the show?"

Wen was shocked. "I didn't watch it."

Paul Wen began to understand that he was the Better Humanity leader in name only. Bob Saunders must have thought the same thing when he said, "Who's running this outfit anyway?"

"You want I should talk to Kopetz?" Stoller asked. "If he's been going behind your back..."

Wen sat down. "Kopetz is coming in for the meeting this Monday. I'll ask...our friends...about the good doctor if I can get up the nerve."

Saunders sat back in his chair, amazed at his good luck. Zimring's plan was all coming together and he hadn't had to do anything. He'd call Ralph tonight.

After Stoller and Saunders left, Paul Wen sat in his chair, holding his hands underneath his chin. It was time for him to retire if they'd let him. He was just a figurehead anyway.

CHAPTER **41**

Saunders called Ralph Zimring that night and told him about Kopetz coming to HQ. "Bob and Ray meet with Paul Wen on the first of next month at noon in the basement," he said. "Kopetz will be there."

Ralph was pleased at how easy it had been. "I'll have Tony Bagdadi try to hack their servers before that meeting. The problem is, we don't know where they are on the World-Net."

Saunders shook his head. "That won't work. The servers are local, air gapped off from the Net."

"Yes, but comms are sent out and received from the basement all over the world, and that's on the Net. You told me that. If we can intercept their comms maybe we can trace them back to those servers and get in."

Saunders shrugged. "I'm no network technician, but I know the Better Humanity servers have an unbreakable firewall around them."

"I'll let Tony Bagdadi work his magic. If he can't get in we'll just have to break into the place and talk to this Kopetz. And maybe nab one of those aliens!"

Saunders grinned. "I'm in. These goddam transformed are an infection of the human race."

* ✸ *

Two weeks later Tony Bagdadi admitted defeat. He and Ralph were sitting at a table in Bagdadi's apartment, drinking Scotch.

"I was able to locate the servers; they aren't air gapped but I can't get in. Whoever Harry Kaine's people are, they're really good. If you want to see that meeting you'll have to breach the walls. You'll be going in blind."

Ralph sighed. He'd been feeling his age after his fight with Saunders. He could tell he was losing his edge.

"Who will you bring along?" Bagdadi asked.

"Just me and Bob, who used to be a demolitions expert."

"A talented guy."

"And a good fighter, one of the best."

The two men sipped their Scotch.

"We've had a guy watching the place for the past ten days. There's no security anywhere around the building. They're hiding in plain sight, Tony. The building has four floors and a basement. The first three floors are just Better Humanity administrative staff; the top floor is for meetings of the top honchos. Staff walk in and out of there all the time, just like a regular office building. They keep a very low profile. But there is a helopad on top of the building."

"Why don't you nab Kopetz before he gets in the building? Then you don't have to break into the basement."

"That's a no-go, Tony. Saunders says that Kopetz takes a helo from the airport and lands right on top of the building. Then a private elevator to the basement right from the roof. The idea is to get to Kopetz but also to see Bob and Ray."

"What's the basement like?"

Ralph took another sip of Scotch. "According to Bob it's fortified. Bio-scanner, absolutely no way to fool it. You step into a booth that leads to a metal door to the comms center. If your DNA doesn't match the booth takes you out."

"Hmmm. So Better Humanity offices are in a nondescript office building, on top of a fortified command and control center with a helopad. Not suspicious at all!"

"Yup. The basement is where the Alexandria group runs the Better Humanity operation. Deep Special Access Program, obviously."

"What's your plan?"

"We don't have one. We wait until we see Kopetz's helo land on the building, then we go in. There's a door that leads to the basement stairs with a keypad. Bob knows the code. After that we wing it. Bob thinks he can blow the booth and the door, but then...who knows?"

"It's not like you, Ralph. Not enough intel."

Ralph shrugged fatalistically. "This might be my last rodeo, Tony. But it will be worth it if I can neutralize Kopetz and Wen. Once the brain is dead, Better Humanity might just go off the rails."

Both men saluted each other and then finished the bottle together.

Ralph walked back to his apartment and showered. For the first time in his life he was nervous before a field mission. He had the feeling he wasn't coming back from this one.

Ralph drove his old car to Alexandria, Virginia on Sunday morning. It would be a fitting end to him, he thought, as he moved toward DC. Fifteen years ago he was still a merc and part of a rogue national security force organized by the Technology Acquisition Consortium. He had driven the same vehicle to Midland, intending at first to take out Max Berglin and dispose of his Cube. Now he was on the other side and at the end of his career, attempting to take out a rogue operation that was threatening the future of humanity.

Ralph met with Bob Saunders at a hotel a mile from the Alexandria group's HQ on Sunday night. They were going in despite the odds and might even be killed. His wound was still sore but his movements were not restricted.

Bob was fatalistic as they sat in the hotel restaurant sipping bourbon. "You convinced me, Zimring. I never thought much about good and bad or right and wrong before. It's why I loved being a mercenary. You fight for the brand, no matter which side it's on. But after you took out Lucas the organization gradually started to go downhill. I began to realize how many of these freaks we were making."

"Yeah. There aren't many of them in Midland but Chicago is infested, and many of the world's big cities. I suppose they are harmless, but they don't have any...self awareness. They're the lowest common denominator a human being can get. And they don't seem to be able think for themselves."

"Yeah. Kinda like a robot army."

"Psalms 84:10: 'Better is one day in your courts than a thousand elsewhere. I had rather be a doorkeeper in the house of my God, than to dwell in the tents of wickedness.'"

"What does that mean?"

"You're a fuckin' barbarian Saunders. It's a saying from the Bible. Better to stay a humble normie and find your purpose in life than to go for fame and bling. I learned that from those kids who do the podcast."

"I like this one better: 'I'd rather spend one day as a lion than a hundred years as a sheep.'"

"Now you're talking, Bob. Il Duce, but pertinent."

"Who the fuck is Il Duce?"

"Mussolini."

Saunders laughed. "You're getting sappy, Zimring."

Ralph finished off his bourbon. "Yeah. Sometimes I get in a mood, but we have a job to do. Tell me how we're going to do this."

"I could probably get past the booth because the scanner will recognize me," Saunders told Ralph. "But I'm not scheduled to be down there when the silver guys come, and the guards will open fire on anybody who comes through the door who isn't authorized to be at that meeting. So we have to blow the booth. Your job is to have my back and take out as many of them as you can."

"We'll be sitting ducks standing in the doorway. Both of us are unauthorized."

"That's right, so be ready to open up." Bob told him that it was only necessary to disable the booth's bio-scanner and the kill-switch mechanism. "If I do it right no one should even notice. That door is thick and it muffles sound."

Bob drew a picture of the booth and the metal door it led to. "The booth's electronics keep the door locked, which pushes out into the room. Once the electronic controls are disabled, we should be able to shove it open."

"The door opens up *into* the room?"

"Yeah. It's a thick solid-metal door."

Ralph felt a little better. "So we're shielded from one side then. We can drop and fire."

Bob drew a picture of the basement control area. "To the right of the door is a big circular command center that is surrounded by bullet-proof plexiglass, ten feet high. Looks like the command deck of a spaceship. That's where the sensitive comm stations and the servers are. Paul Wen meets with...Bob and Ray at a conference table in the middle of the CCC. None of the technicians are down there for the meeting, only the guards, so we don't have to worry about collateral damage."

"OK. The metal door is on the left, the command center is on the right. So all of the fire will come from the center of the room."

"Yeah, and those guys will be looking down the middle for kill shots."

Ralph reached down to the floor. He grabbed a warbag and brought out two thin, lightweight flak jackets and headgear. "These things are made from non-Newtonian material called D450. D450 is designed so that when hit with a bullet, the molecules of the material lock together and become hard enough to stop the bullet. Not only that, it absorbs all of the bullet's energy so you can keep going. Then it gets soft again."

Bob's eyes lit up. "Heard about that stuff but I never tried it."

"Neither have I."

Both men laughed and some of the tension was broken.

Ralph and Bob planned how they would enter the room. "Stay down and spread out as much as possible," Bob said. "I'll go first."

"It's not much of a plan."

"All plans go to hell when the battle starts anyway."

"Yeah. The fog of war."
"A fog of bullets."

CHAPTER 42

At 11:30 on Monday morning the two men were sitting in Ralph's vehicle on a dirt road across the street from the Alexandria building, which was close to the Potomac River at the end of the street. Bob Saunders spotted a helicopter, which landed on the top of the building. "That's it. Kopetz is in there."

The two assailants calmly walked through the unguarded front door in civilian clothes over their flak jackets. One of the women in the lobby recognized Saunders, but he just raised his hand in acknowledgment as the two men turned left and entered a small corridor that led to a metal door to the basement. Saunders punched in the correct key-code and the door slid open to reveal a small staircase. At the bottom Ralph saw a booth encased in clear plastic, which contained a full-body biometric scanner. The kill mechanism was a series of razor-sharp metal blades placed at one-foot intervals from top to bottom of the clear plastic entrance booth.

"Messy," Ralph said. "These guys are paranoid."

"That's no lie." Saunders was carrying a small backpack with C4. He quickly set the explosives in place and the two men walked back up the stairs, punched in the keycode, and closed the door. "We'll be safe here," Saunders said. He calmly pressed the remote to activate the explosive. A soft thud was heard. A building security guy came running into the lobby.

"Everything OK here sir?" he asked Bob. Ralph stood in the doorway to block the guard's view when the door opened.

"I'll let you know if there's anything wrong," Bob told the guard.

The man looked up at Ralph and shrugged. If Saunders said it was OK he was off the hook if something went wrong.

Bob punched in the code and the door opened. Ralph walked calmly down the stairs as Saunders closed the door. They both put on their D450 headgear. A little debris from the destroyed scanner and kill mechanism littered the floor.

"Good job!" Ralph said, admiring the elegance of Saunders' work. "But this is too easy, Bob."

"We're committed now," Saunders said grimly. "The system will notice that the door is unlocked and test the booth. An alarm will sound."

"Fuck it then. Let's go!"

At that moment a siren sounded and Saunders put his shoulder to the door. Both men dropped to the ground as bullets flew over their heads.

Ralph heard cursing and footsteps coming down the stairwell. "Behind you!" he shouted to Saunders, who was on his stomach, firing his Glock 17 automatic. In the split second before he turned back to the staircase Ralph saw Bob run toward a concrete support pillar, firing. A bullet glanced off his vest and another went into his shoulder where the vest didn't cover. Ralph quickly turned toward the staircase as two armed security men came toward him. Ralph waited a split second to see if they would fire. One of them did, and he caught one above the hip, just below his vest (of course). Ralph closed quickly and grabbed both men, smashing their heads against the concrete wall before the pain reaction set in. He could hear gunfire. He raced into the control room, throwing himself to the floor. Bob was still standing behind the concrete pillar. One of the guards was spraying bullets all over the place with his MP-5. A bullet whizzed over Ralph's head and another glanced off his D450 headgear. Ralph was just about to fire when two men suddenly appeared in the room. Each of them was tall, silver-haired, and had big heads and eyes. "STOP!" they commanded.

The three guards stood down, their jaws dropping as they stared at the men, who had appeared out of nowhere. Paul Wen, Manny Stoller, and Anders Kopetz came out of the plexiglass enclosure. Ralph got out his field kit and began sterilizing and patching his wound, stopping the blood flow. The bullet had gone clean through without hitting anything vital; he would be OK.

Bob pushed his Glock 17 out from beyond the pillar and shot one of the silver-haired men. The bullet passed through him and smacked into the concrete wall behind.

One of the guards did the same, with the same results. "Fuck! Who are these guys?"

"Enough!" Paul Wen shouted. "They're just projections!"

"Zimring!" Manny Stoller shouted, seeing Ralph at the entrance. He turned to Saunders. "Are you working with *him* now? You fucking traitor!"

One of the silver-haired men spoke. "We are representatives of the Greater Community."

"What the fuck is that?" Bob asked. He had caught one in the left shoulder where his vest didn't cover (naturally, that's what always happened). He holstered his weapon and began patching up his wound with his field kit now that the firing had stopped.

"Put down your weapons," the projection said.

"You're not even here," a guard said. "You're just a ghost. Make me."

The weapon flew out of the guard's hand and clattered to the floor. The other two guards slowly bent down and carefully placed their weapons gently on the concrete floor. Ralph and Bob stared in awe at the two lifelike projections.

One of the projections spoke to Anders Kopetz. "Dr. Kopetz, you have completed your brief and can now stand down from your activities. Mr. Wen, you may run this organization as you see fit, without further direction from us. We advise you to be wary of those who wish to usurp your position. For the time being your weapons have been disabled."

"What about you two?" Paul Wen asked.

"We have overstayed our visit. There will be no further contact between us and any human being."

"You are leaving us before our mission is complete!" Paul Wen complained.

The two men disappeared.

Anders Kopetz smiled broadly. It had all worked out just as was promised. He turned and walked toward the door, but Ralph caught his arm.

"Not so fast Kopetz. You have some questions to answer."

Saunders stepped toward Paul Wen. "And so do you, Wen."

The two guards picked up their weapons. One of them fired at Bob, the other at Ralph. Nothing happened. A split-second later Ralph returned fire. His weapon was silent. Bob test-fired his Glock at one of the concrete walls. It too refused to function.

Manny Stoller strode angrily toward Bob Saunders, his eyes blazing.

Paul Wen cut him off. "It's over now, Manny. Our guidance counselors have left the building and they're not coming back."

"Tell that to Harry Kaine."

Kopetz laughed.

"What's so funny, Anders?" Paul Wen said, frowning.

"It is over; at least it's over for me." He looked at Stoller and Wen. "If you intend to undergo the procedure, I suggest you do so as quickly as possible."

Stoller frowned. "What does that mean?"

"It means, Manny, that without the technical advice of our two friends it will be harder and harder as time goes on to maintain the integrity of the procedures that create altered humans."

"Nonsense, Anders!" Wen cried. "You know as much as they do!" Paul Wen's face went white as he remembered his conversation with Kopetz a couple of weeks ago. He had dismissed Anders' warnings as exaggerations. Why, millions of people were being successfully transformed every week!

"That is true for the moment. According to our friends, however, human biology is evolving, and the biotech must constantly be tweaked. I don't know how to do that." Kopetz shrugged. "It's been incredible, exciting work for me, but I'm essentially just a technician. Without guidance..."

"Holy shit," Stoller said.

"What do you mean, human biology is evolving?" Bob asked.

"According to 'them' the human race is on some kind of advanced biological evolutionary path," Kopetz replied. "I didn't understand what they were talking about."

"Are those two guys aliens?" Bob asked.

Kopetz looked nervous for a second, then he relaxed. "It doesn't matter now. Those two are from what they call the Greater Community. It's some kind of Local Group of 52 stars."

Ralph was staring at Kopetz.

"They said they are from a star called Altair in the constellation of Aquila, about 17 light years from the sun. Altair is one of the vertices of the Summer Triangle. One of them – I could never tell them apart – showed it to me on a very sophisticated 3D map of the galaxy."

Ralph exploded. "Fuck! That's just what those kids on the podcast said! They claim to be star seeds. Said that humanity was on some kind of spiritual evolution."

Stoller snorted. "What a bunch of bullshit. I've made up my mind. I'm going in tomorrow for the procedure."

Kopetz took out a card and inserted it into his mobile. He handed it to Stoller. "This is your VIP pass to the DC center. Ask for Dr. Janner. She'll get you going."

"Thanks doc. Anybody else?"

Paul Wen's life seemed to pass swiftly through his consciousness. He looked around the control room and wondered what he was doing here. Something occurred to him. "If human biology is evolving, why do we need the transformation procedure?"

Kopetz laughed again. "That's the big joke, Paul, because this evolution – according to our friends – is going to take tens of thousands of years. Those who are, er, impatient undergo the procedure."

Ralph got it. "OK, I understand what those kids on the podcast are saying now. The materialists, like Stoller here, have no spirituality and don't understand that you just come back into a new body after you die. If human biology is advancing, you get a better body each time. And greater...awareness."

"You're a flake, Zimring!" Stoller said. "When you die you're dead. Even if you believe that crap, why keep coming back over and over when you can get a perfect body now?"

It was the standard argument for Transhumanism. Ralph thought about the 122nd street community in Chicago, and the other transformed he'd met. They were all self-indulgent morons. Suddenly he had a huge realization. "I'm playing the long game, Stoller! You're playing the short game."

Ralph was about to explain his new understanding, but realized Stoller wouldn't get it. He saw Paul Wen's face light up.

"The long game!" Wen turned to Ralph. "That's right, isn't it?"

Ralph nodded, still blown away by his realization. He could tell Wen understood. Ralph looked over at Bob Saunders, who was confused. "Sod it. I don't get it Ralph, but I'm with you. No transformation for me."

Two of the guards felt differently. "You got two more of those passes, Dr. Kopetz?"

Kopetz was feeling generous. "Sure, boys." He got out two more of the VIP pass cards and certified them. "You have to get there at 6 a.m. tomorrow. Don't be late or you'll miss out."

Stoller and the two guards walked out of the room and up the staircase. "I don't know what to do," the third guard said.

"I have one more VIP card left, but you only have sixty seconds to make up your mind. I want to get out of here."

The young guard was in an agony of indecision. He had been offered the job with the promise he could become transformed at any time, and it was a comforting option. Now he wasn't so sure.

"Time's up," Kopetz said, and turned to walk out.

"What will you do, Kopetz?" Wen asked.

"I have a little hideaway up in the Appalachians. That's where I'm from. I'm going to do a lot of hiking and observing on the WorldNet. I'll start a VRblog. You won't see me in society anymore."

Ralph nodded. He was glad he wouldn't have to take out the doctor. "All right, Kopetz. But if you get involved again, I'm afraid I'll have to..." Ralph made a slashing motion at his throat.

"That won't be necessary, Mr. Zimring. I'm going to have fun watching."

An understanding passed between Kopetz and the giant. Ralph realized that Kopetz was talking about the inevitable conflict between the transformed and normal humans that must eventually engulf the human race.

Kopetz smiled. It wasn't a nice smile. "I see you understand, Mr. Zimring. You won't need to worry about me. The denouement to this little play will provide all the excitement I desire."

Ralph had a sudden impulse to grab Kopetz by the throat. "Then get the fuck out now, Dr. Kopetz."

Anders Kopetz turned and walked swiftly out of the room, past the blown booth, and up the stairwell. An ambulance had arrived and an EMT was looking at the two guards Ralph had disabled on the stairway.

Ralph looked at the third guard. "A no-decision is still a decision. You did the right thing, kid."

The young guard was sweating in the cool basement. He looked to Paul Wen. "Do I still have a job?

"If you want it. But it won't be under me."

"Watch yourself Wen," Bob said. "You won't be able to walk away like Kopetz did."

Paul Wen was fatalistic. "When your time is up, it's up. I'm tired of this organization. For a while it was exciting...now I know too much."

"That's the problem," Bob said. "You got read in. The only way you get read out is with a bullet. Kopetz was just an independent contractor like me and Manny."

Wen shrugged his shoulders. "I don't care anymore. I'm through here."

"You aren't involved with child trafficking are you Wen?" Bob snarled. "Because if you are I'll kill you right here."

Wen took a step back. "Not me, Bob. I tried to get Lucas to stop but he said he was making too much money."

Ralph could see that Wen was a guy with self awareness who had just gone down the wrong road. "Bob and I are going back to Midland," Ralph suggested. "With your skill set you could probably find something to do there for the good guys."

Paul Wen brightened. "Sure. I'll look you up in a couple of weeks."

But it wasn't to be.

Ralph, Saunders, and the young guard went to a local hospital ER for treatment. Then the guard drove them back to Midland in Ralph's car. They went to Midland East hospital, the same one they had been released from two weeks earlier, because Bob wanted to see Sheila, the ER nurse.

"You two again?" the doctor said when they walked into the ER. "Both of you are crazy." He inspected each of the bullet wounds. "Looks good. You've both lost a little blood but it's nothing serious." After examining the knife wounds the doctor insisted that both men stay off their feet for at least a week. "You both opened up a little; there's some internal bleeding. You need to stay in the IC overnight." This was fine with Saunders, who asked for Nurse Sheila. Ralph had to move into an adjoining single.

After Bob and Ralph got out of the IC, Ralph found a job for the young guard at Berglin Enterprises. Saunders hired out to a PMC in Chicago, and drove down to Midland whenever he could to see his new girlfriend.

Two weeks after they left DC, Ralph heard from Tony Bagdadi that Paul Wen's body had washed up on the Potomac River, downstream from the Alexandria HQ. Cause of death: Two bullets, center mass.

Part V

CHAPTER 43

Four Years later (2043 AD)

Harry Kaine was getting ready for another presentation on his European Tour. He was in Lyon, located at the confluence of the Rhone and Saone rivers, about 290 miles south of Paris. Harry loved the restaurants and the historical landmarks, and especially loved the attention he received while strolling through the city's various venues. Tonight he was in the Vieux Lyon area, the Medieval and Renaissance quarter of the town, with its quaint cobbled streets. A dais had been set up. Harry was to announce the opening of another Better Humanity center! This was his favorite activity. Always charming and magnetic, after years of public speaking his skills had been honed to a fine art. He knew how good he was. Tonight he would wrap them around his finger. The broadcast would, of course, be streamed live all over the WorldNet. Tomorrow normies would flock into the center!

Harry examined himself in the glass window of a little boutique behind the dais, turning around to make sure everything about his appearance was perfect. Yes, he would do nicely.

That afternoon he had been briefed by the stern leadership from Alexandria, who had flown in for the occasion. He had liked it better when Paul Wen was in charge. Poor Paul! Harry never understood why he had committed suicide. Well, sometimes it didn't work out for people; they lost their purpose for living. This thought cheered Harry. He was needed and wanted! Without him A Better Humanity was lost.

Wait, who is that? Standing at the dais now, a half-hour before his speech, he noticed George Chang and Nguyen Tran walking up to the raised platform. Harry noticed that there were three mikes instead of one. He immediately got out his mobile and called Hans

Krueger, the Better Humanity director. "What is this Hans? What are these two normies doing here?"

"Relax Harry!" Hans soothed. "The new trainees are in Lyon to learn from the master."

Harry felt inordinately pleased by this statement. He felt a surge of dopamine in his brain and a rush of well-being flow through his body. That had been happening a lot more lately, and he vaguely wondered why. Harry decided to go with the good feeling. He felt that he should graciously welcome George and Nguyen, so he did.

The event began and he was the star of the show again (naturally), but Harry noticed that the other two spoke much longer than usual. He had recited the statistics he knew from memory; how over 2.5 billion had passed through the centers and become transformed. He glossed over some of the minor failures at many of the Centers; these were mere glitches in which normies had not survived the transformation procedure.

"Look at me!" Harry cried to the crowd, turning around in a circle and showing off his magnificent body. "Join us in perfect health and a long life free of disease and illness!"

Afterward, Harry had to admit that George and Nguyen were excellent. At the WorldNet conference after the meeting, the two trainees were asked why they had not become transformed.

"Oh, that is definitely in my future," George said. "I've been so busy I haven't had time for the procedure."

Nguyen said essentially the same thing.

It was odd, Harry thought. These normies should be eager beavers; the first in line for the procedure! Back in the old days...well, those days were gone now.

When Harry got back to his hotel he felt a little strange. Normally full of energy even after a long day, tonight he felt slightly lethargic. Perhaps it was time for another of the yellow pills. These were called "energizers." Harry didn't understand why he should have to take them, for did not the transformation process guarantee a long, illness-free life? Of course he wasn't sick really, it was just a little feeling of tiredness. Natural, of course.

Harry dismissed these uncomfortable thoughts. Should he call the escort service? No, he didn't feel like it tonight. Harry decided to forego his pill and went to bed. Just before he went to sleep he decided to have a word with Hans tomorrow about some of the "glitches" at the centers. He was getting more and more questions from the public about it.

3037 AD

Li Jing and Li Chun observed the timeline again, as they did now after every one of their Community shifts.

"The temporal manifold is trying to snap back to its original configuration," Li Jing said. "Even though Harry accelerated his progress with Transhumanism and compressed 80 years into ten, the Centers are beginning to fail."

"Just as before."

"That's right. On the old timeline, the failures started around 2120 instead of 2043."

"Harry tried to force it. He went too fast."

The two time techs studied the timeline around the year 2043.

"Harry's compression of the timeline is slowly building temporal pressure."

"Like winding up a spring tighter and tighter."

"That temporal energy will eventually have to be released."

Husband and wife looked at each other. "A timestorm," Li Chun said.

"Like a dam bursting because of too much water pressure."

"And it will flow upstream, into the future!"

"What can we do?"

"More time traveling would be disastrous, and we wouldn't know what to do anyway. We just have to wait it out and hope for the best."

It was a helpless feeling.

2043 AD

Harry called Hans Krueger when he woke up in the morning. He had gotten out of bed feeling energized, which allayed some of his uneasiness because he hadn't taken one of the yellow energizer pills. So, all must be normal.

As he punched in Hans' number code Harry realized with a shock that Hans himself was not transformed. Hans was head of the entire Better Humanity organization! In fact, almost all of the Alexandria group that had succeeded Larry Cernovitz and Paul Wen had yet to undergo the procedure. They were all normies! Why hadn't he recognized this before?

Harry began to feel worried again as Hans Krueger's face appeared on his screen.

"Good job in Lyon last night Harry!" Krueger said encouragingly. Harry felt a wave of euphoria flow through his body. He had never questioned these feelings before; what was different now? Harry felt that Hans was treating him with condescension, as a parent would to a child. Or a favored pet. Harry stared at Hans' chubby, friendly face. Was he being patronized?

"Now now, Harry, there's nothing to get worried about."

Harry felt another wave of well-being go through him. This was the inherent, baseline feeling of the Transhuman body, was it not?

"Of course it is, Harry," Hans said.

Was Krueger reading his mind? Harry wondered. "I have two questions for you, Hans."

"Sure. I'm here for you, Harry."

"First, why haven't you and most of the control group had the procedure? Second, why is it necessary to take these 'enhancement' pills?"

Hans Krueger's face underwent a subtle change. "Ah Harry, that's too bad," he said regretfully.

"Excuse me Hans, but this conversation is becoming uncomfortable for me."

Krueger smiled. "We can fix that Harry!" Another wave of euphoria swept through him. It *did* feel good.

"Yes, it feels good doesn't it?"

An awful thought began to pulse through Harry's consciousness. "You haven't answered my questions, Hans."

"Would you like the truth, Harry, or what we tell everyone?"

Harry gulped. "Uh, the truth of course."

Hans sighed with true regret mixed with vexation. "I'm afraid, Harry, that if I tell you the truth your, er, effectiveness within the organization might be compromised."

"I'll risk it."

Hans smushed his lips together thoughtfully. He spoke softly. "But perhaps I don't wish to."

Harry was getting alarmed. "I'll find out eventually. You might as well tell me now."

Hans leaned back in his chair, resigned at what was to come. He decided to speak bluntly. You never know, Harry might be one of those who don't care. "The enhancers are our best guess to keep the transformed bodies ahead of...the mutating human biology. The truth is, Harry, that myself and the group will not undergo the procedure because...we feel it is too risky."

Harry's face fell and he felt a deep depression that was soon countered by the feeling of well-being. "Is my body being monitored?"

Hans smiled. "But of course Harry! You are the most important figure in our movement." Hans spoke as if Harry was a very valuable commodity. "What does it matter? We must keep you healthy and happy in order to promote the trans- and post-human goal of a glorious future for humanity!"

This platitude had always seemed right to him before. Now it sounded...rehearsed.

"Ahh. Not buying it, are you Harry?"

Harry was confused. "I'm not sure I am."

"Oh dear. That really is too bad."

"Hans, why do I feel suddenly ill?"

Hans smiled with heartfelt admiration and appreciation, as one might for a fallen warrior who has given his all to the cause. "Harry, you have been a great boon to us. I'm truly sorry it has to end like this."

Harry felt very ill now, and very weak. He began to slide off his chair. "What's...what's happening to me?"

"We're shutting you down, Harry. But don't worry: the uncomfortable feelings will soon be over."

Hans watched as Harry slumped to the floor and lay still. There was still time to salvage him; was it worth doing? The decision was his. After several minutes Hans realized that he had done nothing.

The great Harry Kaine was no more.

It was sad. Hans felt real regret, for Harry had been a PR dynamo for years now. But a less-than-motivated Harry Kaine would not be as effective as Nguyen and George.

They would play the pre-recorded message Harry had made last year, when he was thinking of retirement. For the public, Harry Kaine would retire into private life after years of toil as a beloved and selfless public figure.

Of course the Centers would continue, for they were the source of so much money and power. And the human race really did need shepherding. The two silver-haired men had told Anders that humanity's aggressive tendencies needed to be modified, or the race would destroy itself. In Hans' mind the Alexandria Group was truly the savior of humanity.

Hans did wonder about the mutation of human genetics, whose scope went beyond even the Transhumanist technologies. The scientists had no good explanation for it. It will be exciting to see how things play out, he thought.

CHAPTER **44**

3037 AD

A week after their previous examination of the timeline, Li Jing and Li Chun activated the Shifter for a timeline update. The images had been very fuzzy and indistinct for the past several days, indicating that a wildcard event may have occurred. What they saw frightened them.

"Oh my God," Li Jing said to her husband. "Harry Kaine has been killed!" Li Jing saw that a huge temporal wave was sweeping up the timeline like a tsunami, changing events rapidly past the event locus in 2043.

Li Chun was shocked. "But...the conservation of time law should have prevented that!"

"Look for yourself. The dam has broken."

It was true. A huge disturbance of temporal energy was sweeping up the timeline.

Li Chun looked with disgust at their Shifter with its time observation consoles. "Left to itself, the timeline would have gone on as normal."

Husband and wife stared at each other. "*We* are the ones responsible," Li Jing said grimly. "Our original insertion with Pietra 23 must have begun to break down the structure of time."

After further inspection through the time viewer, this was confirmed. "Yes. The time manifold was breached in 2035 with the insertion of Pietra. As more and more people traveled to and from the past, a rip in time was created on this time thread."

"In order to accommodate more and more people, the rip had to get bigger and bigger," Li Jing said.

Li Chun moved the time focus back and forth. "Harry Kaine's death in 2043 is what is causing it. He wasn't supposed to die until 2120!"

"It's coming toward us, from past to future."

"Can you calculate when the timewave will hit us here in 3037?"

"My God. It will be here in about... 137 hours."

"What???"

"It's building momentum, husband. This precarious temporal bubble we're on isn't going to survive. The rip in time is going to blow away our reality. We have to get out now."

"We can't. If we try to go back in time now it will shred the temporal manifold. It would be suicide not just for us, but for everyone downline in the past."

Li Jing was appalled. "What do we tell the Community?"

"The truth."

A somber Li Jing and Li Chun saw the community gather after everyone had eaten. Morale was high; everyone was cheerful after a productive day's work. The children were playing and spirits were buoyant until Li Chun made the announcement and explained what they had seen, and their conclusions.

The community was stunned, particularly Caesar and Pietra.

"All we have built is lost?" Caesar said. "Our lives and our efforts will amount to nothing?"

Li Chun shrugged. "We must accept what fate has in store for us."

Li Jing gave her husband a speaking look that said, 'you're being very harsh.'

"I'm telling it like it is," he replied, consulting his chronometer.

"We have approximately 136 hours until the end of time. My suggestion is that we continue our activities in the fields and with schooling the children. No one knows exactly what will happen." Li Chun paused. That wasn't true, but he didn't want to crush all hope. "Perhaps...perhaps a miracle will occur."

"Why can't we transport the community back in time?" someone asked. "At least we will be able to live out our lives and save our families."

Li Chun looked to his wife. "For two reasons. The first is that we don't have time to insert everyone. There are over 5,000 of us. The second is that the rip in time makes doing more insertions impossible."

Li Jing saw that the children didn't understand, and felt great relief.

Many of the adults were crying openly. Pietra and Caesar, the community leaders, were clearly shocked, and walked over to the Lis and consulted for a time with them. Pietra stepped up to the podium. Her voice breaking, she said, "Please report for duty tomorrow as usual. The work of the community must go on, for the timewave will not hit us for another six days."

Caesar had his arm around Pietra. "Keep the children calm. Our fate is in the hands of God now. Worrying and fretting is pointless. Be strong for your families and for the community."

There was nothing to do but wait for the end. The Community went on as usual in a sort of walking zombie fashion. The adults considered themselves dead men and women walking as they performed their duties, yet the children were unaffected. They played and ran around as if nothing was wrong. The emotionally stronger adults drew some hope from the young ones, but many of the adults stayed indoors, terrified of what was to come.

During the last day no one went out to the fields. Some stayed at home with their families, most went to the Community center. All Li Jing and Li Chun could think about was the final flash of light that would signal their extinction. The two left their stations in the morning and announced that the timewave would hit sometime in the early afternoon.

Caesar and Pietra sat with the others in the open air portion of the Community Center. Their son and daughter, both married now, sat with them and held onto their little children. No one was talking; everyone was too nervous. "It's like Londoners waiting in a bomb shelter for the next Nazi V-2 attack," Caesar whispered to Pietra.

Suddenly the fabric of space began to *twist*. Between the quanta that composed the quarks and the other subatomic particles of matter, a soft light began to shine...

CHAPTER 45

Christian 172 sat in the comm control chair for all of Guardian New York within the imposing Tower of the Overseers. He looked out the window in the direction of what used to be Upper Manhattan. The Tower's systems were automated and collected data and vid on every inhabitant of the city through the brain implants. But the Tower itself was almost a relic now, as Guardian society all the way to Old Harlem had almost completely collapsed. The clones were losing their vitality even though the signal strength from the Tower had been increased to almost unsustainable levels. The new generation of sexless clones had been a complete failure, their life expectancy only a dozen years or so.

A lowering of life force readings had been expected, of course, but the shocking truth was how quickly the clone population had deteriorated! That devil Caesar 13 had told him what would happen before he snuck off into the past...

Christian 172 summoned his assistant Paul 26, now an almost useless, apathetic clone without intelligence or initiative. What good were assistants when there was nothing to do except watch society collapse? "Bring me a ration," Christian 172 ordered. "And check the store inventory."

"Yes sire."

Fifteen minutes later Paul 26 returned with two synthpaks. "There are synthpaks enough for two more months at the current Tower staff levels," Paul 26 informed him, handing him one.

Absently Christian 172 took a bite. "How do we make more?"

Paul 26 scratched his head. "I used to know sire...I have forgotten."

"Turn up the Tower support frequency."

"Sire, it is already at the maximum level."

So there it was then. He was the supreme ruler of...nothing. Why had he not planned better for the future? In his more vital days he was more intelligent, but it had seemed more important then to maintain his power and authority.

Christian 172 rose from his seat at the comm center and walked slowly across the hallway to the ruined Shifter, with its twelve consoles and observation tanks. How unfortunate that he had destroyed the time travel device! His fondest dream for the past two years was to escape into the past. He could then regain his vitality and become a powerful ruler somewhere. He took another bite from his synthpak.

There was always the rebel Shifter, but no way to reach it. Their hovercraft were gone, stolen long ago by the rebel barbarians. He had been too weak for the past ten years to exercise any authority over them. And whose fault was that? The traitor Caesar 13 of course! Christian 172 raged silently at the perfidy of that clone and his twelve time techs. Oh, give me a hovercraft and just one of those old 21st century EMP weapons! He would gladly take them all out, all 5,000 of those so-called "free" humans and the feckless clones who had defected. His body shook as he imagined the death screams of Caesar 13 and that Pietra 23 and all of their filthy children. The fit passed, leaving him even more drained than usual.

Christian 172 stumbled back to his command chair across the hallway from the bank of elevators and stepped over to the transparency on the east wall. He hadn't been outside the Tower in over a year. As he apathetically gazed out the transparency, chewing his food, he noticed a funny rippling on the horizon.

Something was happening out there. Christian 172 widened the transparency to the full. He stood now seemingly out in the air, gazing at the crumbled buildings of the once-great New York City, the mostly empty observation huts, the ugly clone block housing, and the old grid system roadways, completely overgrown with scraggly grass and half-dead bushes and sickly looking trees. He saw a few clones walking out of doors in a straight line, still obeying the mandates from their brain implants.

The disturbance was getting closer. It wasn't a weather pattern, he was certain of that. As the ripple got closer and closer he began to feel a freshness in the air. He could see a little bit now behind the ripple. Why, the deadened landscape had somehow come alive! Tall buildings rose up to the sky, craft floated between them, pedestrians walked in mid-air, and on the ground as well...

Christian 172 saw the ripple rapidly approaching the Tower. He felt a euphoria, an energy he hadn't ever experienced in his life. His dilapidated clone body was surging with life force it could not contain. He lost consciousness, crumbling to the floor of the comm center, his synthpak spilling out of his hand onto the floor.

————— • ✱ • —————

Caesar Jones, holding hands with Pietra, watched as a ripple in the time-space fabric appeared on the horizon. Caesar saw the landscape changing behind it, but it was too far away to make out details. From the old 25th street, where the Community Center had been erected, he could see the fields of grain and vegetables that had been planted from 25th to 30th. Beyond that, on 42nd street, was the imposing black Tower of the Overseers.

"It's getting closer," Pietra said.

Pietra and Caesar saw what Christian 172 was seeing. The air seemed to be highly oxygenated and invigorating. Behind the ripple the entire landscape of New York was changing from lifelessness to something bold and modern; even more vital than the New York of 2035.

"Caesar, look!" Pietra cried. "The Tower of the Overseers! It's gone!"

In the split-second before the ripple struck, Caesar wondered whether this was the end for him. If the timeline was changing would he and Pietra and their children even have been born?

CHAPTER **46**

2043 AD

Harry Kaine would have been disturbed had he known that his death barely created a ripple in the Better Humanity organization. Harry was the keynote speaker at all important international functions, but an army of other speakers like George Chang and Nguyen Tran were necessary for events in the 103 centers worldwide.

Hans Krueger had Harry's body transported to the DC center for analysis. There were twelve central centers in the world. These centers were meticulously monitored. The best scientists did their very best to keep the transformation procedures matched to a changing human biology. The other centers – well, those would struggle along as best they could. Even with their failure rate at 50%, enough successes were achieved to inspire people to come in.

Hans' comm link buzzed: the autopsy report on Harry Kaine had been completed. Shockingly, Harry's body had already begun to fail before his shutdown! Apparently even a fully transformed human body could not survive the unexpected human mutations. Hans sighed. He wanted desperately to become transformed, but it was too risky now. According to the late Paul Wen, Harry had been among the first. He had been carefully monitored by their two friends from the stars, who had personally supervised the procedures. Hans leaned back in his chair. Perhaps that is why the silver-haired men went away: they knew that their bio-enhancements were only temporary. It was too bad. The only question was, how long should he hang on at A Better Humanity?

---•✳•---

In Chicago, Councilperson Shanda Jenks was feeling the effects of the flu. It was a cold November morning (they were always cold in Chicago) and the wind was blowing at 25 mph. She arrived on time for the city council meeting. Mayor Miles Raymond (that freak) was to be questioned. As Miles walked in Shanda blew her nose and sniffled She expected the usual barbs from Raymond about normies and sickness. But today she noticed that the mayor was himself looking less glossy and polished than usual.

"What's the matter Miles?" she asked. "Did you have a bad manicure?"

Raymond looked irritated. He got out a pill case. Shanda noticed that it contained yellow capsules.

"I thought the transformed were always in perfect health," she teased.

"We are. These are just enhancers."

"To make you even more beautiful."

"That's right," he snapped.

Shanda saw that Raymond had (as usual) missed her sarcasm.

"Soon your district will elect one of the transformed," he said.

"After the gerrymandering, of course."

The mayor smiled. "Of course."

Shanda noted the smugness and condescension in his voice. For Miles Raymond, normies were clearly a nuisance now, and he made no attempt to hide his attitude toward her.

When it was her turn to question the mayor about the new city budget (which directed more resources to the transformed communities), Shanda noticed that Raymond's eyes had narrowed, as if he was in pain. "Are you all right, Miles?"

The mayor looked a little shocked and confused. "I don't understand it Shanda. I'm feeling a pain in my side. Most unusual."

"Just normal aches and pains, Miles," she said. "You're over 50."

Shanda saw that the other transformed council members had looks of concern on their faces. "Aches and pains are for normies," Raymond said sternly.

"Apparently not, Mr. Mayor."

What happened next shocked the Council. Mayor Miles Raymond screamed in pain, slumping in his chair. Shanda immediately called 911. An ambulance arrived within one minute (one of the benefits of the transformed city administration, Shanda thought. They were all irritatingly self-centered, but had increased the efficiency and response of social and medical services all over the city).

As the ambulance left the building, Shanda saw looks of complete shock on the faces of the transformed. One of them took out a pill case and swallowed a yellow capsule.

The meeting continued with the deputy mayor answering Council's questions. In half-an-hour the word had come down from the hospital: Miles Raymond had had a routine operation to remove his appendix.

The headline in the *Chicago Bulletin* (a normie newspaper) screamed, "TRANSFORMED MAYOR RUSHED TO HOSPITAL!"

The *Chicago Tribune* (controlled by the transformed) was more measured, headlining, "MAYOR RAYMOND UNDERGOES MINOR OPERATION."

------------ • ✳ • ------------

This incident was one of many reviewed by Hans Krueger in preparation for the bi-annual meeting of the Better Humanity operating group. He ordered a statistical analysis of the procedural failures and deaths in the centers. A statistically valid survey of the world's transformed was undertaken.

Two months after Harry Kaine's demise, the results were examined at a plenary meeting of the seven Better Humanity directors. Each Director represented a continent. The meeting was chaired by Hans Krueger. The Directors took the unusual step of inviting the retired chief scientist, Dr. Anders Kopetz, to the meeting. It was held in the heavily guarded basement at a huge conference table in the bulletproof, walled-off secure area in the Alexandria headquarters building.

Dr. Kopetz was excited, having already digested the report. "The data shows that most of the worldwide transformed population will be dead within the next 24 months," he said.

"You seem pleased with this result, doctor," Hans replied. Even he was appalled by the impending deaths of 2.5 billion humans.

Kopetz smiled. "I have no love for the transformed, but neither do I for normal humans. The human race is...a hopeless barbarism. Our evolution has peaked, and the race will devolve."

"All of us?"

"Of course! History shows that humanity is greedy, anti-social, and selfish. The race has never been able to place the needs of the whole above the desire of individuals for power and influence. Therefore, I am happy to see part of us die away. It's like removing a cancerous tumor from the planet."

There were two transformed on the board. "You are yourself a perfect representation of your critique of humanity," a transformed woman said with disgust.

Kopetz bowed his head in agreement. "Of course! Surely you do not think I exempt myself? I know myself to be a hopeless primitive, encompassing all of the qualities I despise the most in humanity."

Hans thought that Kopetz's cynicism was appalling. There was something twisted about his brutal self-hatred. "The question is, do we shut down the centers or keep them going? Our financial projections show that within six months, if present trends continue, expenses will exceed revenues." Hans looked over the conference table at every participant. "It seems to me pointless to accept the public for a procedure that will soon fail everyone."

"By all means keep them open," Kopetz said calmly. "The more deaths, the sooner the human race will exterminate itself. I regard that as a positive development."

"You're sure that we have only 24 months?" Hans asked.

"At the outside. The Post-human biotech was designed for a static human biology." Kopetz shrugged fatalistically. "Without the help of our friends, we simply cannot keep up with the unexpected mutations in the human population. An entirely unforeseen development, by the way. I don't think even our friends anticipated it."

"What is the source of these biological mutations?" the Asian representative asked. "Some say it is the result of consciousness evolution."

Kopetz scoffed. "That is absurd. Since the turn of the century, human biology has been influenced by powerful viral attacks. That is undoubtedly the source of the biological mutations."

Hans Krueger waved his hand dismissively. "The question is, ladies and gentlemen of the board, what should be done?" Hans asked. "We can simply close the centers and avoid a financial hemorrhage. Or we can let the centers die a natural death."

The European board member argued for an immediate shutdown.

"An immediate shutdown will likely generate an angry backlash from the public," Hans said. "Our very lives may be in danger."

Kopetz laughed. "I am fine either way because no one knows where I live." Anders looked around the room. "But that does not apply to you, ladies and gentlemen of the board. Your physical locations are a matter of public record."

Hans Krueger knew himself to be anti-social. But Kopetz's callousness frightened even him. He spoke to the other board members. "If we fight to the very last, we may be perceived as social heroes and heroines."

Kopetz pinned each board member with his gaze. "All of you are hypocrites, and dishonest as well. Your main concern is for yourselves." Several of the board members squirmed in their seats. Kopetz leaned back in his chair, satisfied.

After several hours of debate, which Dr. Kopetz watched in cynical amusement, it was agreed unanimously (Kopetz didn't have a vote) to let each center fend for itself.

"Prepare your hideaways, ladies and gentlemen," Hans Krueger said at the end.

CHAPTER 47

2045 AD, two years after Harry Kaine's death

Anders Kopetz was right. In less than 24 months, the bodies of all transformed humans began to fail. Many of them simply fell over on the street, or in their homes, or in the market. The hottest new businesses in Chicago were funeral homes and crematoriums. Wide swaths of Chicago, the city with the highest percentage of transformed in the US, became ghost towns. Those normies with more limited means simply moved into neighborhoods that had been resurrected from decay. These residents found pride in their new surroundings, and vowed to keep them in good repair. A new sense of community began to form in the city, as normies no longer felt ostracized.

A similar phenomenon occurred in every city with a Better Humanity center. As in Chicago, there was much belated gratitude toward the transformed. Although self-centered, they had been hard workers who transformed the physical structures and homes in their communities, just as their own bodies had been transformed.

The mutation of human biology is still, even as far up the timeline as I've seen, not completely understood. I'm pretty sure it began right after the turn of the millennium.

Midland Edison settled with Carleton University and old man Taubman for a piece of the magnetic engine pie. Everybody was happy now. The mag engine production had increased substantially, but still not near enough to satisfy demand. Ger is a sought-after speaker all over the world now, but he doesn't have time for it. Ger is an autist and doesn't like publicity. Neither do I.

Ger bought Lab B-103 from Old Man Taubman and resurrected the time viewer. (We could do that now because we were making lots of money.) We saw that the timeline had completely changed. There was no more Guardian society, no more Overseers, and the Tower had disappeared. After we satisfied our curiosity about the future, Ger shut the Time Shifter down. We were way too busy making and installing magnetic engines to fool around with time travel. We were making the future a better place! It was good to know that everything worked out OK for the people upline.

Karen and I bought a house together and stayed in Midland. It's a pretty cosmopolitan place, and about as big as cities get now. After the Pandemic of 2020, people started to move out of the big cities, and some never moved back. Many were afraid of packing together in high-rises and apartment buildings. But Midland has always had a unique culture. Many Chicagoans moved here after the Transhuman debacle.

I have learned to be more assertive; I've had to in order to be with Karen. The mag engine project has given me a passion just like Karen's. We fight sometimes, but what couple doesn't? We do it as equal partners in life.

Karen still works for the *Chronicle*, which has become the most important information source in the Midwest. She's a big wheel there now, a senior editor.

Strangely, Liqao asked Kirra to marry him and she turned him down. Liqao says Kirra has commitment issues. Liqao came to help us with the mag engine project. He now works for us in our tech department. Kirra is working with Standing Bear and John Washakie on indigenous issues. In 2037 the Supreme Court ruled that large swaths of government land actually belongs to the native tribes, giving them complete control over some of the land that was taken from them.

Ger and I sustained a visit from Chief Standing Bear and John Washakie, representing the Native Council of North America. We had expanded production to three factory lines now. The big Arapaho and his friend toured the facilities, demanding to know how the mag engine worked. After they were done Standing Bear gave both of us a big hug that almost killed us. "It is well. The people of earth can become energy independent. The old corporate control structures are breaking down."

The man had an almost sacred presence, and I bowed. "We were glad to be of service." I offered to donate several of the larger, neighborhood-sized mag generators to the Council, but Standing Bear frowned. "That will not be necessary. The days of tribal dependency are over." He brought out a contract and a currency card to pay for the order. "The Council expects expedited delivery, however," he said, winking.

John Washakie laughed. "So that is why we came here in person, you old coot."

Kjirsten and M'basa got married. Karen and M'basa quit the Lightroom podcast after a Mission Accomplished, but Kjirsten continued the Lightroom. She liked to research and follow social trends. M'basa continued his job at Gundarum Electronics, and then received an offer from the Government of Nigeria in their African Cultural Values department. Kjirsten and M'basa are both thinking of moving to Nigeria. Africa is becoming much more important internationally, and African voices are becoming more and more influential as the population of the continent grows.

Lledren is now a successful producer, and is living with the guard from the Alexandria center, who works for Max Berglin. Lledren still works in the soup kitchen part-time, as does his mentor, Karel Friedman.

Ger hit it off well with Tanya, who quit her job at the Taubman Research Center. She works for us now, and has reorganized our company's fly-by-night operation and made it totally professional. She rules us with an iron hand but we like it. One day Edison will hire her as a senior executive, or she'll enter politics. She's quiet but tough.

I think Melanie Fuscaldo from the Interfaith Center was right all along. She always claimed (without evidence) that humanity was undergoing a spiritual transformation. She and Ralph Zimring argue about it all the time. Ralph claims that the mutating of human biology is purely a result of the Pandemic of 2020, and other mutating viruses in later years. Melanie says that these mutations come from a shift in human consciousness. I think they are getting sweet on each other.

And finally, how did this book reach you, dear reader, if it was written in 2045? I sent it back in time using Ger's Time Shifter to a few select people I trust. If you are reading this before 2045, somebody in your time published it. We are too afraid to use Ger's Time Shifter again. Ger disabled it, but we know how to put it back together if necessary.

It is still way too cold in Midland, Illinois. Increased volcanic activity in the 2020s began the process of limiting solar radiation to the earth's surface, along with a decrease in solar radiation. No one knows how long this global cooling period will last, but in places where it has been traditionally hot, around the equator, it's even hotter! Go figure. Melanie Fuscaldo insists that climate change is also related to the change in human consciousness. I wouldn't go that far, but one thing is for sure: the world is changing so fast it's hard to keep up.

The Old Order is gone; it is being replaced by something a lot better.

CHAPTER 48

3057 AD

Caesar and Pietra Jones walked between two skyscrapers under a bright yellow sun, holding hands. The air was refreshing and cool today. They were on their way to pick up their grandson and granddaughter from the Courvall School, where they were discovering their native talents and their life work.

Life was grand but somewhat odd, Caesar reflected.

In his dreams (which occasionally intruded into his daytime awareness) he sometimes lived in a dystopian society where his native intelligence was a liability, and where he had to constantly be on guard from oppressive "authorities." One nightmare in particular was disturbing. He saw the image of a malevolent black tower in midtown, which directed hovercraft that observed his every move. But then he would always wake up from the dream and realize he lived in a vibrant and thriving New York with his wife and their beautiful family. He always discussed his dreams with Pietra, who had similar nightmares. They both agreed that perhaps these images came from a past life, or maybe even a parallel universe.

The couple discovered many others in New York who had similar dreams about a colony of survivors who struggled against long odds, fighting evil forces. But then, had not humanity always had these themes in its collective consciousness? Perhaps the struggle between good and evil was a necessary evolutionary step toward the fulfillment of human potential.

Afterword

The story of how the timeline changed from 2045 to 3057 after Harry Kaine's death is too complicated to include in this memoir, which is already long enough! The moral of this story is that temporal energy is conserved – it always tries to take the path of least resistance – but human intent and human action can change it. How else can Harry's death have changed a dystopian future for the better?

The timeline may change from what I have described here. Past affects future and future affects past. Who knows but that another wildcard like Harry Kaine will appear and completely change the future again? Or that a future civilization might once more intervene in the past and change everything around. One thing is certain: the human race is unpredictable, and time is malleable. Nothing is set in stone!

How can you change your future for the better? Listen not to those who criticize and tear down. Find your passion and go with it. That's a lesson Karen taught me.

Postscript – 2096 AD

Ger and I are old men now. I added this bit later, after the memoir was already written. That's the advantage of having a time device: we can add stuff on anytime we want and send it downline.

In our old age we've learned something about the timeline. It turns out that the original Looking Glass project was right: The future is totally dependent on the consciousness of those who are participating in it. We can see this clearly now through our time viewer.

Around the year 2012 or so, the potential for a positive future was created by humanity. After the collapse of Transhumanism in 2045, this was strengthened. By the year 2095, Ger and I see that all future time threads converge into a single track with a single outcome. The late Melanie Fuscaldo was right after all: The future of humanity is one of awakening and ascension.

The timeline is clear now, we can see without obstructions or fuzziness all the way to the far future.

The final outcome doesn't come for almost 100,000 years though. As the late great Ralph Zimring said, we have to play the long game. It's good to know that the ultimate evolution of humanity is so positive.

From our perspective in 2096, we can see that consciousness changes reality. That might not make sense to some readers, but as events unfold along the timeline, you will see an almost inevitable "push," or temporal momentum, for good outcomes We don't understand how or why this is. it just is. Our time research shows that there really is power to positive thinking. The more we expect a benevolent future, the faster it will arrive.

One final note: the planetary cooling period lasts from about 2030 to 2065 or so, depending on which climate variables are used. Ger and I are proud to say that our magnetic engines keep people warm, and greatly alleviate suffering, during this period.

As the future unfolds you'll see how immense greenhouses, heated and powered by our mag engines, grow food necessary to keep people fed during the coldest period. You'll see how...well, we won't spoil it for you.

It's always darkest just before the dawn. The future is a bright one, so never give up hope!

Author's Note –The Technocratic Wet Dream

For readers in the 2020s

I have tried to be fair and present both sides of the Transhumanist issue, even though I have a strong bias against Transhumanism. The protagonist in the story, Harry Kaine, is gung-ho for Transhumanism. Through Harry and his group I attempt to present both the good side and the bad side of Transhumanism.

My personal belief is that Transhumanism is a part of a dystopian technocratic vision – a techno utopia – for humanity's future. Technocrats and Transhumanists believe that consciousness is merely an epi-phenomenon of the brain. Therefore, if we can just develop enough "smart" algorithms and assemble enough parallel computing power (they say), independent intelligence will magically result.

This idea is behind the technical singularity, which postulates that artificial intelligence will magically become self-booting and self-learning, and far surpass human intelligence.

The other viewpoint is that consciousness exists independent of the body and all physical structure, that human consciousness results from the infusion of the soul into a physical body, and that the most important factor in life is love. This idea has been around since the first caveman, because there is something within every human being that instinctively looks past the biology to a higher understanding of life.

The machine- or artificial-intelligence theory is just as evidence-free as the idea of reincarnation, yet technocrats insist that their materialism is inherently superior.

Technocracy is a rather new idea, put forward in the twentieth century, and is a sort of wet dream of some scientists, technologists, and politicians like Ray Kurzwell (employed by Google), Klaus Schwab, and others, who think that a materialist-based society would be great because it could result in the easy control of the world population. Technocrats look upon themselves as human benefactors who are bringing scientific "enlightenment" to all humans to fulfill their agenda, even if it results in the replacement of humans with artificial general intelligence, or artificial super intelligence within the framework of a technical singularity.

In the book I have tried to present the idea that technocrats seek indefinite life extension because they want physical immortality in the present body (through technologies such as mind uploading and gene editing), and because they aren't self aware enough to understand that their consciousness is already immortal. Anyone with self awareness understands that consciousness enters and exits physical bodies over and over, advancing human biology and awareness each time. Like Harry Kaine in the story, technocrats have only contempt for this idea. Yet they embrace absurd and delusional ideas like the technical singularity, which will supposedly result in a God-like intelligence that all technocrats and Trans-Post-Humanists will be part of, after their minds have been uploaded to a cloud, or to non-biological physical containers.

This calculus leaves out the most important element of life: love, affinity, harmony, and cooperation. The radical beliefs of the technocrats arise, I believe, because these anti-humanists are disconnected from their own souls, and are incapable of feeling the higher emotions.

The new religion of the technocrats is scientism, not science. Scientism (scientific materialism) is a narrative-based, propaganda-oriented pseudo-science with a political agenda that uses whatever data it can muster to convince you to follow the guys in the white lab coats, the new gods of humanity. It's counter-intuitive, the technocrats say, but if you just believe our scientists and follow along, everything will turn out for the best. If you don't follow along, you're just too stupid to understand our great vision. (This is the viewpoint expressed by Harry Kaine in the story.)

Unfortunately, scientism and the technocrats don't tolerate free speech. They attempt to enforce their authoritarian decrees with vaccine passports, lockdowns, gulags, and censorship. The ultimate evolution of a technocratic state is similar to the New York of 3013 AD depicted in the story. Psychopaths like Khan 1, Patton 1, and Christian 172 are the inevitable evolution of technocratic "leaders."

The fundamental schism between a love-based philosophy of consciousness and a technocrat-scientism based philosophy is the inability to be aware of anything past the physical biology, which is itself a creation of the designer of the universe. If you don't understand this, you will when you die. Yes, the materialists are in for a big surprise at the end of life!

Appendices

Appendix A – Child Trafficking

Susan Sarandon, in the movie "Bull Durham," says, "The world is a simple place for those not cursed with self-awareness." Unfortunately this planet is mostly run by people with no self awareness. This is what they promote:

The scope of this problem was reported in 2018 in Westminster, London by The International Tribunal for Natural Justice (ITNJ), as the court convened over a 3-day period to launch their Judicial Commission of Inquiry into Human Trafficking and Child Sex Abuse. During the opening Plenary Session, Chief Counsel Robert David Steele, a former CIA officer, gave perhaps the best summary of the purpose of Tribunal's Judicial Commission.

This is an uncomfortable truth that 99% of the human population on this planet wants to ignore because it's too hard to confront. Our ignorance just allows it to continue. As John Stuart Mill said in 1867, "Bad men need nothing more to compass their ends, than that good men should look on and do nothing."

Testimony from Robert David Steele is here:

https://commission.itnj.org/westminster-seating-april-2018/

Testimony of Robert David Steele from Plenary Seating, Day 1, Session 1, Introductions. Begin at 34:50 for Mr. Steele's presentation.

This website is one of many exposing child trafficking and child sexual abuse. Research this for yourself.

Appendix B – Transhumanism: The New Technocratic Fascism

See the article titled "Klaus Schwab: Great Reset Will 'Lead to a Fusion of Our Physical, Digital and Biological Identity," *Implantable microchips that can read your thoughts,* by Paul Joseph Watson, 16 November, 2020, at

https://summit.news/2020/11/16/klaus-schwab-great-reset-will-lead-to-a-fusion-of-our-physical-digital-and-biological-identity/

From the article:

"Globalist Klaus Schwab made it clear that transhumanism is an integral part of 'The Great Reset' when he said that the fourth industrial revolution would 'lead to a fusion of our physical, digital and biological identity,' which in his book he clarifies is implantable microchips that can read your thoughts."

I can't infringe on the author's copyright, but I encourage you to read the entire article.

Transhumanism begins with cell-altering medicines that use RNA sequences to re-program cells.

See the Moderna website at

https://www.modernatx.com/mrna-technology/mrna-platform-enabling-drug-discovery-development

Moderna calls their mRNA medicines operating systems that execute biological programs or apps.

From the Moderna website:

"Our Operating System

"Recognizing the broad potential of mRNA science, we set out to create an mRNA technology platform that functions very much like an operating system on a computer. It is designed so that it can plug and play interchangeably with different programs. In our case, the "program" or "app" is our mRNA drug – the unique mRNA sequence that codes for a protein."

Moderna calls their biodevices "The Software of Life."

Here's what the company says:

"Utilizing these instruction sets [mRNA instruction sets] gives our investigational mRNA medicines a software-like quality."

In the section "Overcoming Challenges," the company talks about by-passing the immune system to change the protein structure of your body.

The Chemical & Engineering News website describes how mRNA is encapsulated in LNPs, or liquid nano-particles, and inserted into cells. This article is worth a read, and illustrates the process with diagrams.

"Fragile mRNA molecules used in COVID-19 vaccines can't get into cells on their own. They owe their success to lipid nanoparticles (LNPs) that took decades to refine. These LNPs encapsulate mRNA, shield it from destructive enzymes, and shuttle it into cells, where the mRNA is unloaded and used to make proteins."

See https://cen.acs.org/pharmaceuticals/drug-delivery/Without-lipid-shells-mRNA-vaccines/99/i8

For an in-depth discussion of gene editing, see the book *Crack in Creation: Gene Editing and the Unthinkable Power to Control Evolution,* by Jennifer Doudna.

For a excellent summary of Transhumanism, written by Transhumanists, see the Humanity+ Transhumanist FAQ at https://humanityplus.org/philosophy/transhumanist-faq/

Another great website on Transhumanism is https://lifeboat.com/, the website of the Lifeboat Foundation.

Appendix C – Cast of Important Clones

Rebel Clones
>Alan 133 – Alan the Curious, then Alan Percival
>Pietra 23 – Pietra Smith
>Li 355 – Li the Hammer (male)
>Sigmund 78 – Sigmund Drake
>Grace 5 – Grace Drake
>Li 57 – Li Jing (female)
>Li 308 – Li Chun (male)

Tower Temporal Technicians
>Caesar 13 – Caesar Smith
>Tesla 8 – Tesla Jones (female)
>Einstein 3 – Alfredo Einstein

Notes

[1] John F. Kennedy, "Inaugural Address," January 20, 1961, Public Papers of the Presidents: John F. Kennedy, 1961 (Washington, DC, 1961), I.

[2] See Appendix B.

[3] For readers in the 2020s, information on the grand solar minimum, which led to global cooling at an average of 1°C during the period 2030-2065, is available by reading Valentina Zharkov, "Modern Grand Solar Minimum will lead to terrestrial cooling," Taylor & Francis online, at

https://www.tandfonline.com/doi/full/10.1080/23328940.2020.1796243

Scientists in the early 2020s debated whether, after three 11-year solar cycles in which solar activity had been reduced (solar cycles 21-23), that the following 11-year cycles would be stronger. See "Upcoming Grand Solar Minimum Could Wipe Out Global Warming for Decades," October 5, 2020, at

https://www.scienceunderattack.com/blog/2020/10/5/upcoming-grand-solar-minimum-could-wipe-out-global-warming-for-decades-62.

For the opposite view, see Laura Snider, National Center for Atmospheric Research, "New sunspot cycle could be one of the strongest on record, new research predicts," December 7,2020 at https://phys.org/news/2020-12-sunspot-strongest.html

In the early 2020s, the sun's internal overlapping magnetic fields, which determine the sun's activity and thus the amount of solar radiation that reaches the earth, wasn't clearly understood. It turns out that Ms. Zharkrov was right, as you will see during the next three 11-year solar cycles (24-26).

[4] See The End of the Universe, by the author.

[5] Adapted from Dr. Sue Arrigo, "Secrets of the CIA's Global Sex Slave Industry," 2007. Accessed at gailallen.com. I do not vouch for the validity of this information, but sexual slavery (and child harvesting) certainly does exist. (See Appendix A.)

[6] Much of the information in this scene taken from a fantastic website called Lifeboat, at https://lifeboat.com/ex/transhumanist.technologies and https://lifeboat.com/ex/cybernetics.upgrades

[7] See *The Old Soul*, by the author.

[8] "All Our yesterdays," Star Trek: The Original Series, Season 3 Episode 23.

[9] This story is from the website

http://www.indigenouspeople.net/Hopination

https://www.manataka.org/page239.html.

[10] Diagram is an author created image in Photoshop, adapted from Harvard's SITN site, article titled, "CRISPR – A game changing genetic engineering technique" at http://sitn.hms.harvard.edu/flash/2014/crispr-a-game-changing-genetic-engineering-technique/

[11] The information for this scene was taken from several excellent online articles, primarily from "What Is CRISPR?" by Aparna Vidyasagar at Live Science. https://www.livescience.com/58790-crispr-explained.html.

[12] See the great Lifeboat site, lifeboat.com, article titled "AI Shield," at https://lifeboat.com/ex/aishield, from which some of this info was taken.

[13] *Id.*

[14] See "A beginners guide to A.I. superintelligence and 'the singularity,'" Luke Dormehl, the Digital Trends website, https://www.digitaltrends.com/cool-tech/what-is-the-singularity-ai/

[15] Much of the info in this section comes from a website called humanityplus.org, "Transhumanist FAQ," at https://humanityplus.org/philosophy/transhumanist-faq/

[16] See Transhumanist FAQ, "What is Uploading?" at Humanity+,

https://humanityplus.org/transhumanism/transhumanist-faq/

[17] See "These tiny robots could be disease-fighting machines inside the human body," from NBC News at https://www.nbcnews.com/mach/science/these-tiny-robots-could-be-disease-fighting-machines-inside-body-ncna861451

[18] *Id.*

[19] See "What is a posthuman?" at

https://humanityplus.org/transhumanism/transhumanist-faq/

[20] See "Will transhuman technologies make us inhuman?" at

https://humanityplus.org/philosophy/transhumanist-faq/

[21] *Id.*

[22] See "Won't things like uploading, cryonics, and AI fail because they can't preserve or create the soul?"

at https://humanityplus.org/philosophy/transhumanist-faq/

[23] "There is a possibility that a scientist who is very much involved his whole life [with computers], then the next life…[he would be reborn in a computer], same process! [Laughter.] Then this machine which is half-human and half-machine has been reincarnated." From "Dreams Precede Everything," by Sanjin Đumišić, at https://sanjindumisic.com/dalai-lama-on-reincarnation-and-artificial-intelligence/

[24] I adapted this beautiful story for this scene. The original is found on an awesome website called http://www.native-art-in-canada.com/creationstory.html, from an article titled "The Ojibwa Creation Story – This is the Creation Story of My Childhood."

[25] See *Tesla's Lost Notebook*, by the author.

[26] The solution was discovered by Li Chun, who led a massive shopping contingent to a 21st century Sams Club after the attack on the Tower. The group used the still active and maintained Shifter, and all three of Caesar 13's currency cards, to make the purchases. That incident led to an embarrassing confrontation by a hotheaded Li Hammer with a Midland Police black and white at the abandoned warehouse. This story, in which Li Hammer almost got arrested by a Midland Police officer who questioned why groceries were being delivered to a filthy warehouse, will not be chronicled here. Only that Caesar 13's three apartments were rented out for twelve months. The former rebel clones had only lived there for two months. So ten months of occupancy were still available in 2035, as long as no overlapping in 2035 occurred. The colony in 3016 could escape back to 2035 in an emergency.

[27] See *Tesla's Lost Notebook*, by the author.

[28] See "NASA Vegetation Index: Globe Continues Rapid Greening Trend, Sahara Alone Shrinks 700,000 Sq Km!", Feb 21, 2021, at https://www.climatedepot.com/2021/02/25/nasa-vegetation-index-globe-continues-rapid-greening-trend-sahara-alone-shrinks-700000-sq-km-2/

[29] *Id.*